Graphing Technology Guide

Benjamin N. Levy
and
Laurel Technical Services

HOUGHTON MIFFLIN COMPANY   Boston   New York

Sponsoring Editor: Christine B. Hoag
Senior Associate Editor: Maureen Brooks
Managing Editor: Catherine B. Cantin
Assistant Editor: Carolyn Johnson
Supervising Editor: Karen Carter
Associate Project Editor: Rachel D'Angelo Wimberly
Editorial Assistant: Caroline Lipscomb
Production Supervisor: Lisa Merrill
Art Supervisor: Gary Crespo
Marketing Manager: Charles Cavaliere
Marketing Associate: Ros Kane
Marketing Assistant: Kate Burden Thomas

Printed in the United States of America.

International Standard Book Number: 0-669-41756-4

23456789-CS 00 99 98 97

Contents

Detailed Contents
Texas Instruments Calculators

Chapter Key

x = 1   Texas Instruments TI-81
x = 2   Texas Instruments TI-82
x = 3   Texas Instruments TI-83
x = 4   Texas Instruments TI-85
x = 5   Texas Instruments TI-92

Detailed Contents
Casio, Sharp, and Hewlett Packard Calculators

Chapter Key

x = 6   Casio fx-7700GE/9700GE
x = 7   Casio CFX-9800G
x = 8   Casio CFX-9850G
x = 9   Sharp EL-9200/9300
x = 10 Hewlett Packard HP 48G
x = 11 Hewlett Packard HP 38G

Preface

The *Graphing Technology Guide* provides step-by-step, keystroke-level calculator commands and instructions for working through exercises in your textbook. Graphing calculator screens and technology tips are included throughout. The *Guide* will help you to become familiar with your calculator's capabilities and learn how to use it as a tool to learn mathematics. However, it does not replace the instruction manual that comes with the calculator. Refer to that manual to learn how to use additional capabilities of your calculator.

The *Graphing Technology Guide* contains parallel topics for each calculator model. Locate the discussion for your calculator in the appropriate column of numbers on the detailed table of contents. See pages iv-v for the Texas Instruments calculators and pages vi-vii for the Casio, Sharp, and Hewlett Packard calculators.

Please note that different typefaces are used to distinguish keystrokes that you press from the rest of the text. Thus, MATH and ENTER will represent the labels on your calculator's keys.

Calculators have function keys that assume different behavior in different contexts. To clarify the effect expected from depressing a function key, we sometimes write F1 [COMMAND], where F1 names the key and [COMMAND] represents its corresponding functions in the current menu. In Chapters 10 and 11, since the HP 48G and the HP 38G have six white function keys, we write [COMMAND] to represent the function key below the menu item COMMAND.

For convenience, when you are asked to type a number, say 345.67, we shall express the keystrokes as 345.67, without any spaces between the individual keys, instead of writing 3 4 5 . 6 7.

The calculator manufacturers--Texas Instruments, Casio, Sharp, and Hewlett Packard-- have all been cooperative through the writing of this guide, and the authors gratefully acknowledge their helpfulness.

Chapter 1

Texas Instruments TI-81

1.1   Getting started with the TI-81

1.1.1 Basics: Press the ON key to begin using your TI-81 calculator. If you need to adjust the display contrast, first press 2nd, then press and hold ▲ (the *up* arrow key) to increase the contrast or ▼ (the *down* arrow key) to decrease the contrast. As you press and hold ▲ or ▼, an integer between 0 (lightest) and 9 (darkest) appears in the upper right corner of the display. When you have finished with the calculator, turn it off to conserve battery power by pressing 2nd and then OFF.

Check the TI-81's settings by pressing MODE. If necessary, use the arrow keys to move the blinking cursor to a setting you want to change. Press ENTER to select a new setting. To start with, select the options along the left side of the MODE menu as illustrated in Figure 1.1: normal display, floating decimals, radian measure, function graphs, connected lines, sequential plotting, grid off, and rectangular coordinates. Details on alternative options will be given later in this guide. For now, leave the MODE menu by pressing CLEAR.

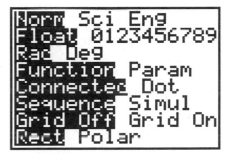

Figure 1.1: MODE menu

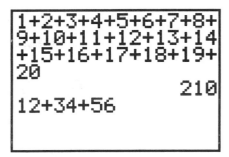

Figure 1.2: Home screen

1.1.2 Editing: One advantage of the TI-81 is that up to 8 lines are visible at one time, so you can *see* a long calculation. For example, type this sum (see Figure 1.2):

$$1 + 2 + 3 + 4 + 5 + 6 + 7 + 8 + 9 + 10 + 11 + 12 + 13 + 14 + 15 + 16 + 17 + 18 + 19 + 20$$

Then press ENTER to see the answer, too.

Often we do not notice a mistake until we see how unreasonable an answer is. The TI-81 permits you to re-display an entire calculation, edit it easily, then execute the *corrected* calculation.

Suppose you had typed 12 + 34 + 56 as in Figure 1.2 but had *not* yet pressed ENTER, when you realize that 34 should have been 74. Simply press ◀ (the *left* arrow key) as many times as necessary to move the blinking cursor left to 3, then type 7 to write over it. On the other hand, if 34 should have been 384, move the cursor back to 4, press INS (the cursor changes to a blinking underline) and then type 8 (inserts at the cursor position and other characters are pushed to the right). If the 34 should have been 3 only, move the cursor to 4 and press DEL to delete it.

Even if you had pressed ENTER, you may still edit the previous expression. Press 2nd and then ENTRY to *recall* the last expression that was entered. Now you can change it. If you have not pressed any key since the last ENTER, you can recall the previous expression by pressing ▲.

Technology Tip: When you need to evaluate a formula for different values of a variable, use the editing feature to simplify the process. For example, suppose you want to find the balance in an investment account if there is now $5000 in the account and interest is compounded annually at the rate of 8.5%. The formula for the balance is $P\left(1+\frac{r}{n}\right)^{nt}$, where P = principal, r = rate of interest (expressed as a decimal), n = number of times interest is compounded each year, and t = number of years. In our example, this becomes $5000(1+.085)^{t}$. Here are the keystrokes for finding the balance after $t = 3, 5,$ and 10 years.

Years	Keystrokes	Balance
3	5000 (1 + .085) ^ 3 ENTER	$6386.45
5	▲ ◀ 5 ENTER	$7518.28
10	▲ ◀ 10 ENTER	$11,304.92

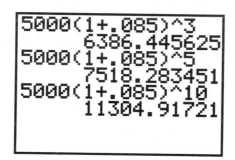

Figure 1.3: Editing expressions

Then to find the balance from the same initial investment but after 5 years when the annual interest rate is 7.5%, press these keys to change the last calculation above: ▲ ◀ DEL ◀ 5 ◀ ◀ ◀ ◀ ◀ 7 ENTER.

1.1.3 Key Functions: Most keys on the TI-81 offer access to more than one function, just as the keys on a computer keyboard can produce more than one letter ("g" and "G") or even quite different characters ("5" and "%"). The primary function of a key is indicated on the key itself, and you access that function by a simple press on the key.

To access the *second* function indicated to the *left* above a key, first press 2nd (the cursor changes to a blinking ↑) and *then* press the key. For example, to calculate $\sqrt{25}$, press 2nd $\sqrt{}$ 25 ENTER.

When you want to use a letter or other character printed to the *right* above a key, first press **ALPHA** (the cursor changes to a blinking **A**) and then the key. For example, to use the letter K in a formula, press **ALPHA K**. If you need several letters in a row, press **2nd A-LOCK**, which is like CAPS LOCK on a computer keyboard, and then press all the letters you want. Remember to press **ALPHA** when you are finished and want to restore the keys to their primary functions.

1.1.4 Order of Operations: The TI-81 performs calculations according to the standard algebraic rules. Working outwards from inner parentheses, calculations are performed from left to right. Powers and roots are evaluated first, followed by multiplications and divisions, and then additions and subtractions.

Note that the TI-81 distinguishes between *subtraction* and the *negative sign*. If you wish to enter a negative number, it is necessary to use the (-) key. For example, you would evaluate $-5-(4\cdot-3)$ by pressing (-) 5 − (4 × (-) 3) **ENTER** to get 7.

Enter these expressions to practice using your TI-81.

Expression	Keystrokes	Display
$7-5\cdot3$	7 − 5 × 3 ENTER	-8
$(7-5)\cdot3$	(7 − 5) × 3 ENTER	6
$20-10^2$	120 − 10 x² ENTER	20
$(120-10)^2$	(120 − 10) x² ENTER	12100
$\dfrac{24}{2^3}$	24 ÷ 2 ^ 3 ENTER	3
$\left(\dfrac{24}{2}\right)^3$	(24 ÷ 2) ^ 3 ENTER	1728
$(7--5)\cdot-3$	(7 − (-) 5) × (-) 3 ENTER	-36

1.1.5 Algebraic Expressions and Memory: Your calculator can evaluate expressions such as $\dfrac{N(N+1)}{2}$ *after* you have entered a value for N. Suppose you want $N = 200$. Press 200 STO ► N ENTER to store the value 200 in memory location N. (The STO ► key prepares the TI-81 for an alphabetical entry, so it is *not* necessary to press ALPHA also.) Whenever you use N in an expression, the calculator will substitute the value 200 until you make a change by storing *another* number in N. Next enter the expression $\dfrac{N(N+1)}{2}$ by typing

ALPHA N (ALPHA N + 1) ÷ 2 ENTER. For $N = 200$, you will find that $\dfrac{N(N+1)}{2} = 20100$.

The contents of any memory location may be revealed by typing just its letter name and then ENTER. And the TI-81 retains memorized values even when it is turned off, so long as its batteries are good.

1.1.6 Repeated Operations with ANS: The result of your *last* calculation is always stored in memory location ANS and replaces any previous result. This makes it easy to use the answer from one computation in another computation. For example, press 30 + 15 ENTER so that 45 is the last result displayed. Then press 2nd ANS ÷ 9 ENTER and get 5 because $\frac{45}{9} = 5$.

With a function like division, you press the ÷ key *after* you enter an argument. For such functions, whenever you would start a new calculation with the previous answer followed by pressing the function key, you may press just the function key. So instead of 2nd ANS ÷ 9 in the previous example, you could have pressed simply ÷ 9 to achieve the same result. This technique also works for these functions: + - × x² ^ x⁻¹.

Here is a situation where this is especially useful. Suppose a person makes $5.85 per hour and you are asked to calculate earnings for a day, a week, and a year. Execute the given keystrokes to find the person's incomes during these periods (results are shown in Figure 1.4):

Pay period	Keystrokes	Earnings
8-hour day	5.85 × 8 ENTER	$46.80
5-day week	× 5 ENTER	$234
52-week year	× 52 ENTER	$12,168

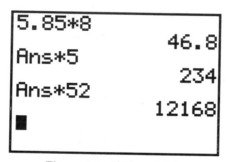

Figure 1.4: ANS variable

1.1.7 The MATH Menu: Operators and functions associated with a scientific calculator are available either immediately from the keys of the TI-81 or by 2nd keys. You have direct key access to common arithmetic operations (x², 2nd √ , x⁻¹, ^, 2nd ABS), trigonometric functions (SIN, COS, TAN) and their inverses (2nd SIN⁻¹, 2nd COS⁻¹, 2nd TAN⁻¹), exponential and logarithmic functions (LOG, 2nd 10ˣ, LN, 2nd eˣ), and a famous constant (2nd π).

A significant difference between the TI-81 and many scientific calculators is that the TI-81 requires the argument of a function *after* the function, as you would see a formula written in your textbook. For example, on the TI-81 you calculate $\sqrt{16}$ by pressing the keys 2nd √ 16 in that order.

Here are keystrokes for basic mathematical operations. Try them for practice on your TI-81.

Expression	Keystrokes	Display
$\sqrt{3^2+4^2}$	2nd $\sqrt{\ }$ (3 x² + 4 x²) ENTER	5
$2\frac{1}{3}$	2 + 3 x⁻¹ ENTER	2.333333333
$|-5|$	2nd ABS (-) 5 ENTER	5
$\log 200$	LOG 200 ENTER	2.301029996
$2.34 \cdot 10^5$	2.34 × 2nd 10ˣ 5 ENTER	234000

Additional mathematical operations and functions are available from the MATH menu (Figure 1.5). Press MATH to see the various options. You will learn in your mathematics textbook how to apply many of them. As an example, calculate $\sqrt[3]{7}$ by pressing MATH and then *either* 4 *or* ▼ ▼ ▼ ENTER; finally press 7 ENTER to see 1.912931183. To leave the MATH menu and take no other action, press 2nd QUIT or just CLEAR.

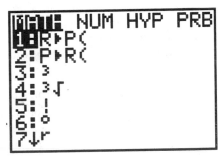

Figure 1.5: MATH menu

The *factorial* of a non-negative integer is the *product* of *all* the integers from 1 up to the given integer. The symbol for factorial is the exclamation point. So 4! (pronounced *four factorial*) is $1 \cdot 2 \cdot 3 \cdot 4 = 24$. You will learn more about applications of factorials in your textbook, but for now use the TI-81 to calculate 4! Press these keystrokes: 4 MATH 5 ENTER *or* 4 MATH ▼ ▼ ▼ ▼ ENTER ENTER.

1.2   Functions and Graphs

1.2.1 Evaluating Functions: Suppose you receive a monthly salary of $1975 plus a commission of 10% of sales. Let x = your sales in dollars; then your wages W in dollars are given by the equation $W = 1975 + .10x$. If your January sales were $2230 and your February sales were $1865, what was your income during those months?

Here's how to use your TI-81 to perform this task. Press the Y= key at the top of the calculator to display the function editing screen (Figure 1.6). You may enter as many as four different functions for the TI-81 to use at one time. If there is already a function Y_1, press ▲ or ▼ as many times as necessary to move the cursor to Y_1 and then press CLEAR to delete whatever was there. Then enter the expression $1975 + .10x$ by pressing these keys: 1975 + .10 XⱵT. (The XⱵT key lets you enter the variable x easily without having to use the ALPHA key.) Now press 2nd QUIT to return to the main calculations screen.

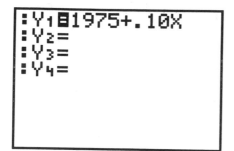

Figure 1.6: Y= screen

Figure 1.7: Evaluating a function

Assign the value 2230 to the variable x by these keystrokes (see Figure 1.7): 2230 STO ▶ XⱵT ENTER. Next press the following keystrokes to evaluate Y_1 and find January's wages: 2nd Y-VARS 1 ENTER. Repeat these steps to find the February wages. Each time the TI-81 evaluates the function Y_1, it uses the *current* value of x.

Technology Tip: The TI-81 does not require multiplication to be expressed between variables, so xxx means x^3. It is often easier to press two or three x's together than to search for the square key or the cube operation. Of course, expressed multiplication is also not required between a constant and a variable. Hence to enter $2x^3 + 3x^2 - 4x + 5$ in the TI-81, you might save keystrokes and press just these keys: 2 XⱵT XⱵT XⱵT + 3 XⱵT XⱵT - 4 XⱵT + 5.

1.2.2 Functions in a Graph Window: Once you have entered a function in the Y= screen of the TI-81, just press GRAPH to see its graph. The ability to draw a graph contributes substantially to our ability to solve problems.

For example, here is how to graph $y = -x^3 + 4x$. First press Y= and delete anything that may be there by moving with the arrow keys to Y_1 or to any of the other lines and pressing CLEAR wherever necessary. Then, with the cursor on the top line Y_1, press (-) XⱵT MATH 3 + 4 XⱵT to enter the function (as in Figure 1.8). Now press GRAPH and the TI-81 changes to a window with the graph of $y = -x^3 + 4x$.

Your graph window may look like the one in Figure 1.9 or it may be different. Since the graph of $y = -x^3 + 4x$ extends infinitely far left and right and also infinitely far up and down, the TI-81 can display only a piece of the actual graph. This displayed rectangular part is called a *viewing rectangle*. You can easily change the viewing rectangle to enhance your investigation of a graph.

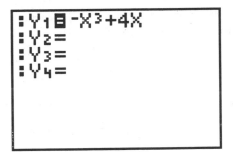

Figure 1.8: Y= screen

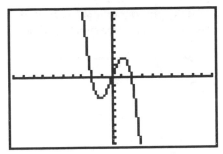

Figure 1.9: Graph of $y = -x^3 + 4x$

The viewing rectangle in Figure 1.9 shows the part of the graph that extends horizontally from -10 to 10 and vertically from -10 to 10. Press RANGE to see information about your viewing rectangle. Figure 1.10 shows the RANGE screen that corresponds to the viewing rectangle in Figure 1.9. This is the *standard* viewing rectangle for the TI-81.

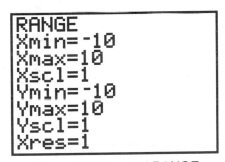

Figure 1.10: Standard RANGE

The variables Xmin and Xmax are the minimum and maximum x-values of the viewing rectangle; Ymin and Ymax are its minimum and maximum y-values.

Xscl and Yscl set the spacing between tick marks on the axes.

Xres is an integer from 1 to 8 that controls the resolution of the plot and also the speed of plotting. When Xres = 1, the calculator evaluates the function and plots a point 96 times along the x-axis. When Xres = 2, the calculator evaluates and plots at every *second* point, 48 times along the x-axis. Keep Xres = 1 to have the best resolution for your graphs.

Use the arrow keys ▲ and ▼ to move up and down from one line to another in this list; pressing the ENTER key will move down the list. Press CLEAR to delete the current value and then enter a new value. You may also edit the entry as you would edit an expression. Remember that a minimum *must* be less than the corresponding maximum or the TI-81 will issue an error message. Also, remember to use the (-) key, not

− (which is subtraction), when you want to enter a negative value. The following figures show different RANGE screens and the corresponding viewing rectangle for each one.

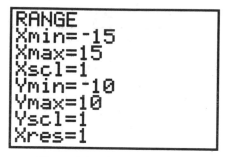

Figure 1.11: Square window

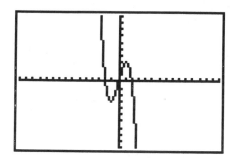

Figure 1.12: Graph of $y = -x^3 + 4x$

To set the range quickly to standard values (see Figure 1.10), press ZOOM 6. To set the viewing rectangle quickly to a square (Figure 1.11), press ZOOM 5. More information about square windows is presented later in Section 1.2.4.

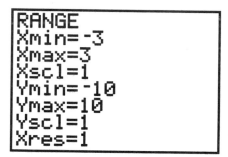

Figure 1.13: Custom window

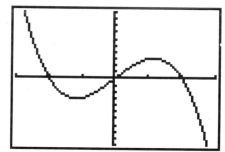

Figure 1.14: Graph of $y = -x^3 + 4x$

Sometimes you may wish to display grid points corresponding to tick marks on the axes. In the MODE menu (Figure 1.1), use arrow keys to move the blinking cursor to Grid On, then press ENTER and 2nd QUIT GRAPH. Figure 1.15 shows the same graph as in Figure 1.14 but with the grid turned on. In general, you'll want the grid turned *off*, so do that now by pressing MODE, use the arrow keys to move the blinking cursor to Grid Off, and press ENTER.

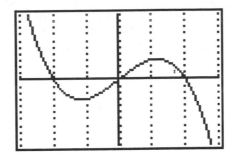

Figure 1.15: Grid turned on for $y = -x^3 + 4x$

1.2.3 Piecewise-Defined Functions: The greatest integer function, written [[x]], gives the greatest *integer* less than or equal to a number *x*. On the TI-81, the greatest integer function is called Int and is located under the NUM sub-menu of the MATH menu (see Figure 1.5). So calculate [[6.78]] = 6 by pressing MATH ▶ 4 6.78 ENTER.

To graph y = [[x]], go in the Y= menu, move beside Y_1 and press CLEAR MATH ▶ 4 XIT GRAPH. Figure 1.16 shows this graph in a viewing rectangle from -5 to 5 in both directions.

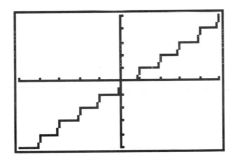

Figure 1.16: **Connected** graph of $y = [[x]]$

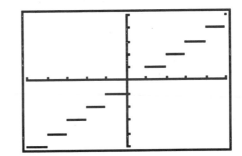

Figure 1.17: **Dot** graph of $y = [[x]]$

The true graph of the greatest integer function is a step graph, like the one in Figure 1.17. For the graph of y = [[x]], a segment should *not* be drawn between every pair of successive points. You can change from Connected line to Dot graph on the TI-81 by opening the MODE menu. Move the cursor down to the fifth line; select whichever graph type you require; press ENTER to put it into effect and GRAPH to see the result.

You should also change to Dot graph when plotting a piecewise-defined function. For example, to plot the graph of $f(x) = \begin{cases} x^2 + 1, & x < 0 \\ x - 1, & x \geq 0 \end{cases}$, enter the expression $(x^2 + 1)(x < 0) + (x - 1)(x \geq 0)$ somewhere in your Y= list by pressing (XIT x² + 1) (XIT 2nd TEST 5 0) + (XIT − 1) (XIT 2nd TEST 4 0). Then change the mode to Dot graph and press GRAPH.

1.2.4 Graphing a Circle: Here is a useful technique for graphs that are not functions, but that can be "split" into a top part and a bottom part, or into multiple parts. Suppose you wish to graph the circle whose equation is $x^2 + y^2 = 36$. First solve for y and get an equation for the top semicircle, $y = \sqrt{36 - x^2}$, and for the bottom semicircle, $y = -\sqrt{36 - x^2}$. Then graph the two semicircles simultaneously.

The keystrokes to draw this circle's graph follow. Enter $\sqrt{36 - x^2}$ as Y₁ and $-\sqrt{36 - x^2}$ as Y₂ (see Figure 1.18) by pressing Y= CLEAR 2nd $\sqrt{\ }$ (36 - X|T x²) ENTER CLEAR (-) 2nd $\sqrt{\ }$ (36 - X|T x²). Then press GRAPH to draw them both.

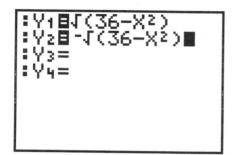

Figure 1.18: Two semicircles

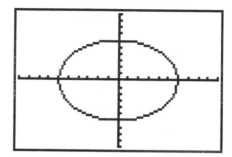

Figure 1.19: Circle's graph - standard view

If your range were set to the standard viewing rectangle, your graph would look like Figure 1.19. Now this does *not* look like a circle, because the units along the axes are not the same. This is where the square viewing rectangle is important. Press ZOOM 5 and see a graph that appears more circular.

Technology Tip: Another way to get a square graph is to change the range variables so that the value of Ymax - Ymin is $\frac{2}{3}$ times Xmax - Xmin. For example, see the RANGE in Figure 1.20 and the corresponding graph in Figure 1.21. The method works because the dimensions of the TI-81's display are such that the ratio of vertical to horizontal is $\frac{2}{3}$.

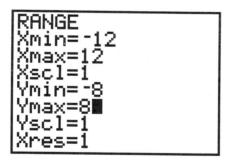

Figure 1.20: $\frac{\text{vertical}}{\text{horizontal}} = \frac{16}{24} = \frac{2}{3}$

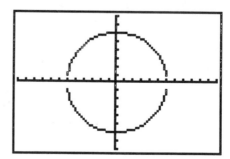

Figure 1.21: A "square" circle

The two semicircles in Figure 1.21 do not meet because of an idiosyncrasy in the way the TI-81 plots a graph.

Back when you entered $\sqrt{36-x^2}$ as Y_1 and $-\sqrt{36-x^2}$ as Y_2, you could have entered $-Y_1$ as Y_2 and saved some keystrokes. Try this by going back to the Y= menu and pressing the arrow key to move the cursor down to Y_2. Then press CLEAR (-) 2nd Y-VARS 1. The graph should be just as it was before.

1.2.5 TRACE: Graph $y = -x^3 + 4x$ in the standard viewing rectangle. Press any of the arrow keys ▲ ▼ ◀ ▶ and see the cursor move from the center of the viewing rectangle. The coordinates of the cursor's location are displayed at the bottom of the screen, as in Figure 1.22, in floating decimal format. This cursor is called a *free-moving cursor* because it can move from dot to dot *anywhere* in the graph window.

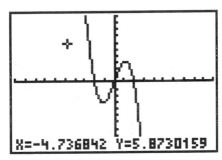

Figure 1.22: Free moving cursor

Remove the free-moving cursor and its coordinates from the window by pressing GRAPH or ENTER. If you press GRAPH, the next time you press an arrow key the free-moving cursor will appear again from the center of the viewing rectangle. If you press ENTER, the cursor will reappear at the same point you left it.

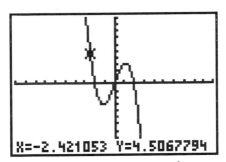

Figure 1.23: Trace on $y = -x^3 + 4x$

Press TRACE to enable the left ◀ and right ▶ arrow keys to move the cursor along the function. The cursor is no longer free-moving, but is now constrained to the function. The coordinates that are displayed belong to points on the function's graph, so the y-coordinate is the calculated value of the function at the corresponding x-coordinate.

Now plot a second function, $y = -.25x$, along with $y = -x^3 + 4x$. Press Y= and enter $-.25x$ for Y_2, then press GRAPH.

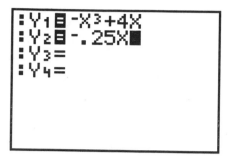

Figure 1.24: Two functions

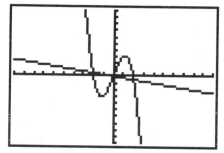

Figure 1.25: $y = -x^3 + 4x$ and $y = -.25x$

Note in Figure 1.24 that the equal signs next to Y_1 and Y_2 are *both* highlighted. This means *both* functions will be graphed. In the Y= screen, move the cursor directly on top of the equal sign next to Y_1 and press ENTER. This equal sign should no longer be highlighted (see Figure 1.26). Now press GRAPH and see that only Y_2 is plotted (Figure 1.27).

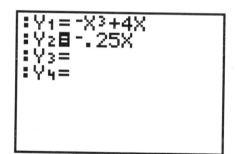

Figure 1.26: Y= screen with only Y_2 active

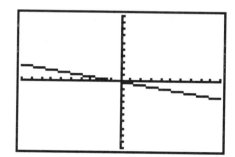

Figure 1.27: Graph of $y = -.25x$

So up to 4 different functions may be stored in the Y= list and any combination of them may be graphed simultaneously. You can make a function active or inactive for graphing by pressing ENTER on its equal sign to highlight (activate) or remove the highlight (deactivate). Go back to the Y= screen and do what is needed in order to graph Y_1 but not Y_2.

Now activate Y_2 again so that both graphs are plotted. Press TRACE and the cursor appears first on the graph of $y = -x^3 + 4x$ because it is higher up in the Y= list. Press the up ▲ or down ▼ arrow key to move the cursor vertically to the graph of $y = -.25x$. Next press the right and left arrow keys to trace along the graph

TI-81 Graphics Calculator

of $y = -.25x$. When more than one function is plotted, you can move the trace cursor vertically from one graph to another in this way.

Technology Tip: By the way, trace along the graph of $y = -.25x$ and press and hold either ◄ or ►. Eventually you will reach the left or right edge of the window. Keep pressing the arrow key and the TI-81 will allow you to continue the trace by panning the viewing rectangle. Check the RANGE screen to see that Xmin and Xmax are automatically updated.

The TI-81's display has 96 horizontal columns of pixels and 64 vertical rows. So when you trace a curve across a graph window, you are actually moving from Xmin to Xmax in 95 equal jumps, each called Δx. You would calculate the size of each jump to be $\Delta x = \dfrac{Xmax - Xmin}{95}$. Sometimes you may want the jumps to be friendly numbers like .1 or .25 so that, when you trace along the curve, the x-coordinates will be incremented by such a convenient amount. Just set your viewing rectangle for a particular increment Δx by making Xmax = Xmin + 95·Δx. For example, if you want Xmin = -5 and Δx = .3, set Xmax = -5 + 95·.3 = 23.5. Likewise, set Ymax = Ymin + 63·Δy if you want the vertical increment to be some special Δy.

To center your window around a particular point, say (h, k), and also have a certain Δx, set Xmin = h - 47·Δx and Xmax = h + 48·Δx. Likewise, make Ymin = k - 31·Δy and Ymax = k + 32·Δy. For example, to center a window around the origin, (0, 0), with both horizontal and vertical increments of .25, set the range so that Xmin = 0 - 47·.25 = -11.75, Xmax = 0 + 48·.25 = 12, Ymin = 0 - 31·.25 = -7.75, and Ymax = 0 + 32·.25 = 8.

See the benefit by first plotting $y = x^2 + 2x + 1$ in a standard graphing window. Trace near its y-intercept, which is (0, 1), and move towards its x-intercept, which is (-1, 0). Then change to a viewing rectangle from -9 to 10 horizontally and from -6 to 6.6 vertically, and trace again near the intercepts.

1.2.6 ZOOM: Plot again the two graphs, for $y = -x^3 + 4x$ and for $y = -.25x$. There appears to be an intersection near $x = 2$. The TI-81 provides several ways to enlarge the view around this point. You can change the viewing rectangle directly by pressing RANGE and editing the values of Xmin, Xmax, Ymin, and Ymax. Figure 1.29 shows a new viewing rectangle for the range displayed in Figure 1.28. Trace has been turned on and the coordinates of a point on $y = -x^3 + 4x$ that is close to the intersection are displayed.

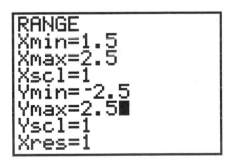

Figure 1.28: New RANGE

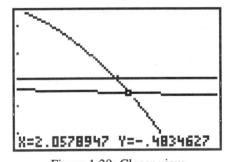

Figure 1.29: Closer view

A more efficient method for enlarging the view is to draw a new viewing rectangle with the cursor. Start again with a graph of the two functions $y = -x^3 + 4x$ and $y = -.25x$ in a standard viewing rectangle (press ZOOM 6 for the standard window, from -10 to 10 along both axes).

Now imagine a small rectangular box around the intersection point, near $x = 2$. Press ZOOM 1 (Figure 1.30) to draw a box to define this new viewing rectangle. Use the arrow keys to move the cursor, whose coordinates are displayed at the bottom of the window, to one corner of the new viewing rectangle you imagine.

Figure 1.30: ZOOM menu

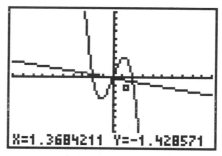

Figure 1.31: One corner selected

Press ENTER to fix the corner where you have moved the cursor; it changes shape and becomes a blinking square (Figure 1.31). Use the arrow keys again to move the cursor to the diagonally opposite corner of the new rectangle (Figure 1.32). If this box looks all right to you, press ENTER. The rectangular area you have enclosed will now enlarge to fill the graph window (Figure 1.33).

You may cancel the zoom any time *before* you press this last ENTER. Press 2nd QUIT to cancel the zoom and return to the home screen, or select another screen by pressing GRAPH or ZOOM and also cancel the zoom.

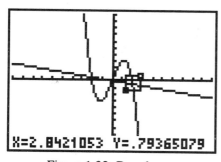

Figure 1.32: Box drawn

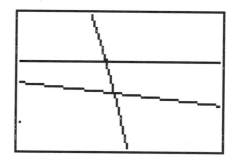

Figure 1.33: New viewing rectangle

You can also gain a quick magnification of the graph around the cursor's location. Return once more to the standard range for the graph of the two functions $y = -x^3 + 4x$ and $y = -.25x$. Press ZOOM 2 and then press arrow keys to move the cursor as close as you can to the point of intersection near $x = 2$ (see Figure 1.34).

TI-81 Graphics Calculator

Then press ENTER and the calculator draws a magnified graph, centered at the cursor's position (Figure 1.35). The range variables are changed to reflect this new viewing rectangle. Look in the RANGE menu to check.

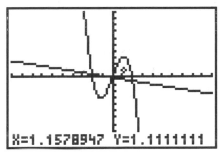

Figure 1.34: Before a zoom in

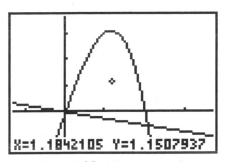

Figure 1.35: After a zoom in

As you see in the ZOOM menu (Figure 1.30), the TI-81 can Zoom In (press ZOOM 2) or Zoom Out (press ZOOM 3). Zoom out to see a larger view of the graph, centered at the cursor position. You can change the horizontal and vertical scale of the magnification by pressing ZOOM 4 (see Figure 1.36) and editing XFact and YFact, the horizontal and vertical magnification factors.

The default zoom factor is 4 in both directions. It is not necessary for XFact and YFact to be equal. Sometimes, you may prefer to zoom in one direction only, so the other factor should be set to 1. As usual, press 2nd QUIT to leave the ZOOM menu.

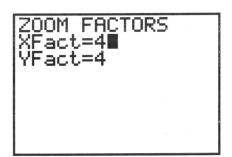

Figure 1.36: Set zoom factors

Technology Tip: If you should zoom in too much and lose the curve, zoom back to the standard viewing rectangle and start over.

1.2.7 Relative Minimums and Maximums: Graph $y = -x^3 + 4x$ once again in the standard viewing rectangle (Figure 1.9). This function appears to have a relative minimum near $x = -1$ and a relative maximum near $x = 1$. You may zoom and trace to approximate these extreme values.

First trace along the curve near the local minimum. Notice by how much the x-values and y-values change as you move from point to point. Trace along the curve until the y-coordinate is as *small* as you can get it, so that you are as close as possible to the local minimum, and zoom in (press ZOOM 2 or use a zoom box). Now trace again along the curve and, as you move from point to point, see that the coordinates change by smaller amounts than before. Keep zooming and tracing until you find the coordinates of the local minimum point as accurately as you need them, approximately (-1.15, -3.08).

Follow a similar procedure to find the local maximum. Trace along the curve until the y-coordinate is as *great* as you can get it, so that you are as close as possible to the local maximum, and zoom in. The local maximum point on the graph of $y = -x^3 + 4x$ is approximately (1.15, 3.08).

1.3   Solving Equations and Inequalities

1.3.1 Intercepts and Intersections: Tracing and zooming are also used to locate an x-intercept of a graph, where a curve crosses the x-axis. For example, the graph of $y = x^3 - 8x$ crosses the x-axis three times (see Figure 1.37). After tracing over to the x-intercept point that is furthest to the left, zoom in (Figure 1.38). Continue this process until you have located all three intercepts with as much accuracy as you need. The three x-intercepts of $y = x^3 - 8x$ are approximately -2.828, 0, and 2.828.

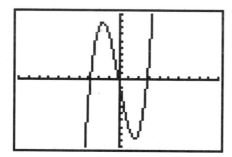

Figure 1.37: Graph of $y = x^3 - 8x$

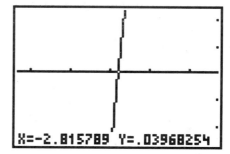

Figure 1.38: An x-intercept of $y = x^3 - 8x$

Technology Tip: As you zoom in, you may also wish to change the spacing between tick marks on the x-axis so that the viewing rectangle shows scale marks near the intercept point. Then the accuracy of your approximation will be such that the error is less than the distance between two tick marks. Change the x-scale on the TI-81 from the RANGE menu. Move the cursor down to Xscl and enter an appropriate value.

TRACE and ZOOM are especially important for locating the intersection points of two graphs, say the graphs of $y = -x^3 + 4x$ and $y = -.25x$. Trace along one of the graphs until you arrive close to an intersection point. Then press ◮ or ◿ to jump to the other graph. Notice that the x-coordinate does not change, but the y-coordinate is likely to be different (see Figures 1.39 and 1.40).

When the two y-coordinates are as close as they can get, you have come as close as you now can to the point of intersection. So zoom in around the intersection point, then trace again until the two y-coordinates are as close as possible. Continue this process until you have located the point of intersection with as much accuracy as necessary.

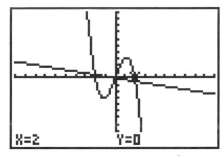

Figure 1.39: Trace on $y = -x^3 + 4x$

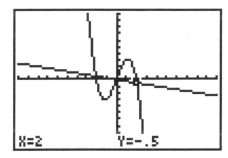

Figure 1.40: Trace on $y = -.25x$

1.3.2 Solving Equations by Graphing: Suppose you need to solve the equation $24x^3 - 36x + 17 = 0$. First graph $y = 24x^3 - 36x + 17$ in a window large enough to exhibit *all* its x-intercepts, corresponding to all its roots. Then use trace and zoom to locate each one. In fact, this equation has just one solution, approximately $x = -1.414$.

Remember that when an equation has more than one x-intercept, it may be necessary to change the viewing rectangle a few times to locate all of them.

Technology Tip: To solve an equation like $24x^3 + 17 = 36x$, you may first transform it into standard form, $24x^3 - 36x + 17 = 0$, and proceed as above. However, you may also graph the *two* functions $y = 24x^3 + 17$ and $y = 36x$, then zoom and trace to locate their point of intersection.

1.3.3 Solving Systems by Graphing: The solutions to a system of equations correspond to the points of intersection of their graphs (Figure 1.41). For example, to solve the system $y = x^2 - 3x - 4$ and $y = x^3 + 3x^2 - 2x - 1$, first graph them together. Then zoom and trace to locate their point of intersection, approximately (-2.17, 7.25).

You must judge whether the two current y-coordinates are sufficiently close for $x = -2.17$ or whether you should continue to zoom and trace to improve the approximation.

The solutions of the system of two equations $y = x^3 + 3x^2 - 2x - 1$ and $y = x^2 - 3x - 4$ correspond to the solutions of the single equation $x^3 + 3x^2 - 2x - 1 = x^2 - 3x - 4$, which simplifies to $x^3 + 2x^2 + x + 3 = 0$. So you may also graph $y = x^3 + 2x^2 + x + 3$ and find its x-intercepts to solve the system.

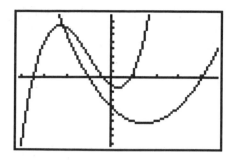

Figure 1.41: Solving a system of equations

1.3.4 Solving Inequalities by Graphing: Consider the inequality $1-\dfrac{3x}{2} \geq x-4$. To solve it with your TI-81,

graph the two functions $y = 1-\dfrac{3x}{2}$ and $y = x-4$ (Figure 1.42). First locate their point of intersection, at $x =$

2. The inequality is true when the graph of $y = 1-\dfrac{3x}{2}$ lies *above* the graph of $y = x-4$, and that occurs for

$x < 2$. So the solution is the half-line $x \leq 2$, or $(-\infty, 2]$.

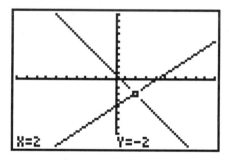

Figure 1.42: Solving $1-\dfrac{3x}{2} \geq x-4$

The TI-81 is capable of shading the region above or below a graph or between two graphs. For example, to graph $y \geq x^2 -1$, first graph the function $y = x^2 -1$ as Y_1. Then press 2nd DRAW 7 2nd Y-VARS 1 ALPHA , 10 ALPHA , 2) ENTER (see Figure 1.43). These keystrokes instruct the TI-81 to shade the region *above* $y = x^2 -1$ and *below* $y = 10$ (chosen because this is the greatest y-value in the graph window) with shading resolution value of 2. The result is shown in Figure 1.44.

To clear the shading, press 2nd DRAW 1 ENTER.

Now use shading to solve the previous inequality, $1 - \dfrac{3x}{2} \geq x - 4$. The function whose graph forms the lower boundary is named *first* in the SHADE command (see Figure 1.45). To enter this in your TI-81, press these keys: 2nd DRAW 7 x|T - 4 ALPHA , 1 - 3 x|T ÷ 2 ALPHA , 2) ENTER (Figure 1.46). The shading extends left from $x = 2$, hence the solution to $1 - \dfrac{3x}{2} \geq x - 4$ is the half-line $x \leq 2$, or $(-\infty, 2]$.

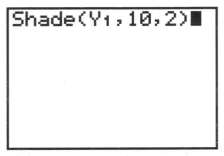

Figure 1.43: DRAW Shade

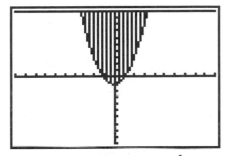

Figure 1.44: Graph of $y \geq x^2 - 1$

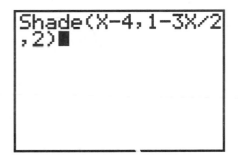

Figure 1.45: DRAW Shade command

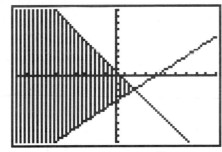

Figure 1.46: Graph of $1 - \dfrac{3x}{2} \geq x - 4$

More information about the DRAW menu is in the TI-81 manual.

1.4   Trigonometry

1.4.1 Degrees and Radians: The trigonometric functions can be applied to angles measured either in radians or degrees, but you should take care that the TI-81 is configured for whichever measure you need. Press MODE to see the current settings. Press ▼ twice and move down to the third line of the mode menu where angle measure is selected. Then press ◀ or ▶ to move between the displayed options. When the blinking cursor is on the measure you want, press ENTER to select it. Then press 2nd QUIT to leave the mode menu.

It's a good idea to check the angle measure setting before executing a calculation that depends on a particular measure. You may change a mode setting at any time and not interfere with pending calculations. Try the following keystrokes to see this in action.

Expression	Keystrokes	Display
$\sin 45°$	MODE ▼ ▼ ▶ ENTER	
	2nd QUIT SIN 45 ENTER	.7071067812
$\sin \pi°$	SIN 2nd π ENTER	.0548036651
$\sin \pi$	SIN 2nd π MODE ▼ ▼	
	ENTER 2nd QUIT ENTER	0
$\sin 45$	SIN 45 ENTER	.8509035245
$\sin \frac{\pi}{6}$	SIN (2nd π ÷ 6) ENTER	.5

The first line of keystrokes sets the TI-81 in degree mode and calculates the sine of 45 *degrees*. While the calculator is still in degree mode, the second line of keystrokes calculates the sine of π *degrees*, 3.1415°. The third line changes to radian mode just before calculating the sine of π *radians*. The fourth line calculates the sine of 45 *radians* (the calculator is already in radian mode).

The TI-81 makes it possible to mix degrees and radians in a calculation. Execute these keystrokes to calculate $\tan 45° + \sin \frac{\pi}{6}$ as shown in Figure 1.47: TAN 45 MATH 6 + SIN (2nd π ÷ 6) MATH 7 ENTER. Do you get 1.5 whether your calculator is set *either* in degree mode *or* in radian mode?

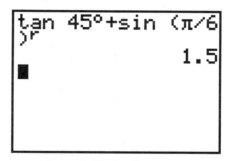

Figure 1.47: Angle measure

1.4.2 Graphs of Trigonometric Functions: When you graph a trigonometric function, you need to pay careful attention to the choice of graph window. For example, graph $y = \dfrac{\sin 30x}{30}$ in the standard viewing rectangle. Trace along the curve to see where it is. Zoom in to a better window, or use the period and amplitude to establish better RANGE values.

Technology Tip: Since $\pi \approx 3.1$, set Xmin = 0 and Xmax = 6.3 to cover the interval from 0 to 2π.

Next graph $y = \tan x$ in the standard window. The TI-81 plots consecutive points and then connects them with a segment, so the graph is not exactly what you should expect. You may wish to change from Connected line to Dot graph (see Section 1.2.3) when you plot the tangent function.

1.5   Scatter Plots

1.5.1 Entering Data: This table shows total prize money (in millions of dollars) awarded at the Indianapolis 500 race from 1981 to 1989. (*Source:* Indianapolis Motor Speedway Hall of Fame.)

Year	1981	1982	1983	1984	1985	1986	1987	1988	1989
Prize ($ million)	$1.61	$2.07	$2.41	$2.80	$3.27	$4.00	$4.49	$5.03	$5.72

We'll now use the TI-81 to construct a scatter plot that represents these points and to find a linear model that approximates the given data.

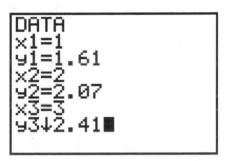

Figure 1.48: Entering data points

Before entering the data, press 2nd STAT ◀ 2 ENTER to clear away any previous data.

Note that you can select a sub-menu from the STAT menu by pressing either ◀ or ▶. It is a bit easier to press ◀ once than to press ▶ twice to get to the DATA sub-menu.

Now press 2nd STAT ◀ 1 to prepare to input data from the table. Instead of entering the full year 198x, enter only x. Here are the keystrokes for the first three years: 1 ENTER 1.61 ENTER 2 ENTER 2.07 ENTER 3 ENTER 2.41 ENTER and so on (see Figure 1.48). Continue to enter all the given data. Press 2nd QUIT when you have finished.

You may edit statistical data in the same way you edit expressions in the home screen. Move the cursor to the x or y value for any data point you wish to change, then type the correction. To insert or delete statistical data, move the cursor over the = next to the x or y value for any data point you wish to add or delete. Press INS and a new data point is created; press DEL and the data point is deleted.

1.5.2 Plotting Data: Once all the data points have been entered, press 2nd STAT ▶ 2 ENTER to draw a scatter plot. Your viewing rectangle is important, so you may wish to change the RANGE first to improve the view of the data. If you change the RANGE *after* drawing the scatter plot, you will have to create the plot again by pressing 2nd STAT ▶ 2 ENTER once more. Figure 1.49 shows the scatter plot for the range in Figure 1.50.

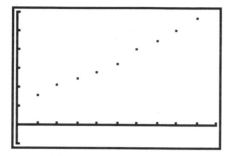

Figure 1.49: Scatter plot

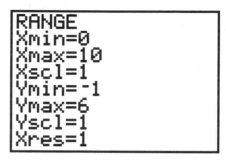

Figure 1.50: Range for scatter plot

1.5.3 Regression Line: The TI-81 calculates the slope and y-intercept for the line that best fits all the data. After the data points have been entered, press 2nd STAT 2 ENTER to calculate a linear regression model. As you see in Figure 1.51, the TI-81 names the y-intercept a and calls the slope b. The number r (between -1 and 1) is called the *correlation coefficient* and measures the goodness of fit of the linear regression equation with the data. The closer |r| is to 1, the better the fit; the closer |r| is to 0, the worse the fit.

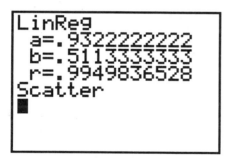

Figure 1.51: Linear regression model

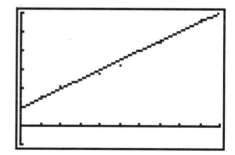

Figure 1.52: Linear regression line

Graph the line $y = a + bx$ by pressing Y=, inactivating any existing functions, moving to a free line or clearing one, then pressing VARS ▶ ▶ 4 GRAPH. Redraw the scatter plot to see how well this line fits with your data points (see Figure 1.52).

1.5.4 Exponential Growth Model: After data points have been entered, press 2nd STAT ▶ 4 ENTER to calculate an exponential growth model $y = a \cdot b^x$ for the data.

TI-81 Graphics Calculator

1.6.1 Making a Matrix: The TI-81 can display and use three different matrices, each with up to six rows and

up to six columns. Here's how to create this 3×4 matrix $\begin{bmatrix} 1 & -4 & 3 & 5 \\ -1 & 3 & -1 & -3 \\ 2 & 0 & -4 & 6 \end{bmatrix}$ in your calculator.

Press MATRX to see the matrix menu (Figure 1.53); then press ▶ to switch to the matrix EDIT menu (Figure 1.54). Whenever you enter the matrix EDIT menu, the cursor starts at the top matrix. Move to another matrix by repeatedly pressing ▼ . For now, just press ENTER to edit matrix [A].

Figure 1.53: MATRX menu

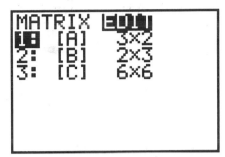

Figure 1.54: Matrix EDIT menu

You may now change the dimensions of matrix [A] to 3×4 by pressing 3 ENTER 4 ENTER. Simply press ENTER or an arrow key to accept an existing dimension. As you change the dimensions, a small black rectangle appears at the top of the TI-81 screen, as in Figure 1.55. This rectangle represents the matrix and shows its size and the element where the cursor is positioned.

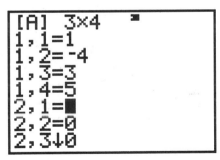

Figure 1.55: Editing the 2nd row, 1st column element

Use ▲ and ▼ to move the cursor to a matrix element you want to change, and watch the white mark move in the black rectangle to show the cursor's location within the matrix. At the bottom of the screen in Figure

1.55, there is ↓ instead of = to indicate that more elements are below, off the screen. Go to them by pressing ▼ as many times as necessary. The white mark in the black rectangle indicates that the cursor in Figure 1.55 is currently on the element in the second row and first column. Continue to enter all the elements of matrix [A].

Leave the matrix [A] editing screen by pressing 2nd QUIT and return to the home screen.

1.6.2 Matrix Math: From the home screen you can perform many calculations with matrices. First, let's see matrix [A] itself by pressing 2nd [A] ENTER (Figure 1.56).

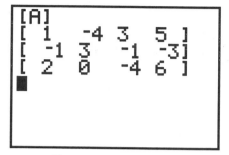

Figure 1.56: Matrix [A]

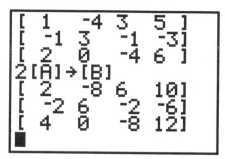

Figure 1.57: Matrix [B]

Calculate the scalar multiplication 2[A] by pressing 2 2nd [A] ENTER. To replace matrix [B] by 2[A], press 2 2nd [A] STO ▸ 2nd [B] ENTER, or if you do this immediately after calculating 2[A], press only STO ▸ 2nd [B] ENTER (see Figure 1.57). Press MATRX ▶ 2 to verify that the dimensions of matrix [B] have been changed automatically to reflect this new value.

Add the two matrices [A] and [B] by pressing 2nd [A] + 2nd [B] ENTER. Subtraction is similar.

Now set the dimensions of matrix [C] to 2×3 and enter this as [C]: $\begin{bmatrix} 2 & 0 & 3 \\ 1 & -5 & -1 \end{bmatrix}$. For matrix multiplication of [C] by [A], press 2nd [C] × 2nd [A] ENTER. If you tried to multiply [A] by [C], your TI-81 would signal an error because the dimensions of the two matrices do not permit multiplication this way.

The *transpose* of a matrix [A] is another matrix with the rows and columns interchanged. The symbol for the transpose of [A] is $[A]^T$. To calculate $[A]^T$, press 2nd [A] MATRX 6 ENTER.

1.6.3 Row Operations: Here are the keystrokes necessary to perform elementary row operations on a matrix. Your textbook provides more careful explanation of the elementary row operations and their uses.

To interchange the second and third rows of the matrix [A] that was defined above, press MATRX 1 2nd [A] ALPHA , 2 ALPHA , 3) ENTER (see Figure 1.58). The format of this command is RowSwap(*matrix, row1, row2*).

To add row 2 and row 3 and store the results in row 3, press MATRX 2 2nd [A] ALPHA , 2 ALPHA , 3) ENTER. The format of this command is Row+(*matrix, row1, row2*).

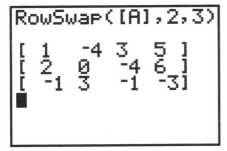

Figure 1.58: Swap rows 2 and 3

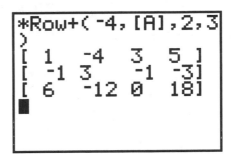

Figure 1.59: Add -4 times row 2 to row 3

To multiply row 2 by -4 and *store* the results in row 2, thereby replacing row 2 with new values, press MATRX 3 (-) 4 ALPHA , 2nd [A] ALPHA , 2) ENTER. The format of this command is *Row(*scalar, matrix, row*).

To multiply row 2 by -4 and *add* the results to row 3, thereby replacing row 3 with new values, press MATRX 4 (-) 4 ALPHA , 2nd [A] ALPHA , 2 ALPHA , 3) ENTER (see Figure 1.59). The format of this command is *Row+(*scalar, matrix, row1, row2*).

Technology Tip: It is important to remember that your TI-81 does *not* store a matrix obtained as the result of any row operations. So when you need to perform several row operations in succession, it is a good idea to store the result of each one in a temporary place. You may wish to use matrix [C] to hold such intermediate results.

For example, use elementary row operations to solve this system of linear equations: $\begin{cases} x - 2y + 3z = 9 \\ -x + 3y = -4 \\ 2x - 5y + 5z = 17 \end{cases}$.

First enter this *augmented matrix* as [A] in your TI-81: $\begin{bmatrix} 1 & -2 & 3 & 9 \\ -1 & 3 & 0 & -4 \\ 2 & -5 & 5 & 17 \end{bmatrix}$. Next store this matrix in [C]

(press 2nd [A] STO ▶ 2nd [C] ENTER) so you may keep the original in case you need to recall it.

Here are the row operations and their associated keystrokes. At each step, the result is stored in [C] and replaces the previous matrix [C]. The solution is shown in Figure 1.60.

Row Operation	Keystrokes
Row+([C], 1, 2)	MATRX 2 2nd [C] ALPHA , 1 ALPHA , 2) STO ▶ 2nd [C] ENTER
*Row+(-2, [C], 1, 3)	MATRX 4 (-) 2 ALPHA , 2nd [C] ALPHA , 1 ALPHA , 3) STO ▶ 2nd [C] ENTER
Row+([C], 2, 3)	MATRX 2 2nd [C] ALPHA , 2 ALPHA , 3) STO ▶ 2nd [C] ENTER
*Row(½, [C], 3)	MATRX 3 1 ÷ 2 ALPHA , 2nd [C] ALPHA , 3) STO ▶ 2nd [C] ENTER

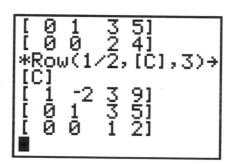

Figure 1.60: Final matrix after row operations

Thus $z = 2$, so $y = -1$ and $x = 1$.

1.6.4 Determinants and Inverses: Enter this 3×3 square matrix as [A]: $\begin{bmatrix} 1 & -2 & 3 \\ -1 & 3 & 0 \\ 2 & -5 & 5 \end{bmatrix}$. To calculate its de-

terminant, $\begin{vmatrix} 1 & -2 & 3 \\ -1 & 3 & 0 \\ 2 & -5 & 5 \end{vmatrix}$, press MATRX 5 2nd [A] ENTER. You should find that $|[A]| = 2$, as shown in

Figure 1.59.

Since the determinant of matrix [A] is not zero, it has an inverse, $[A]^{-1}$. Press 2nd [A] x⁻¹ ENTER to calculate the inverse of matrix [A], also shown in Figure 1.61.

Now let's solve a system of linear equations by matrix inversion. Once more, consider $\begin{cases} x - 2y + 3z = 9 \\ -x + 3y = -4 \\ 2x - 5y + 5z = 17 \end{cases}$.

TI-81 Graphics Calculator

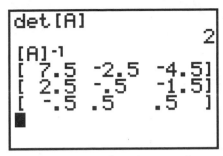

Figure 1.61: $|[A]|$ and $[A]^{-1}$

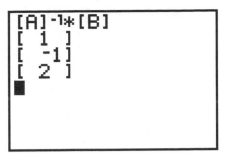

Figure 1.62: Solution matrix

The coefficient matrix for this system is the matrix $\begin{bmatrix} 1 & -2 & 3 \\ -1 & 3 & 0 \\ 2 & -5 & 5 \end{bmatrix}$ that was entered in the previous example. If

necessary, enter it again as [A] in your TI-81. Enter the matrix $\begin{bmatrix} 9 \\ -4 \\ 17 \end{bmatrix}$ as [B]. Then press 2nd [A] x^{-1} × 2nd [B]

ENTER to calculate the solution matrix (Figure 1.62). The solutions are still $x = 1$, $y = -1$, and $z = 2$.

1.7   Sequences

1.7.1 Iteration with the ANS Key: The ANS feature permits you to perform *iteration*, the process of evaluating a function repeatedly. As an example, calculate $\dfrac{n-1}{3}$ for $n = 27$. Then calculate $\dfrac{n-1}{3}$ for $n =$ the answer to the previous calculation. Continue to use each answer as n in the *next* calculation. Here are keystrokes to accomplish this iteration on the TI-81 calculator (see the results in Figure 1.63). Notice that when you use ANS in place of n in a formula, it is sufficient to press ENTER to continue an iteration.

Iteration	Keystrokes	Display
1	27 ENTER	27
2	(ANS - 1) ÷ 3 ENTER	8.666666667
3	ENTER	2.555555556
4	ENTER	.5185185185
5	ENTER	-.1604938272

Press ENTER several more times and see what happens with this iteration. You may wish to try it again with a different starting value.

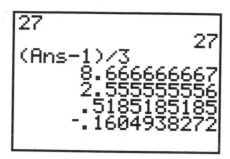

Figure 1.63: Iteration

1.7.2 Arithmetic and Geometric Sequences: Use iteration with the ANS variable to determine the n-th term of a sequence. For example, find the 18th term of an *arithmetic* sequence whose first term is 7 and whose common difference is 4. Enter the first term 7, then start the progression with the recursion formula, ANS + 4 ENTER. This yields the 2nd term, so press ENTER sixteen more times to find the 18th term. For a *geometric* sequence whose common ratio is 4, start the progression with ANS × 4 ENTER.

Of course, you could also use the *explicit* formula for the n-th term of an arithmetic sequence, $t_n = a + (n-1)d$. First enter values for the variables a, d, and n, then evaluate the formula by pressing ALPHA A + (ALPHA N - 1) ALPHA D ENTER. For a geometric sequence whose n-th term is given by $t_n = a \cdot r^{n-1}$, enter values for the variables a, r, and n, then evaluate the formula by pressing ALPHA A ALPHA R ^ (ALPHA N - 1) ENTER.

1.8   Parametric and Polar Graphs

1.8.1 Graphing Parametric Equations: The TI-81 plots parametric equations as easily as it plots functions. Just use the MODE menu (Figure 1.1), go to the fourth line from the top, and change the setting from Function to Param. Be sure, if the independent parameter is an angle measure, that MODE is set to whichever you need, Rad or Deg.

For example, here are the keystrokes needed to graph the parametric equations $x = \cos^3 t$ and $y = \sin^3 t$. First check that angles are currently being measured in radians. Change to parametric mode and press Y= to examine the new parametric equation menu (Figure 1.64). Enter the two parametric equations by pressing (COS X|T) ^ 3 ENTER (SIN X|T) ^ 3 ENTER. Now, when you press the variable key X|T, you get a T because the calculator is in parametric mode.

Figure 1.64: Parametric Y= menu

Figure 1.65: Parametric RANGE menu

Also look at the new RANGE menu (Figure 1.65). In the standard window, the values of T go from 0 to 2π in steps of $\frac{\pi}{30}$ =.1047, with the view from -10 to 10 in both directions. But here the viewing rectangle has been changed to extend from -2 to 2 in both directions. Press GRAPH to see the parametric graph (Figure 1.66).

You may ZOOM and TRACE along parametric graphs just as you did with function graphs. As you trace along this graph, notice that the cursor moves in the *counterclockwise* direction as T increases.

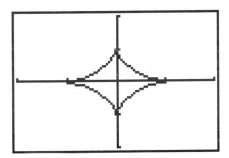

Figure 1.66: Parametric graph of $x = \cos^3 t$ and $y = \sin^3 t$

1.8.2 Rectangular-Polar Coordinate Conversion: The MATH menu (Figure 1.5) provides functions for converting between rectangular and polar coordinate systems. These functions use the current MODE settings, so it is a good idea to check the default angle measure before any conversion. Of course, you may use the MATH menu to override the current angle measure setting, as explained in Section 1.4.1. For the following examples, the TI-81 is set to radian measure.

Given rectangular coordinates $(x, y) = (4, -3)$, convert *from* these rectangular coordinates *to* polar coordinates (r, θ) by pressing MATH 1 4 ALPHA , (-) 3) ENTER. The value of r is displayed; press ALPHA θ ENTER to display the value of θ.

Suppose $(r, \theta) = (3, \pi)$. To convert *from* these polar coordinates *to* rectangular coordinates (x, y), press MATH 2 3 ALPHA , 2nd π) ENTER. The *x*-coordinate is displayed; press ALPHA Y ENTER to display the *y*-coordinate.

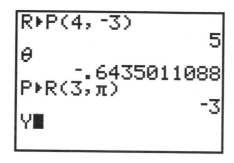

Figure 1.67: Coordinate conversions

1.8.3 Graphing Polar Equations: The TI-81 graphs a polar function when you write it in parametric form. For example, to graph $r = 4 \sin \theta$, enter the parametric equations $x = (4 \sin t) \cos t$ and $y = (4 \sin t) \sin t$. While you are changing MODE to parametric, also change the last MODE menu item to Polar (see below for the reason). Choose a good viewing rectangle and an appropriate range for the parameter *t*. In Figure 1.68, the graphing window is square and extends from -6 to 6 horizontally and from -4 to 4 vertically.

Figure 1.68 shows *rectangular* coordinates of the cursor's location on the graph. When you change MODE to Polar, you are able to trace along a parametric curve and see the *polar* coordinates of the cursor's location.

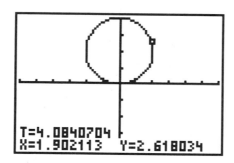

Figure 1.68: Polar graph of $r = 4 \sin \theta$

1.9 Probability

1.9.1 Random Numbers: The command Rand generates a number between 0 and 1. You will find this command in the PRB (probability) sub-menu of the MATH menu. Press MATH ◀ 1 ENTER to generate a random number. Press ▲ ENTER to generate another number; keep pressing ▲ ENTER to generate more of them.

If you need a random number between, say, 0 and 10, then press 10 MATH ◀ 1 ENTER. To get a random number between 5 and 15, press 5 + 10 MATH ◀ 1 ENTER.

Note that you can select a sub-menu from the MATH menu by pressing either ◀ or ▶. It is easier to press ◀ once than to press ▶ three times to get to the PRB sub-menu.

1.9.2 Permutations and Combinations: To calculate the number of *permutations* of 12 objects taken 7 at a time, $_{12}P_7$, press 12 MATH ◀ 2 7 ENTER. Then $_{12}P_7 = 3,991,680$, as shown in Figure 1.69.

For the number of *combinations* of 12 objects taken 7 at a time, $_{12}C_7$, press 12 MATH ◀ 3 7 ENTER. So $_{12}C_7 = 792$.

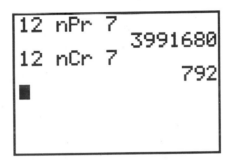

Figure 1.69: $_{12}P_7$ and $_{12}C_7$

1.9.3 Probability of Winning: A state lottery is configured so that each player chooses six different numbers from 1 to 40. If these six numbers match the six numbers drawn by the State Lottery Commission, the player wins the top prize. There are $_{40}C_6$ ways for the six numbers to be drawn. If you purchase a single lottery ticket, your probability of winning is 1 in $_{40}C_6$. Press 1 ÷ 40 MATH ◀ 3 6 ENTER to calculate your chances, but don't be disappointed.

1.10   Programming

1.10.1 Entering a Program: The TI-81 is a programmable calculator that can store sequences of commands for later replay. Here's an example to show you how to enter a useful program that solves quadratic equations by the quadratic formula.

Press PRGM to access the programming menu. The TI-81 has space for up to 37 programs, each named by a number or letter. If a program area is empty, there will be nothing to the right of its name in the PRGM window. You may ERASE a program area to make one clear by pressing ▶ ▶ and then the number or letter of the program. When you see a clear program area, press ▶ or ◀ as many times as necessary to move the cursor to EDIT; then press the key corresponding to its number or letter. For example, to edit program 5, press 5; to edit program B, press ALPHA B.

For convenience, the cursor is now a blinking **A**, indicating that the calculator is set to receive alphabetic characters. Enter a descriptive title of up to eight characters, letters or numerals, and end by pressing ENTER. Let's call this program QUADRAT.

In the program, each line begins with a colon : supplied automatically by the calculator. Any command you could enter directly in the TI-81's home screen can be entered as a line in a program. There are also special programming commands.

Enter the program QUADRAT by pressing the keystrokes given in the listing below.

Program Line	*Keystrokes*
: Disp "ENTER A"	PRGM ▶ 1 2nd A-LOCK " E N T E R ⌴ A " ENTER

displays the words *Enter A* on the TI-81 screen

| : Input A | PRGM ▶ 2 ALPHA A ENTER |

waits for you to input a value that will be assigned to the variable *A*

: Disp "ENTER B"	PRGM ▶ 1 2nd A-LOCK " E N T E R ⌴ B " ENTER
: Input B	PRGM ▶ 2 ALPHA B ENTER
: Disp "ENTER C"	PRGM ▶ 1 2nd A-LOCK " E N T E R ⌴ C " ENTER
: Input C	PRGM ▶ 2 ALPHA C ENTER
: B^2-4AC → D	ALPHA B x^2 - 4 ALPHA A ALPHA C STO▶ D ENTER

calculates the discriminant and stores its value as *D*

| : If D<0 | PRGM 3 ALPHA D 2nd TEST 5 0 ENTER |

tests to see if the discriminant is negative

| : Goto 1 | PRGM 2 1 ENTER |

in case the discriminant is negative, jumps to the line Lbl 1 below;
if the discriminant is not negative, continues on to the next line

| : If D=0 | PRGM 3 ALPHA D 2nd TEST 5 0 ENTER |

tests to see if the discriminant is zero

: Goto 2   PRGM 2 1 ENTER

    in case the discriminant is zero, jumps to the line Lbl 2 below;
    if the discriminant is not zero, continues on to the next line

: Disp "TWO REAL   PRGM ▶ 1 2nd A-LOCK " T W O ␣ R E A L ␣ R O O T S "
    ROOTS"   ENTER

: (-B+√D)/(2A) → M   ((-) ALPHA B + 2nd √ ALPHA D) ÷ (2 ALPHA A) STO▶ M
                                        ENTER

    calculates one root and stores it as M

: Disp M   PRGM ▶ 1 ALPHA M ENTER

    displays one root

: (-B-√D)/(2A) → N   ((-) ALPHA B - 2nd √ ALPHA D) ÷ (2 ALPHA A) STO▶ N
                                          ENTER

: Disp N   PRGM ▶ 1 ALPHA N ENTER

: End   PRGM 7 ENTER

    stops program execution

: Lbl 1   PRGM 1 1 ENTER

    jumping point for the Goto command above

: Disp "COMPLEX   PRGM ▶ 1 2nd A-LOCK " C O M P L E X ␣ R O O T S "
    ROOTS"   ENTER

    displays a message in case the roots are complex numbers

: Disp "REAL PART"   PRGM ▶ 1 2nd A-LOCK " R E A L ␣ P A R T " ENTER

: -B/(2A) → R   (-) ALPHA B ÷ (2 ALPHA A) STO▶ R ENTER

    calculates the real part $\dfrac{-b}{2a}$ of the complex roots

: Disp R   PRGM ▶ 1 ALPHA R ENTER

: Disp "IMAGINARY   PRGM ▶ 1 2nd A-LOCK " I M A G I N A R Y ␣ P A R T "
    PART"   ENTER

| : √-D/(2A) → I | 2nd √ (-) ALPHA D ÷ (2 ALPHA A) STO▶ I ENTER |

calculates the imaginary part $\dfrac{\sqrt{-D}}{2a}$ of the complex roots;

since $D < 0$, we must use $-D$ as the radicand

: Disp I	PRGM ▶ 1 ALPHA I ENTER
: End	PRGM 7 ENTER
: Lbl 2	PRGM 1 2 ENTER
: Disp "DOUBLE ROOT"	PRGM ▶ 1 2nd A-LOCK " D O U B L E ⌴ R O O T " ENTER

displays a message in case there is a double root

| : -B/(2A) → M | (-) ALPHA B ÷ (2 ALPHA A) STO▶ M ENTER |

the quadratic formula reduces to $\dfrac{-b}{2a}$ when $D = 0$

| : Disp M | PRGM ▶ 1 ALPHA M ENTER |
| : End | PRGM 7 |

When you have finished, press 2nd QUIT to leave the program editor.

1.10.2 Executing a Program: To execute the program just entered, press PRGM and then the number or letter that it was named. If you have forgotten its name, use the arrow keys to move through the program listing to find its description QUADRAT. Then press ENTER to select this program and ENTER again to execute it.

The program has been written to prompt you for values of the coefficients a, b, and c in a quadratic equation $ax^2 + bx + c = 0$. Input a value, then press ENTER to continue the program.

If you need to interrupt a program during execution, press ON.

The instruction manual for your TI-81 gives detailed information about programming. Refer to it to learn more about programming and how to use other features of your calculator.

1.11 Differentiation

1.11.1 Limits: Suppose you need to find this limit: $\lim\limits_{x \to 0} \dfrac{\sin 4x}{x}$. Plot the graph of $f(x) = \dfrac{\sin 4x}{x}$ in a conven-
ient viewing rectangle (Figure 1.70) that contains the point where the function appears to intersect the line $x = 0$ (because you want the limit as $x \to 0$). Your graph should support the conclusion that $\lim\limits_{x \to 0} \dfrac{\sin 4x}{x} = 4$.

To test the reasonableness of the conclusion that $\lim\limits_{x \to \infty} \dfrac{2x-1}{x+1} = 2$, evaluate the function $f(x) = \dfrac{2x-1}{x+1}$ for several large positive values of x (since you want the limit as $x \to \infty$). For example, evaluate $f(100)$, $f(1000)$, and $f(10,000)$. Another way to test the reasonableness of this result is to examine the graph of $f(x) = \dfrac{2x-1}{x+1}$ in a viewing rectangle that extends over large values of x. See, as in Figure 1.71 (where the viewing rectangle extends horizontally from −20 to 100), whether the graph is asymptotic to the horizontal line $y = 2$ (enter $\dfrac{2x-1}{x+1}$ for Y_1 and 2 for Y_2).

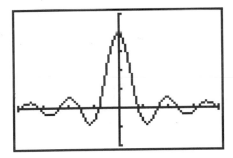

Figure 1.70: Checking $\lim\limits_{x \to 0} \dfrac{\sin 4x}{x} = 4$

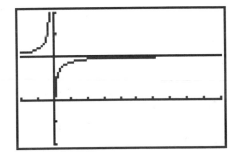

Figure 1.71: Checking $\lim\limits_{x \to \infty} \dfrac{2x-1}{x+1} = 2$

1.11.2 Numerical Derivatives: The derivative of a function f at x can be defined as the limit of the slopes of secant lines, so $f'(x) = \lim\limits_{\Delta x \to 0} \dfrac{f(x + \Delta x) - f(x - \Delta x)}{2\Delta x}$. And for small values of Δx, the expression $\dfrac{f(x + \Delta x) - f(x - \Delta x)}{2\Delta x}$ gives a good approximation to the limit.

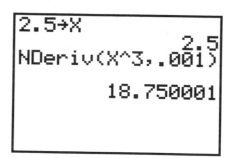

Figure 1.72: Using NDeriv(

The TI-81 has a function, NDeriv(in the MATH menu to calculate the *symmetric difference*, $\dfrac{f(x+\Delta x) - f(x-\Delta x)}{2\Delta x}$. So, to find a numerical approximation to $f'(2.5)$ when $f(x) = x^3$ and with $\Delta x = 0.001$, first save 2.5 for x, then press MATH 8 X|T ∧ 3 ALPHA , .001) ENTER as shown in Figure 1.72.

Technology Tip: It is sometimes helpful to plot both a function and its derivative together. In Figure 1.74, the function $f(x) = \dfrac{5x-2}{x^2+1}$ and its numerical derivative (actually, an approximation to the derivative given by the symmetric difference) are graphed. You can duplicate this graph by first entering $\dfrac{5x-2}{x^2+1}$ for Y_1 and then entering its numerical derivative for Y_2 by pressing MATH 8 2nd Y-VARS 1 ALPHA, .001) as you see in Figure 1.73.

:Y₁ ...	(graph)

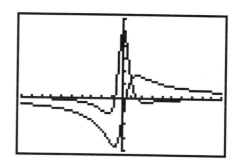

Figure 1.73: Entering $f(x)$ and $f'(x)$ Figure 1.74: Graphs of $f(x)$ and $f'(x)$

Technology Tip: To approximate the *second* derivative $f''(x)$ of a function $y = f(x)$ or to plot the second derivative, first enter the expression for Y_1 and its derivative for Y_2 as above. Then enter the second derivative for Y_3 by pressing MATH 8 2nd Y-VARS 1 2 , X,T,θ , X,T,θ).

1.11.3 Newton's Method: With your TI-81, you may iterate using Newton's method to find the zeros of a function. Recall that Newton's Method determines each successive approximation by the formula

$$x_{n+1} = x_n - \frac{f(x_n)}{f'(x_n)}.$$

As an example of the technique, consider $f(x) = 2x^3 + x^2 - x + 1$. Enter this function as Y_1. A look at its graph suggests that it has a zero near $x = -1$, so start the iteration by going to the home screen and storing -1 as x (see figure 1.75). Then press these keystrokes: X|T – 2nd Y-VARS 1 ÷ MATH 8 2nd Y-VARS 1 ALPHA , .001) STO► X|T. Press ENTER repeatedly until two successive approximations differ by less than some pre-determined value, say 0.0001. Note that each time you press ENTER, the TI-81 will use its *current* value of x, and that value is changing as you continue the iteration.

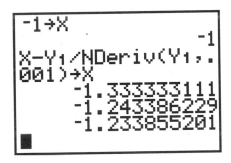

Figure 1.75: Newton's method

Technology Tip: Newton's Method is sensitive to your seed value for x, so look carefully at the function's graph to make a good first estimate. Also, remember that the method sometimes fails to converge!

You may want to save the Newton's Method formula as a short program. See your calculator's manual for further information on programming the TI-81.

1.12   Integration

It is not possible to integrate using the TI-81 without writing a program.

Chapter 2

Texas Instruments TI-82

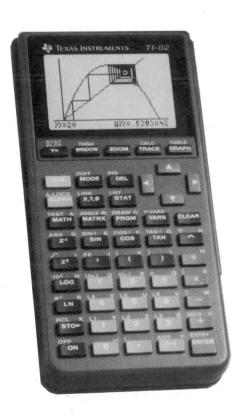

2.1   Getting started with the TI-82

2.1.1 Basics: Press the ON key to begin using your TI-82 calculator. If you need to adjust the display contrast, first press 2nd, then press and hold (the *up* arrow key) to increase the contrast or (the *down* arrow key) to decrease the contrast. As you press and hold or , an integer between 0 (lightest) and 9 (darkest) appears in the upper right corner of the display. When you have finished with the calculator, turn it off to conserve battery power by pressing 2nd and then OFF.

Check the TI-82's settings by pressing MODE. If necessary, use the arrow keys to move the blinking cursor to a setting you want to change. Press ENTER to select a new setting. To start with, select the options along the left side of the MODE menu as illustrated in Figure 2.1: normal display, floating decimals, radian measure, function graphs, connected lines, sequential plotting, and full screen display. Details on alternative options will be given later in this guide. For now, leave the MODE menu by pressing CLEAR.

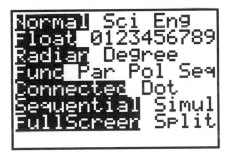

Figure 2.1: MODE menu

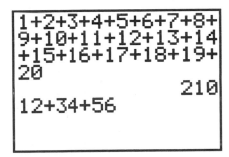

Figure 2.2: Home screen

2.1.2 Editing: One advantage of the TI-82 is that up to 8 lines are visible at one time, so you can *see* a long calculation. For example, type this sum (see Figure 2.2):

$$1 + 2 + 3 + 4 + 5 + 6 + 7 + 8 + 9 + 10 + 11 + 12 + 13 + 14 + 15 + 16 + 17 + 18 + 19 + 20$$

Then press ENTER to see the answer, too.

Often we do not notice a mistake until we see how unreasonable an answer is. The TI-82 permits you to re-display an entire calculation, edit it easily, then execute the *corrected* calculation.

Suppose you had typed 12 + 34 + 56 as in Figure 2.2 but had *not* yet pressed ENTER, when you realize that 34 should have been 74. Simply press ◀ (the *left* arrow key) as many times as necessary to move the blinking cursor left to 3, then type 7 to write over it. On the other hand, if 34 should have been 384, move the cursor back to 4, press 2nd INS (the cursor changes to a blinking underline) and then type 8 (inserts at the cursor position and other characters are pushed to the right). If the 34 should have been 3 only, move the cursor to 4 and press DEL to delete it.

Even if you had pressed ENTER, you may still edit the previous expression. Press 2nd and then ENTRY to *recall* the last expression that was entered. Now you can change it. In fact, the TI-82 retains many prior entries in a "last entry" storage area. Press 2nd ENTRY repeatedly until the previous line you want replaces the current line.

Technology Tip: When you need to evaluate a formula for different values of a variable, use the editing feature to simplify the process. For example, suppose you want to find the balance in an investment account if there is now \$5000 in the account and interest is compounded annually at the rate of 8.5%. The formula for the balance is $P\left(1+\frac{r}{n}\right)^{nt}$, where P = principal, r = rate of interest (expressed as a decimal), n = number of times interest is compounded each year, and t = number of years. In our example, this becomes $5000(1+.085)^t$. Here are the keystrokes for finding the balance after $t = 3, 5,$ and 10 years.

Years	Keystrokes	Balance
3	5000 (1 + .085) ^ 3 ENTER	\$6386.45
5	2nd ENTRY ◀ 5 ENTER	\$7518.28
10	2nd ENTRY ◀ 10 ENTER	\$11,304.92

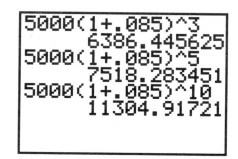

Figure 2.3: Editing expressions

Then to find the balance from the same initial investment but after 5 years when the annual interest rate is 7.5%, press these keys to change the last calculation above: 2nd ENTRY ◀ DEL ◀ 5 ◀ ◀ ◀ ◀ ◀ 7 ENTER.

2.1.3 Key Functions: Most keys on the TI-82 offer access to more than one function, just as the keys on a computer keyboard can produce more than one letter ("g" and "G") or even quite different characters ("5" and "%"). The primary function of a key is indicated on the key itself, and you access that function by a simple press on the key.

To access the *second* function indicated to the *left* above a key, first press 2nd (the cursor changes to a blinking ↑) and *then* press the key. For example, to calculate $\sqrt{25}$, press 2nd √ 25 ENTER.

When you want to use a letter or other character printed to the *right* above a key, first press ALPHA (the cursor changes to a blinking **A**) and then the key. For example, to use the letter K in a formula, press ALPHA K. If you need several letters in a row, press 2nd A-LOCK, which is like CAPS LOCK on a computer keyboard, and then press all the letters you want. Remember to press ALPHA when you are finished and want to restore the keys to their primary functions.

2.1.4 Order of Operations: The TI-82 performs calculations according to the standard algebraic rules. Working outwards from inner parentheses, calculations are performed from left to right. Powers and roots are evaluated first, followed by multiplications and divisions, and then additions and subtractions.

Note that the TI-82 distinguishes between *subtraction* and the *negative sign*. If you wish to enter a negative number, it is necessary to use the (-) key. For example, you would evaluate $-5 - (4 \cdot -3)$ by pressing (-) 5 − (4 × (-) 3) ENTER to get 7.

Enter these expressions to practice using your TI-82.

Expression	Keystrokes	Display
$7 - 5 \cdot 3$	7 − 5 × 3 ENTER	-8
$(7-5) \cdot 3$	(7 − 5) × 3 ENTER	6
$20 - 10^2$	120 − 10 x² ENTER	20
$(120-10)^2$	(120 − 10) x² ENTER	12100
$\dfrac{24}{2^3}$	24 ÷ 2 ^ 3 ENTER	3
$\left(\dfrac{24}{2}\right)^3$	(24 ÷ 2) ^ 3 ENTER	1728
$(7--5) \cdot -3$	(7 − (-) 5) × (-) 3 ENTER	-36

2.1.5 Algebraic Expressions and Memory: Your calculator can evaluate expressions such as $\dfrac{N(N+1)}{2}$ *after* you have entered a value for *N*. Suppose you want $N = 200$. Press 200 STO▶ ALPHA N ENTER to store the value 200 in memory location *N*. Whenever you use *N* in an expression, the calculator will substitute the value 200 until you make a change by storing *another* number in *N*. Next enter the expression $\dfrac{N(N+1)}{2}$ by typing ALPHA N (ALPHA N + 1) ÷ 2 ENTER. For $N = 200$, you will find that $\dfrac{N(N+1)}{2} = 20100$.

The contents of any memory location may be revealed by typing just its letter name and then ENTER. And the TI-82 retains memorized values even when it is turned off, so long as its batteries are good.

2.1.6 Repeated Operations with ANS: The result of your *last* calculation is always stored in memory location ANS and replaces any previous result. This makes it easy to use the answer from one computation in another computation. For example, press 30 + 15 ENTER so that 45 is the last result displayed. Then press 2nd ANS ÷ 9 ENTER and get 5 because $\frac{45}{9} = 5$.

With a function like division, you press the ÷ key *after* you enter an argument. For such functions, whenever you would start a new calculation with the previous answer followed by pressing the function key, you may press just the function key. So instead of 2nd ANS ÷ 9 in the previous example, you could have pressed simply ÷ 9 to achieve the same result. This technique also works for these functions: + − × x² ^ x⁻¹.

Here is a situation where this is especially useful. Suppose a person makes $5.85 per hour and you are asked to calculate earnings for a day, a week, and a year. Execute the given keystrokes to find the person's incomes during these periods (results are shown in Figure 2.4):

Pay period	Keystrokes	Earnings
8-hour day	5.85 × 8 ENTER	$46.80
5-day week	× 5 ENTER	$234
52-week year	× 52 ENTER	$12,168

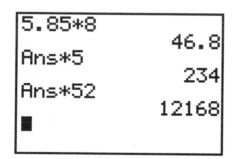

Figure 2.4: ANS variable

2.1.7 The MATH Menu: Operators and functions associated with a scientific calculator are available either immediately from the keys of the TI-82 or by 2nd keys. You have direct key access to common arithmetic operations (x², 2nd √ , x⁻¹, ^, 2nd ABS), trigonometric functions (SIN, COS, TAN) and their inverses (2nd SIN⁻¹, 2nd COS⁻¹, 2nd TAN⁻¹), exponential and logarithmic functions (LOG, 2nd 10ˣ, LN, 2nd eˣ), and a famous constant (2nd π).

A significant difference between the TI-82 and many scientific calculators is that the TI-82 requires the argument of a function *after* the function, as you would see a formula written in your textbook. For example, on the TI-82 you calculate $\sqrt{16}$ by pressing the keys 2nd √ 16 in that order.

Here are keystrokes for basic mathematical operations. Try them for practice on your TI-82.

Expression	Keystrokes	Display
$\sqrt{3^2+4^2}$	2nd √ (3 x² + 4 x²) ENTER	5
$2\frac{1}{3}$	2 + 3 x⁻¹ ENTER	2.333333333
$|-5|$	2nd ABS (-) 5 ENTER	5
$\log 200$	LOG 200 ENTER	2.301029996
$2.34 \cdot 10^5$	2.34 × 2nd 10ˣ 5 ENTER	234000

Additional mathematical operations and functions are available from the MATH menu (Figure 2.5). Press MATH to see the various options. You will learn in your mathematics textbook how to apply many of them. As an example, calculate $\sqrt[3]{7}$ by pressing MATH and then *either* 4 *or* ▼ ▼ ▼ ENTER; finally press 7 ENTER to see 1.912931183. To leave the MATH menu and take no other action, press 2nd QUIT or just CLEAR.

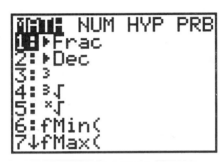

Figure 2.5: MATH menu

The *factorial* of a non-negative integer is the *product* of *all* the integers from 1 up to the given integer. The symbol for factorial is the exclamation point. So 4! (pronounced *four factorial*) is $1 \cdot 2 \cdot 3 \cdot 4 = 24$. You will learn more about applications of factorials in your textbook, but for now use the TI-82 to calculate 4! The factorial command is located in the MATH menu's PRB sub-menu. To compute 4!, press these keystrokes: 4 MATH ◀ 4 ENTER *or* 4 MATH ◀ ▼ ▼ ▼ ENTER ENTER.

Note that you can select a sub-menu from the MATH menu by pressing either ◀ or ▶. It is easier to press ◀ once than to press ▶ three times to get to the PRB sub-menu.

2.2   Functions and Graphs

2.2.1 Evaluating Functions: Suppose you receive a monthly salary of $1975 plus a commission of 10% of sales. Let x = your sales in dollars; then your wages W in dollars are given by the equation $W = 1975 + .10x$.

If your January sales were \$2230 and your February sales were \$1865, what was your income during those months?

Here's how to use your TI-82 to perform this task. Press the Y= key at the top of the calculator to display the function editing screen (Figure 2.6). You may enter as many as ten different functions for the TI-82 to use at one time. If there is already a function Y_1, press 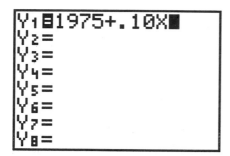 or ▼ as many times as necessary to move the cursor to Y_1 and then press CLEAR to delete whatever was there. Then enter the expression $1975 + .10x$ by pressing these keys: 1975 + .10 X,T,θ. (The X,T,θ key lets you enter the variable X easily without having to use the ALPHA key.) Now press 2nd QUIT to return to the main calculations screen.

Figure 2.6: Y= screen

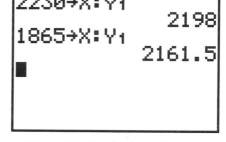

Figure 2.7: Evaluating a function

Assign the value 2230 to the variable x by these keystrokes (see Figure 2.7): 2230 STO ▸ X,T,θ. Then press 2nd : to allow another expression to be entered on the same command line. Next press the following keystrokes to evaluate Y_1 and find January's wages: 2nd Y-VARS 1 1 ENTER.

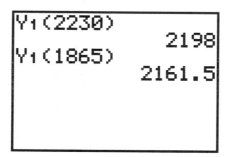

Figure 2.8: Function notation

It is not necessary to repeat all these steps to find the February wages. Simply press 2nd ENTRY to recall the entire previous line and change 2230 to 1865. Each time the TI-82 evaluates the function Y_1, it uses the *current* value of x.

Like your textbook, the TI-82 uses standard function notation. So to evaluate $Y_1(2230)$ when $Y_1(x) = 1975 + .10x$, press 2nd Y-VARS 1 1 (2230) ENTER (see Figure 2.8). Then to evaluate $Y_1(1865)$, press 2nd ENTRY to recall the last line and change 2230 to 1865.

You may also have the TI-82 make a table of values for the function. Press 2nd TblSet to set up the table (Figure 2.9). Move the blinking cursor onto Ask beside Indpnt:, then press ENTER. This configuration permits you to input values for x one at a time. Now press 2nd TABLE, enter 2230 in the x column, and press ENTER (see Figure 2.10). Continue to enter additional values for x and the calculator automatically completes the table with corresponding values of Y_1. Press 2nd QUIT to leave the TABLE screen.

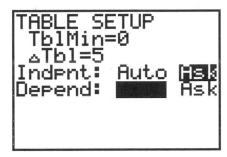

Figure 2.9: TblSet screen

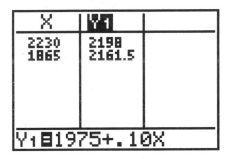

Figure 2.10: Table of values

For a table containing values for x = 1, 2, 3, 4, 5, and so on, set TblMin = 1 to start at x = 1, ΔTbl = 1 to increment in steps of 1, and both Indpnt and Depend to Auto.

Technology Tip: The TI-82 does not require multiplication to be expressed between variables, so xxx means x^3. It is often easier to press two or three x's together than to search for the square key or the cube operation. Of course, expressed multiplication is also not required between a constant and a variable. Hence to enter $2x^3 + 3x^2 - 4x + 5$ in the TI-82, you might save keystrokes and press just these keys: 2 X,T,θ X,T,θ X,T,θ + 3 X,T,θ X,T,θ - 4 X,T,θ + 5.

2.2.2 Functions in a Graph Window: Once you have entered a function in the Y= screen of the TI-82, just press GRAPH to see its graph. The ability to draw a graph contributes substantially to our ability to solve problems.

For example, here is how to graph $y = -x^3 + 4x$. First press Y= and delete anything that may be there by moving with the arrow keys to Y_1 or to any of the other lines and pressing CLEAR wherever necessary. Then, with the cursor on the top line Y_1, press (-) X,T,θ MATH 3 + 4 X,T,θ to enter the function (as in Figure 2.11). Now press GRAPH and the TI-82 changes to a window with the graph of $y = -x^3 + 4x$.

While the TI-82 is calculating coordinates for a plot, it displays a busy indicator at the top right of the graph window.

Technology Tip: If you would like to see a function in the Y= menu and its graph in a graph window, both at the same time, open the MODE menu, move the cursor down to the last line, and select Split screen. Your TI-82's screen is now divided horizontally (see Figure 2.11), with an upper graph window and a lower window that can display the home screen or an editing screen. The split screen is also useful when you need to do some calculations as you trace along a graph. For now, restore the TI-82 to FullScreen.

Your graph window may look like the one in Figure 2.12 or it may be different. Since the graph of $y = -x^3 + 4x$ extends infinitely far left and right and also infinitely far up and down, the TI-82 can display only a piece of the actual graph. This displayed rectangular part is called a *viewing rectangle*. You can easily change the viewing rectangle to enhance your investigation of a graph.

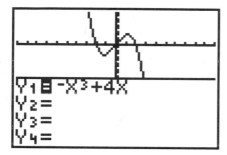

Figure 2.11: Split screen: Y= below

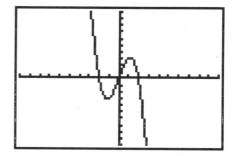

Figure 2.12: Graph of $y = -x^3 + 4x$

The viewing rectangle in Figure 2.12 shows the part of the graph that extends horizontally from -10 to 10 and vertically from -10 to 10. Press WINDOW to see information about your viewing rectangle. Figure 2.13 shows the WINDOW screen that corresponds to the viewing rectangle in Figure 2.12. This is the *standard* viewing rectangle for the TI-82.

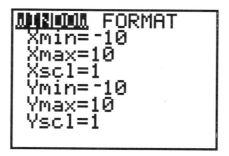

Figure 2.13: Standard WINDOW

The variables Xmin and Xmax are the minimum and maximum *x*-values of the viewing rectangle; Ymin and Ymax are its minimum and maximum *y*-values.

Xscl and Yscl set the spacing between tick marks on the axes.

Use the arrow keys and to move up and down from one line to another in this list; pressing the ENTER key will move down the list. Press CLEAR to delete the current value and then enter a new value. You may also edit the entry as you would edit an expression. Remember that a minimum *must* be less than the corresponding maximum or the TI-82 will issue an error message. Also, remember to use the (-) key, not − (which is subtraction), when you want to enter a negative value. Figures 2.12-13, 2.14-15, and 2.16-17 show different WINDOW screens and the corresponding viewing rectangle for each one.

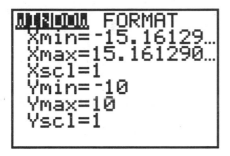

Figure 2.14: Square window

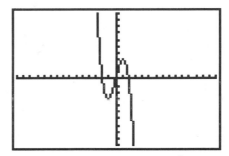

Figure 2.15: Graph of $y = -x^3 + 4x$

To set the range quickly to standard values (see Figure 2.13), press ZOOM 6. To set the viewing rectangle quickly to a square (Figure 2.14), press ZOOM 5. More information about square windows is presented later in Section 2.2.4.

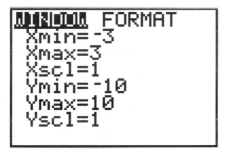

Figure 2.16: Custom window

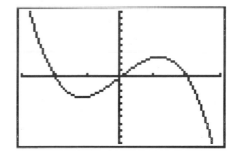

Figure 2.17: Graph of $y = -x^3 + 4x$

Sometimes you may wish to display grid points corresponding to tick marks on the axes. This and other graph format options may be changed by pressing WINDOW ▶ to display the FORMAT menu (Figure 2.18). Use arrow keys to move the blinking cursor to GridOn; press ENTER and then GRAPH to redraw the graph. Figure 2.19 shows the same graph as in Figure 2.17 but with the grid turned on. In general, you'll want the grid turned *off*, so do that now by pressing WINDOW ▶, use the arrow keys to move the blinking cursor to GridOff, and press ENTER and CLEAR.

Figure 2.18: WINDOW FORMAT menu

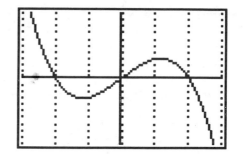

Figure 2.19: Grid turned on for $y = -x^3 + 4x$

2.2.3 Piecewise-Defined Functions: The greatest integer function, written [[x]], gives the greatest *integer* less than or equal to a number x. On the TI-82, the greatest integer function is called Int and is located under the NUM sub-menu of the MATH menu (see Figure 2.5). So calculate [[6.78]] = 6 by pressing MATH ▶ 4 6.78 ENTER.

To graph $y = [[x]]$, go in the Y= menu, move beside Y_1 and press CLEAR MATH ▶ 4 X,T,θ GRAPH. Figure 2.20 shows this graph in a viewing rectangle from -5 to 5 in both directions.

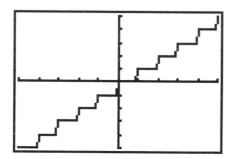

Figure 2.20: Connected graph of $y = [[x]]$

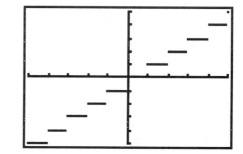

Figure 2.21: Dot graph of $y = [[x]]$

The true graph of the greatest integer function is a step graph, like the one in Figure 2.21. For the graph of $y = [[x]]$, a segment should *not* be drawn between every pair of successive points. You can change from Connected line to Dot graph on the TI-82 by opening the MODE menu. Move the cursor down to the fifth line; select whichever graph type you require; press ENTER to put it into effect and GRAPH to see the result.

You should also change to Dot graph when plotting a piecewise-defined function. For example, to plot the graph of $f(x) = \begin{cases} x^2 + 1, & x < 0 \\ x - 1, & x \geq 0 \end{cases}$, enter the expression $(x^2 + 1)(x < 0) + (x - 1)(x \geq 0)$ somewhere in your Y=

list by pressing (X,T,θ x² + 1) (X,T,θ 2nd TEST 5 0) + (X,T,θ − 1) (X,T,θ 2nd TEST 4 0). Then change the mode to Dot graph and press GRAPH.

2.2.4 Graphing a Circle: Here is a useful technique for graphs that are not functions, but that can be "split" into a top part and a bottom part, or into multiple parts. Suppose you wish to graph the circle whose equation is $x^2 + y^2 = 36$. First solve for y and get an equation for the top semicircle, $y = \sqrt{36 - x^2}$, and for the bottom semicircle, $y = -\sqrt{36 - x^2}$. Then graph the two semicircles simultaneously.

The keystrokes to draw this circle's graph follow. Enter $\sqrt{36 - x^2}$ as Y_1 and $-\sqrt{36 - x^2}$ as Y_2 (see Figure 2.22) by pressing Y= CLEAR 2nd √ (36 - X,T,θ x²) ENTER CLEAR (-) 2nd √ (36 - X,T,θ x²). Then press GRAPH to draw them both.

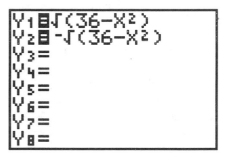
Figure 2.22: Two semicircles

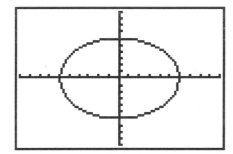
Figure 2.23: Circle's graph - standard view

If your range were set to the standard viewing rectangle, your graph would look like Figure 2.23. Now this does *not* look like a circle, because the units along the axes are not the same. This is where the square viewing rectangle is important. Press ZOOM 5 and see a graph that appears more circular.

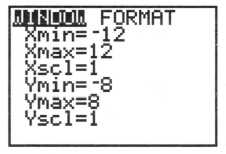
Figure 2.24: $\frac{\text{vertical}}{\text{horizontal}} = \frac{16}{24} = \frac{2}{3}$

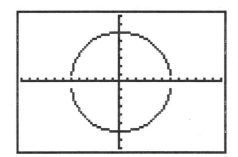
Figure 2.25: A "square" circle

Technology Tip: Another way to get a square graph is to change the range variables so that the value of Ymax - Ymin is approximately $\frac{2}{3}$ times Xmax - Xmin. For example, see the WINDOW in Figure 2.24 and the corresponding graph in Figure 2.25. The method works because the dimensions of the TI-82's display are such that the ratio of vertical to horizontal is approximately $\frac{2}{3}$.

The two semicircles in Figure 2.25 do not meet because of an idiosyncrasy in the way the TI-82 plots a graph.

Back when you entered $\sqrt{36-x^2}$ as Y_1 and $-\sqrt{36-x^2}$ as Y_2, you could have entered -Y_1 as Y_2 and saved some keystrokes. Try this by going back to the $Y=$ menu and pressing the arrow key to move the cursor down to Y_2. Then press CLEAR (-) 2nd Y-VARS 1 1. The graph should be just as it was before.

2.2.5 TRACE:

Graph $y = -x^3 + 4x$ in the standard viewing rectangle. Press any of the arrow keys ▲ ▼ ◀ ▶ and see the cursor move from the center of the viewing rectangle. The coordinates of the cursor's location are displayed at the bottom of the screen, as in Figure 2.26, in floating decimal format. This cursor is called a *free-moving cursor* because it can move from dot to dot *anywhere* in the graph window.

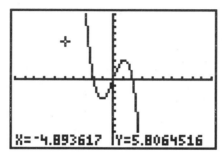

Figure 2.26: Free-moving cursor

Remove the free-moving cursor and its coordinates from the window by pressing GRAPH, CLEAR, or ENTER. Press an arrow key again and the free-moving cursor will reappear at the same point you left it.

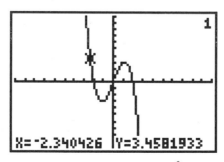

Figure 2.27: Trace on $y = -x^3 + 4x$

Press TRACE to enable the left ◀ and right ▶ arrow keys to move the cursor along the function. The cursor is no longer free-moving, but is now constrained to the function. The coordinates that are displayed belong to points on the function's graph, so the y-coordinate is the calculated value of the function at the corresponding x-coordinate.

Now plot a second function, $y = -.25x$, along with $y = -x^3 + 4x$. Press Y= and enter $-.25x$ for Y_2, then press GRAPH.

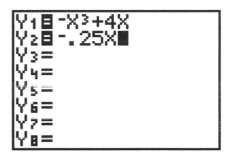

Figure 2.28: Two functions

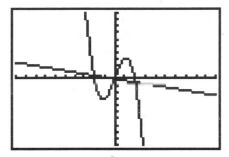

Figure 2.29: $y = -x^3 + 4x$ and $y = -.25x$

Note in Figure 2.28 that the equal signs next to Y_1 and Y_2 are *both* highlighted. This means *both* functions will be graphed. In the Y= screen, move the cursor directly on top of the equal sign next to Y_1 and press ENTER. This equal sign should no longer be highlighted (see Figure 2.30). Now press GRAPH and see that only Y_2 is plotted (Figure 2.31).

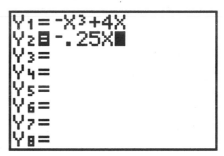

Figure 2.30: Y= screen with only Y_2 active

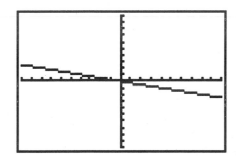

Figure 2.31: Graph of $y = -.25x$

Many different functions may be stored in the Y= list and any combination of them may be graphed simultaneously. You can make a function active or inactive for graphing by pressing ENTER on its equal sign to highlight (activate) or remove the highlight (deactivate). Go back to the Y= screen and do what is needed in order to graph Y_1 but not Y_2.

Now activate Y_2 again so that both graphs are plotted. Press **TRACE** and the cursor appears first on the graph of $y = -x^3 + 4x$ because it is higher up in the **Y=** list. You know that the cursor is on this function, Y_1, because of the numeral 1 that is displayed in the upper right corner of the window (see Figure 2.27). Press the up ▲ or down ▼ arrow key to move the cursor vertically to the graph of $y = -.25x$. Now the numeral 2 is displayed in the top right corner of the window. Next press the right and left arrow keys to trace along the graph of $y = -.25x$. When more than one function is plotted, you can move the trace cursor vertically from one graph to another in this way.

Technology Tip: By the way, trace along the graph of $y = -.25x$ and press and hold either ◀ or ▶. Eventually you will reach the left or right edge of the window. Keep pressing the arrow key and the TI-82 will allow you to continue the trace by panning the viewing rectangle. Check the **WINDOW** screen to see that **Xmin** and **Xmax** are automatically updated.

The TI-82's display has 95 horizontal columns of pixels and 63 vertical rows. So when you trace a curve across a graph window, you are actually moving from **Xmin** to **Xmax** in 94 equal jumps, each called Δx. You would calculate the size of each jump to be $\Delta x = \dfrac{\text{Xmax} - \text{Xmin}}{94}$. Sometimes you may want the jumps to be friendly numbers like .1 or .25 so that, when you trace along the curve, the x-coordinates will be incremented by such a convenient amount. Just set your viewing rectangle for a particular increment Δx by making **Xmax** = **Xmin** + 94·Δx. For example, if you want **Xmin** = -5 and Δx = .3, set **Xmax** = -5 + 94·.3 = 23.2. Likewise, set **Ymax** = **Ymin** + 62·Δy if you want the vertical increment to be some special Δy.

To center your window around a particular point, say (h, k), and also have a certain Δx, set **Xmin** = h - 47·Δx and **Xmax** = h + 47·Δx. Likewise, make **Ymin** = k - 31·Δy and **Ymax** = k + 31·Δy. For example, to center a window around the origin, (0, 0), with both horizontal and vertical increments of .25, set the range so that **Xmin** = 0 - 47·.25 = -11.75, **Xmax** = 0 + 47·.25 = 11.75, **Ymin** = 0 - 31·.25 = -7.75, and **Ymax** = 0 + 31·.25 = 7.75.

See the benefit by first plotting $y = x^2 + 2x + 1$ in a standard graphing window. Trace near its y-intercept, which is (0, 1), and move towards its x-intercept, which is (-1, 0). Then press **ZOOM 4** and trace again near the intercepts.

2.2.6 ZOOM: Plot again the two graphs, for $y = -x^3 + 4x$ and for $y = -.25x$. There appears to be an intersection near $x = 2$. The TI-82 provides several ways to enlarge the view around this point. You can change the viewing rectangle directly by pressing **WINDOW** and editing the values of **Xmin**, **Xmax**, **Ymin**, and **Ymax**. Figure 2.33 shows a new viewing rectangle for the range displayed in Figure 2.32. The cursor has been moved near the point of intersection; move your cursor closer to get the best approximation possible for the coordinates of the intersection.

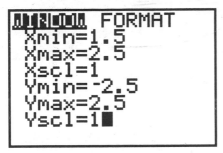

Figure 2.32: New WINDOW

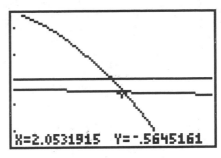

Figure 2.33: Closer view

A more efficient method for enlarging the view is to draw a new viewing rectangle with the cursor. Start again with a graph of the two functions $y = -x^3 + 4x$ and $y = -.25x$ in a standard viewing rectangle (press ZOOM 6 for the standard window, from -10 to 10 along both axes).

Now imagine a small rectangular box around the intersection point, near $x = 2$. Press ZOOM 1 (Figure 2.34) to draw a box to define this new viewing rectangle. Use the arrow keys to move the cursor, whose coordinates are displayed at the bottom of the window, to one corner of the new viewing rectangle you imagine.

Figure 2.34: ZOOM menu

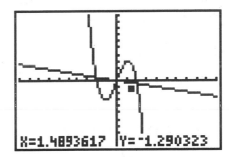

Figure 2.35: One corner selected

Press ENTER to fix the corner where you have moved the cursor; it changes shape and becomes a blinking square (Figure 2.35). Use the arrow keys again to move the cursor to the diagonally opposite corner of the new rectangle (Figure 2.36). If this box looks all right to you, press ENTER. The rectangular area you have enclosed will now enlarge to fill the graph window (Figure 2.37).

You may interrupt the zoom any time *before* you press this last ENTER. Press ZOOM once more and start over. Press CLEAR or GRAPH to cancel the zoom, or press 2nd QUIT to cancel the zoom and return to the home screen.

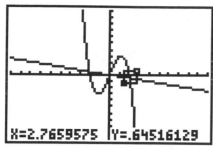

Figure 2.36: Box drawn

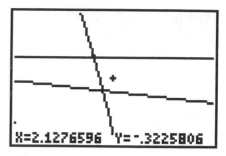

Figure 2.37: New viewing rectangle

You can also gain a quick magnification of the graph around the cursor's location. Return once more to the standard window for the graph of the two functions $y = -x^3 + 4x$ and $y = -.25x$. Press ZOOM 2 and then press arrow keys to move the cursor as close as you can to the point of intersection near $x = 2$ (see Figure 2.38). Then press ENTER and the calculator draws a magnified graph, centered at the cursor's position (Figure 2.39). The range variables are changed to reflect this new viewing rectangle. Look in the WINDOW menu to verify this.

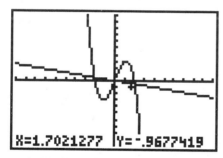

Figure 2.38: Before a zoom in

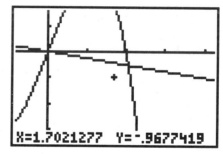

Figure 2.39: After a zoom in

As you see in the ZOOM menu (Figure 2.34), the TI-82 can Zoom In (press ZOOM 2) or Zoom Out (press ZOOM 3). Zoom out to see a larger view of the graph, centered at the cursor position. You can change the horizontal and vertical scale of the magnification by pressing ZOOM ▶ 4 (see Figure 2.41) and editing XFact and YFact, the horizontal and vertical magnification factors.

The default zoom factor is 4 in both directions. It is not necessary for XFact and YFact to be equal. Sometimes, you may prefer to zoom in one direction only, so the other factor should be set to 1. As usual, press 2nd QUIT to leave the ZOOM menu.

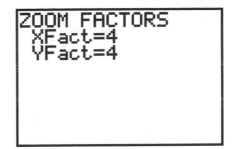

Figure 2.40: ZOOM MEMORY menu   Figure 2.41: ZOOM MEMORY SetFactors...

Technology Tip: The TI-82 remembers the window it displayed before a zoom. So if you should zoom in too much and lose the curve, press ZOOM ▶ 1 to go back to the window before. If you want to execute a series of zooms but then return to a particular window, press ZOOM ▶ 2 to store the current window's dimensions. Later, press ZOOM ▶ 3 to recall the stored window.

2.2.7 Relative Minimums and Maximums: Graph $y = -x^3 + 4x$ once again in the standard viewing rectangle (Figure 2.12). This function appears to have a relative minimum near $x = -1$ and a relative maximum near $x = 1$. You may zoom and trace to approximate these extreme values.

First trace along the curve near the local minimum. Notice by how much the x-values and y-values change as you move from point to point. Trace along the curve until the y coordinate is as *small* as you can get it, so that you are as close as possible to the local minimum, and zoom in (press ZOOM 2 or use a zoom box). Now trace again along the curve and, as you move from point to point, see that the coordinates change by smaller amounts than before. Keep zooming and tracing until you find the coordinates of the local minimum point as accurately as you need them, approximately (-1.15, -3.08).

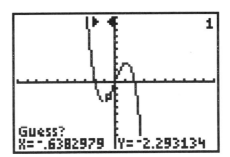

Figure 2.42: CALCULATE menu   Figure 2.43: Finding a minimum

Follow a similar procedure to find the relative maximum. Trace along the curve until the y-coordinate is as *great* as you can get it, so that you are as close as possible to the relative maximum, and zoom in. The local maximum point on the graph of $y = -x^3 + 4x$ is approximately (1.15, 3.08).

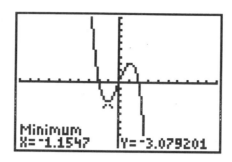

Figure 2.44: Relative minimum on $y = -x^3 + 4x$

The TI-82 automates the search for relative minimum and relative maximum points. Press **2nd CALC** to display the CALCULATE menu (Figure 2.42). Choose **3** to calculate the minimum value of the function and **4** for the maximum. You will be prompted to trace the cursor along the graph first to a point *left* of the minimum/maximum (press **ENTER** to set this *lower bound*). Then move to a point *right* of the minimum/maximum and set an *upper bound* (as in Figure 2.43) and press **ENTER**. Note the two arrows marking the lower and upper bounds at the top of the display.

Next move the cursor along the graph between the two bounds and as close to the minimum/maximum as you can; this serves as a *guess* for the TI-82 to start its search. Good choices for the lower bound, upper bound, and guess can help the calculator work more efficiently and quickly. Press **ENTER** and the coordinates of the relative minimum/maximum point will be displayed (see Figure 2.44).

2.3   Solving Equations and Inequalities

2.3.1 Intercepts and Intersections: Tracing and zooming are also used to locate an x-intercept of a graph, where a curve crosses the x-axis. For example, the graph of $y = x^3 - 8x$ crosses the x-axis three times (see Figure 2.45). After tracing over to the x-intercept point that is furthest to the left, zoom in (Figure 2.46). Continue this process until you have located all three intercepts with as much accuracy as you need. The three x-intercepts of $y = x^3 - 8x$ are approximately -2.828, 0, and 2.828.

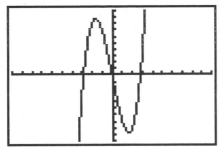

Figure 2.45: Graph of $y = x^3 - 8x$

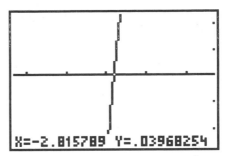

Figure 2.46: An x-intercept of $y = x^3 - 8x$

Technology Tip: As you zoom in, you may also wish to change the spacing between tick marks on the x-axis so that the viewing rectangle shows scale marks near the intercept point. Then the accuracy of your approximation will be such that the error is less than the distance between two tick marks. Change the x-scale on the TI-82 from the **WINDOW** menu. Move the cursor down to **Xscl** and enter an appropriate value.

The x-intercept of a function's graph is a *root* of the equation $f(x) = 0$. So press **2nd CALC** to display the **CALCULATE** menu (Figure 2.42) and choose **2** to find a root of this function. Set a lower bound, upper bound, and guess as described above in Section 2.2.7.

TRACE and **ZOOM** are especially important for locating the intersection points of two graphs, say the graphs of $y = -x^3 + 4x$ and $y = -.25x$. Trace along one of the graphs until you arrive close to an intersection point. Then press ▲ or ▼ to jump to the other graph. Notice that the x-coordinate does not change, but the y-coordinate is likely to be different (see Figures 2.47 and 2.48).

When the two y-coordinates are as close as they can get, you have come as close as you now can to the point of intersection. So zoom in around the intersection point, then trace again until the two y-coordinates are as close as possible. Continue this process until you have located the point of intersection with as much accuracy as necessary.

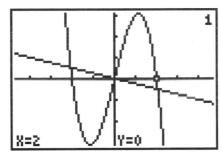

Figure 2.47: Trace on $y = -x^3 + 4x$

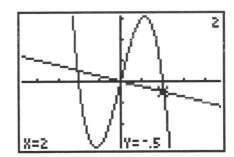

Figure 2.48: Trace on $y = -.25x$

You can also find the point of intersection of two graphs by pressing 2nd CALC 5. Trace with the cursor first along one graph near the intersection and press ENTER; then trace with the cursor along the other graph and press ENTER. Marks + are placed on the graphs at these points. Finally, move the cursor near the point of intersection and press ENTER again. Coordinates of the intersection will be displayed at the bottom of the window.

2.3.2 Solving Equations by Graphing: Suppose you need to solve the equation $24x^3 - 36x + 17 = 0$. First graph $y = 24x^3 - 36x + 17$ in a window large enough to exhibit *all* its x-intercepts, corresponding to all its roots. Then use trace and zoom, or the TI-82's root finder, to locate each one. In fact, this equation has just one solution, approximately $x = -1.414$.

Remember that when an equation has more than one x-intercept, it may be necessary to change the viewing rectangle a few times to locate all of them.

Technology Tip: To solve an equation like $24x^3 + 17 = 36x$, you may first transform it into standard form, $24x^3 - 36x + 17 = 0$, and proceed as above. However, you may also graph the *two* functions $y = 24x^3 + 17$ and $y = 36x$, then zoom and trace to locate their point of intersection.

2.3.3 Solving Systems by Graphing: The solutions to a system of equations correspond to the points of intersection of their graphs (Figure 2.49). For example, to solve the system $y = x^2 - 3x - 4$ and $y = x^3 + 3x^2 - 2x - 1$, first graph them together. Then zoom and trace, or use the intersect option in the CALC menu, to locate their point of intersection, approximately (-2.17, 7.25).

You must judge whether the two current y-coordinates are sufficiently close for $x = -2.17$ or whether you should continue to zoom and trace to improve the approximation.

The solutions of the system of two equations $y = x^3 + 3x^2 - 2x - 1$ and $y = x^2 - 3x - 4$ correspond to the solutions of the single equation $x^3 + 3x^2 - 2x - 1 = x^2 - 3x - 4$, which simplifies to $x^3 + 2x^2 + x + 3 = 0$. So you may also graph $y = x^3 + 2x^2 + x + 3$ and find its x-intercepts to solve the system.

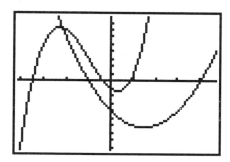

Figure 2.49: Solving a system of equations

2.3.4 Solving Inequalities by Graphing: Consider the inequality $1 - \dfrac{3x}{2} \geq x - 4$. To solve it with your TI-82, graph the two functions $y = 1 - \dfrac{3x}{2}$ and $y = x - 4$ (Figure 2.50). First locate their point of intersection, at $x =$ 2. The inequality is true when the graph of $y = 1 - \dfrac{3x}{2}$ lies *above* the graph of $y = x - 4$, and that occurs for x < 2. So the solution is the half-line $x \leq 2$, or $(-\infty, 2]$.

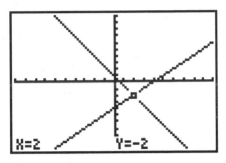

Figure 2.50: Solving $1 - \dfrac{3x}{2} \geq x - 4$

The TI-82 is capable of shading the region above or below a graph or between two graphs. For example, to graph $y \geq x^2 - 1$, first graph the function $y = x^2 - 1$ as Y_1. Then press 2nd DRAW 7 2nd Y-VARS 1 1 , 10 , 2) ENTER (see Figure 2.51). These keystrokes instruct the TI-82 to shade the region *above* $y = x^2 - 1$ and *below* $y = 10$ (chosen because this is the greatest y-value in the graph window) with shading resolution value of 2. The result is shown in Figure 2.52.

To clear the shading, press 2nd DRAW 1.

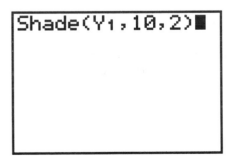

Figure 2.51: DRAW Shade

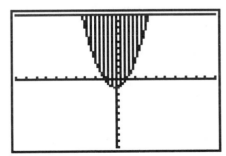

Figure 2.52: Graph of $y \geq x^2 - 1$

Now use shading to solve the previous inequality, $1 - \dfrac{3x}{2} \geq x - 4$. The function whose graph forms the lower boundary is named *first* in the SHADE command (see Figure 2.53). To enter this in your TI-82, press these keys: 2nd DRAW 7 X,T,θ - 4 , 1 - 3 X,T,θ ÷ 2 , 2) ENTER (Figure 2.54). The shading extends left from $x = 2$, hence the solution to $1 - \dfrac{3x}{2} \geq x - 4$ is the half-line $x \leq 2$, or $(-\infty, 2]$.

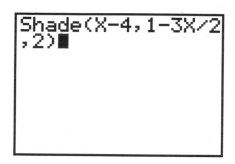

Figure 2.53: DRAW Shade command

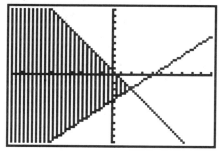

Figure 2.54: Graph of $1 - \dfrac{3x}{2} \geq x - 4$

More information about the DRAW menu is in the TI-82 manual.

2.4   Trigonometry

2.4.1 Degrees and Radians: The trigonometric functions can be applied to angles measured either in radians or degrees, but you should take care that the TI-82 is configured for whichever measure you need. Press MODE to see the current settings. Press ▼ twice and move down to the third line of the mode menu where angle measure is selected. Then press ◀ or ▶ to move between the displayed options. When the blinking cursor is on the measure you want, press ENTER to select it. Then press CLEAR or 2nd QUIT to leave the mode menu.

It's a good idea to check the angle measure setting before executing a calculation that depends on a particular measure. You may change a mode setting at any time and not interfere with pending calculations. Try the following keystrokes to see this in action.

Expression	Keystrokes	Display
$\sin 45°$	MODE ▼ ▼ ▶ ENTER	
	2nd QUIT SIN 45 ENTER	.7071067812
$\sin \pi°$	SIN 2nd π ENTER	.0548036651

$\sin \pi$	SIN 2nd π MODE ▼ ▼	
	ENTER 2nd QUIT ENTER	0
$\sin 45$	SIN 45 ENTER	.8509035245
$\sin \frac{\pi}{6}$	SIN (2nd π ÷ 6) ENTER	.5

The first line of keystrokes sets the TI-82 in degree mode and calculates the sine of 45 *degrees*. While the calculator is still in degree mode, the second line of keystrokes calculates the sine of π *degrees*, 3.1415°. The third line changes to radian mode just before calculating the sine of π *radians*. The fourth line calculates the sine of 45 *radians* (the calculator is already in radian mode).

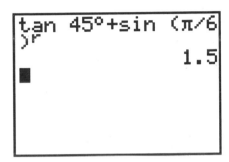

Figure 2.55: Angle measure

The TI-82 makes it possible to mix degrees and radians in a calculation. Execute these keystrokes to calculate $\tan 45° + \sin \frac{\pi}{6}$ as shown in Figure 2.55: TAN 45 2nd ANGLE 1 + SIN (2nd π ÷ 6) 2nd ANGLE 3 ENTER. Do you get 1.5 whether your calculator is set *either* in degree mode *or* in radian mode?

2.4.2 Graphs of Trigonometric Functions: When you graph a trigonometric function, you need to pay careful attention to the choice of graph window. For example, graph $y = \dfrac{\sin 30x}{30}$ in the standard viewing rectangle. Trace along the curve to see where it is. Zoom in to a better window, or use the period and amplitude to establish better WINDOW values.

Technology Tip: Since $\pi \approx 3.1$, set Xmin = 0 and Xmax = 6.3 to cover the interval from 0 to 2π.

Next graph $y = \tan x$ in the standard window first, then press ZOOM 7 to change to a special window for trigonometric functions in which the horizontal increment is $\frac{\pi}{24}$. The TI-82 plots consecutive points and then connects them with a segment, so the graph is not exactly what you should expect. You may wish to change from Connected line to Dot graph (see Section 2.2.3) when you plot the tangent function.

2.5   Scatter Plots

2.5.1 Entering Data: This table shows total prize money (in millions of dollars) awarded at the Indianapolis 500 race from 1981 to 1989. (*Source:* Indianapolis Motor Speedway Hall of Fame.)

Year	1981	1982	1983	1984	1985	1986	1987	1988	1989
Prize ($ million)	$1.61	$2.07	$2.41	$2.80	$3.27	$4.00	$4.49	$5.03	$5.72

We'll now use the TI-82 to construct a scatter plot that represents these points and to find a linear model that approximates the given data.

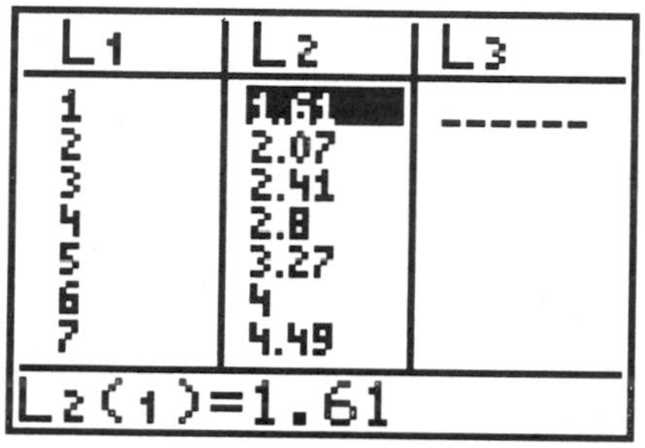

Figure 2.56: Entering data points

The TI-82 holds data in up to six lists. Before entering this new data, press STAT 4 2nd L1 , 2nd L2 , 2nd L3 , 2nd L4 , 2nd L5 , 2nd L6 ENTER to clear all data lists.

Now press STAT 1 to reach the list editor. Instead of entering the full year 198x, enter only x. Here are the keystrokes for the first three years: 1 ENTER 2 ENTER 3 ENTER and so on, then press ▶ to move to the first element of the next list and press 1.61 ENTER 2.07 ENTER 2.41 ENTER and so on (see Figure 2.56). Press 2nd QUIT when you have finished.

You may edit statistical data in the same way you edit expressions in the home screen. Move the cursor to any value you wish to change, then type the correction. To insert or delete data, move the cursor over the data point you wish to add or delete. Press INS and a new data point is created; press DEL and the data point is deleted.

2.5.2 Plotting Data: Once all the data points have been entered, press 2nd STAT PLOT 1 to display the Plot1 screen. Press ENTER to turn Plot1 on, select the other options shown in Figure 2.57, and press GRAPH. Figure 2.58 shows this plot in a window from 0 to 10 in both directions. You may now press TRACE to move from data point to data point.

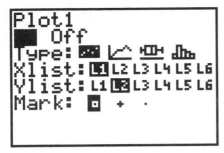

Figure 2.57: Plot1 menu

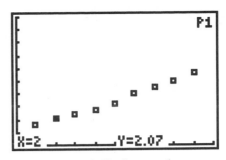

Figure 2.58: Scatter plot

To draw the scatter plot in a window adjusted automatically to include all the data you entered, press ZOOM 9 *[ZoomStat]*.

When you no longer want to see the scatter plot, press 2nd $\frac{STAT}{PLOT}$ 1, move the cursor to OFF, and press ENTER. The TI-82 still retains all the data you entered.

2.5.3 Regression Line:

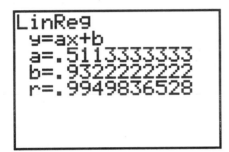

Figure 2.59: Linear regression model

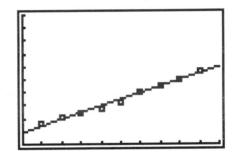

Figure 2.60: Linear regression line

The TI-82 calculates the slope and *y*-intercept for the line that best fits all the data. After the data points have been entered, press STAT ▶ 5 ENTER to calculate a linear regression model. As you see in Figure 2.59, the TI-82 names the slope a and calls the *y*-intercept b. The number *r* (between -1 and 1) is called the *correlation coefficient* and measures the goodness of fit of the linear regression equation with the data. The closer |r| is to 1, the better the fit; the closer |r| is to 0, the worse the fit.

Turn Plot1 on again, if it is not currently displayed. Graph the regression line $y = ax + b$ by pressing Y=, in-activating any existing functions, moving to a free line or clearing one, then pressing VARS 5 ▶ ▶ 7 GRAPH. See how well this line fits with your data points (Figure 2.60).

2.5.4 Exponential Growth Model: After data points have been entered, press STAT ▶ ALPHA A ENTER to calculate an exponential growth model $y = a \cdot b^x$ for the data.

2.6 Matrices

2.6.1 Making a Matrix: The TI-82 can display and use five different matrices. Here's how to create this 3×4 matrix $\begin{bmatrix} 1 & -4 & 3 & 5 \\ -1 & 3 & -1 & -3 \\ 2 & 0 & -4 & 6 \end{bmatrix}$ in your calculator.

Press MATRX to see the matrix menu (Figure 2.61); then press ▶ ▶ or just ◀ to switch to the matrix EDIT menu (Figure 2.62). Whenever you enter the matrix EDIT menu, the cursor starts at the top matrix. Move to another matrix by repeatedly pressing ▼. For now, press ENTER to edit matrix [A].

You may now change the dimensions of matrix [A] to 3×4 by pressing 3 ENTER 4 ENTER. Simply press ENTER or an arrow key to accept an existing dimension. The matrix shown in the window changes in size to reflect a changed dimension.

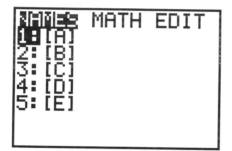

Figure 2.61: MATRX menu

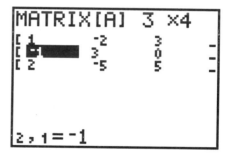

Figure 2.62: Editing a matrix

Use the arrow keys or ENTER to move the cursor to a matrix element you want to change. At the right edge of the screen in Figure 2.62, there are dashes to indicate more columns than are shown. Go to them by pressing ▶ as many times as necessary. The ordered pair at the bottom left of the screen show the cursor's current location within the matrix. The element in the second row and first column in Figure 2.62 is currently highlighted, so the ordered pair at the bottom of the window is 2, . Continue to enter all the elements of matrix [A].

Leave the matrix [A] editing screen by pressing 2nd QUIT and return to the home screen.

2.6.2 Matrix Math: From the home screen you can perform many calculations with matrices. First, let's see matrix [A] itself by pressing MATRX 1 ENTER (Figure 2.63).

Calculate the scalar multiplication 2[A] by pressing 2 MATRX 1 ENTER. To replace matrix [B] by 2[A], press 2 MATRX 1 STO► MATRX 2 ENTER, or if you do this immediately after calculating 2[A], press only STO► MATRX 2 ENTER (see Figure 2.64). Press MATRX ▶ ▶ 2 to verify that the dimensions of matrix [B] have been changed automatically to reflect this new value.

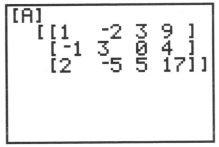

Figure 2.63: Matrix [A]

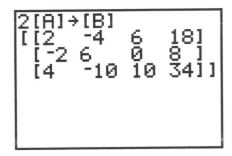

Figure 2.64: Matrix [B]

Add the two matrices [A] and [B] by pressing MATRX 1 + MATRX 2 ENTER. Subtraction is similar.

Now set the dimensions of matrix [C] to 2×3 and enter this as [C]: $\begin{bmatrix} 2 & 0 & 3 \\ 1 & -5 & -1 \end{bmatrix}$. For matrix multiplication of [C] by [A], press MATRX 3 × MATRX 1 ENTER. If you tried to multiply [A] by [C], your TI-82 would signal an error because the dimensions of the two matrices do not permit multiplication this way.

The *transpose* of a matrix [A] is another matrix with the rows and columns interchanged. The symbol for the transpose of [A] is $[A]^T$. To calculate $[A]^T$, press MATRX 1 MATRX ▶ 2 ENTER.

2.6.3 Row Operations: Here are the keystrokes necessary to perform elementary row operations on a matrix. Your textbook provides more careful explanation of the elementary row operations and their uses.

To interchange the second and third rows of the matrix [A] that was defined above, press MATRX ▶ 8 MATRX 1 , 2 , 3) ENTER (see Figure 2.65). The format of this command is rowSwap(*matrix, row1, row2*).

To add row 2 and row 3 and store the results in row 3, press MATRX ▶ 9 MATRX 1 , 2 , 3) ENTER. The format of this command is row+(*matrix, row1, row2*).

To multiply row 2 by -4 and *store* the results in row 2, thereby replacing row 2 with new values, press MATRX ▶ 0 (-) 4 , MATRX 1 , 2) ENTER. The format of this command is *row(*scalar, matrix, row*).

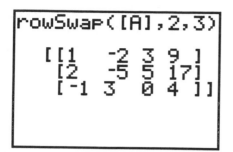

Figure 2.65: Swap rows 2 and 3

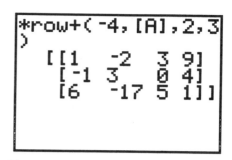

Figure 2.66: Add -4 times row 2 to row 3

To multiply row 2 by -4 and *add* the results to row 3, thereby replacing row 3 with new values, press MATRX ▶ ALPHA A (-) 4 , MATRX 1 , 2 , 3) ENTER (see Figure 2.66). The format of this command is *row+(*scalar, matrix, row1, row2*).

Technology Tip: It is important to remember that your TI-82 does *not* store a matrix obtained as the result of any row operations. So when you need to perform several row operations in succession, it is a good idea to store the result of each one in a temporary place. You may wish to use matrix [E] to hold such intermediate results.

For example, use elementary row operations to solve this system of linear equations: $\begin{cases} x - 2y + 3z = 9 \\ -x + 3y = -4 \\ 2x - 5y + 5z = 17 \end{cases}$.

First enter this *augmented matrix* as [A] in your TI-82: $\begin{bmatrix} 1 & -2 & 3 & 9 \\ -1 & 3 & 0 & -4 \\ 2 & -5 & 5 & 17 \end{bmatrix}$. Next store this matrix in [E] (press MATRX 1 STO ▶ MATRX 5 ENTER) so you may keep the original in case you need to recall it.

Here are the row operations and their associated keystrokes. At each step, the result is stored in [E] and replaces the previous matrix [E]. The solution is shown in Figure 2.67.

Row Operation	Keystrokes
row+([E], 1, 2)	MATRX ▶ 9 MATRX 5 , 1 , 2) STO ▶ MATRX 5 ENTER
*row+(-2, [E], 1, 3)	MATRX ▶ ALPHA A (-) 2 , MATRX 5 , 1 , 3) STO ▶ MATRX 5 ENTER
row+([E], 2, 3)	MATRX ▶ 9 MATRX 5 , 2 , 3) STO ▶ MATRX 5 ENTER

*row(½, [E], 3)   MATRX ▶ 0 1 ÷ 2 , MATRX 5 , 3)
                      STO ► MATRX 5 ENTER

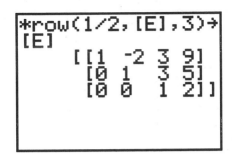

Figure 2.67: Final matrix after row operations

Thus $z = 2$, so $y = -1$ and $x = 1$.

2.6.4 Determinants and Inverses: Enter this 3×3 square matrix as [A]: $\begin{bmatrix} 1 & -2 & 3 \\ -1 & 3 & 0 \\ 2 & -5 & 5 \end{bmatrix}$. To calculate its de-

terminant, $\begin{vmatrix} 1 & -2 & 3 \\ -1 & 3 & 0 \\ 2 & -5 & 5 \end{vmatrix}$, press MATRX ▶ 1 MATRX 1 ENTER. You should find that $|[A]| = 2$, as shown

in Figure 2.68.

Since the determinant of matrix [A] is not zero, it has an inverse, [A]⁻¹. Press MATRX 1 x⁻¹ ENTER to calculate the inverse of matrix [A], also shown in Figure 2.68.

Now let's solve a system of linear equations by matrix inversion. Once more, consider $\begin{cases} x - 2y + 3z = 9 \\ -x + 3y = -4 \\ 2x - 5y + 5z = 17 \end{cases}$.

The coefficient matrix for this system is the matrix $\begin{bmatrix} 1 & -2 & 3 \\ -1 & 3 & 0 \\ 2 & -5 & 5 \end{bmatrix}$ that was entered in the previous example.

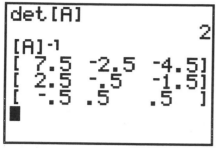

Figure 2.68: $|[A]|$ and $[A]^{-1}$

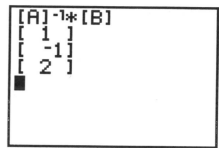

Figure 2.69: Solution matrix

If necessary, enter it again as [A] in your TI-82. Enter the matrix $\begin{bmatrix} 9 \\ -4 \\ 17 \end{bmatrix}$ as [B]. Then press **MATRX 1 x^{-1} $\times$**

MATRX 2 ENTER to calculate the solution matrix (Figure 2.69). The solutions are still $x = 1$, $y = -1$, and $z = 2$.

2.7 Sequences

*2.7.1 Iteration with the **ANS** Key:* The ANS feature permits you to perform *iteration*, the process of evaluating a function repeatedly. As an example, calculate $\dfrac{n-1}{3}$ for $n = 27$. Then calculate $\dfrac{n-1}{3}$ for $n =$ the answer to the previous calculation. Continue to use each answer as n in the *next* calculation. Here are keystrokes to accomplish this iteration on the TI-82 calculator (see the results in Figure 2.70). Notice that when you use ANS in place of n in a formula, it is sufficient to press **ENTER** to continue an iteration.

Iteration	Keystrokes	Display
1	27 ENTER	27
2	(2nd ANS - 1) ÷ 3 ENTER	8.666666667
3	ENTER	2.555555556
4	ENTER	.5185185185
5	ENTER	-.1604938272

TI-82 Graphics Calculator

Press ENTER several more times and see what happens with this iteration. You may wish to try it again with a different starting value.

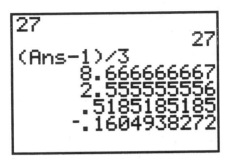

Figure 2.70: Iteration

2.7.2 Arithmetic and Geometric Sequences: Use iteration with the ANS variable to determine the *n*-th term of a sequence. Enter the first term 7, then start the progression with the recursion formula, 2nd ANS + 4 ENTER. This yields the 2nd term, so press ENTER sixteen more times to find the 18th term. For a *geometric* sequence whose common ratio is 4, start the progression with 2nd ANS × 4 ENTER.

You can also define the sequence recursively with the TI-82 by selecting Seq in the MODE menu (see Figure 2.1). Once again, let's find the 18th term of an *arithmetic* sequence whose first term is 7 and whose common difference is 4. Press MODE ▼ ▼ ▼ ▶ ▶ ▶ ENTER 2nd QUIT. Then press Y= to edit either of the TI-82's two sequences, u_n and v_n. Make $u_n = u_{n-1} + 4$ by pressing 2nd u_{n-1} + 4. Now make $u_1 = 7$ by pressing WINDOW and setting U*n*Start = 7 and *n*Start = 1 (because the first term is u_1 where $n = 1$). Press 2nd QUIT to leave this menu and return to the home screen. To find the 18th term of this sequence, calculate u_{18} by pressing 2nd Y-VARS 4 1 (18) ENTER (see Figure 2.71).

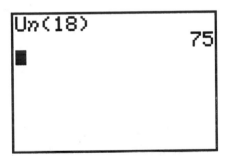

Figure 2.71: Sequence mode

Of course, you could use the *explicit* formula for the *n*-th term of an arithmetic sequence, $t_n = a + (n-1)d$. First enter values for the variables a, d, and n, then evaluate the formula by pressing ALPHA A + (ALPHA N - 1) ALPHA D ENTER. For a geometric sequence whose *n*-th term is given by $t_n = a \cdot r^{n-1}$, enter values for the variables a, r, and n, then evaluate the formula by pressing ALPHA A ALPHA R ^ (ALPHA N - 1) ENTER.

To use the explicit formula in Seq MODE, make $u_n = 7 + (n-1) \cdot 4$ by pressing Y= and then 7 + (2nd n - 1) × 4. Once more, calculate u_{18} by pressing 2nd Y-VARS 4 1 (18) ENTER.

There are more instructions for using sequence mode in the TI-82 manual.

2.7.3 Sums of Sequences: Calculate the sum $\sum_{n=1}^{12} 4(0.3)^n$ on the TI-82 by pressing 2nd LIST ▶ 5 2nd LIST 5 4 × .3 ^ ALPHA N , ALPHA N , 1 , 12 , 1) ENTER. You should get 1.714284803. The format of this command is sum seq(*expression, variable, begin, end, increment*).

2.8   Parametric and Polar Graphs

2.8.1 Graphing Parametric Equations: The TI-82 plots up to six pairs of parametric equations as easily as it plots functions. Just use the MODE menu (Figure 2.1), go to the fourth line from the top, and change the setting to Par. Be sure, if the independent parameter is an angle measure, that MODE is set to whichever you need, Rad or Deg.

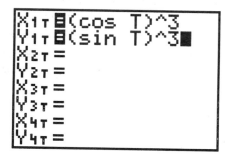

Figure 2.72: Parametric Y= menu

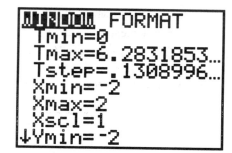

Figure 2.73: Parametric WINDOW menu

For example, here are the keystrokes needed to graph the parametric equations $x = \cos^3 t$ and $y = \sin^3 t$. First check that angles are currently being measured in radians. Change to parametric mode and press Y= to examine the new parametric equation menu (Figure 2.72). Enter the two parametric equations for X1T and Y1T by pressing (COS X,T,θ) ^ 3 ENTER (SIN X,T,θ) ^ 3 ENTER. Now, when you press the variable key X,T,θ, you get a T because the calculator is in parametric mode.

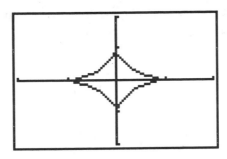

Figure 2.74: Parametric graph of $x = \cos^3 t$ and $y = \sin^3 t$

Also look at the new **WINDOW** menu (Figure 2.73). In the standard window, the values of **T** go from 0 to 2π in steps of $\frac{\pi}{24} = .1309$, with the view from -10 to 10 in both directions. But here the viewing rectangle has been changed to extend from -2 to 2 in both directions. Press **GRAPH** to see the parametric graph (Figure 2.74).

You may **ZOOM** and **TRACE** along parametric graphs just as you did with function graphs. As you trace along this graph, notice that the cursor moves in the *counterclockwise* direction as **T** increases.

2.8.2 Rectangular-Polar Coordinate Conversion: The 2nd **ANGLE** menu provides functions for converting between rectangular and polar coordinate systems. These functions use the current **MODE** settings, so it is a good idea to check the default angle measure before any conversion. Of course, you may use the **MATH** menu to override the current angle measure setting, as explained in Section 2.4.1. For the following examples, the TI-82 is set to radian measure.

Given rectangular coordinates $(x, y) = (4, -3)$, convert *from* these rectangular coordinates *to* polar coordinates (r, θ) by pressing 2nd **ANGLE** 5 4 , (-) 3) **ENTER** to display the value of r. The value of θ is displayed after you press 2nd **ANGLE** 6 4 , (-) 3) **ENTER**.

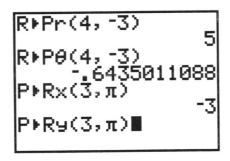

Figure 2.75: Coordinate conversions

Suppose $(r, \theta) = (3, \pi)$. To convert *from* these polar coordinates *to* rectangular coordinates (x, y), press 2nd ANGLE 7 3 , 2nd π) for the *x*-coordinate; next press 2nd ANGLE 8 3 , 2nd π) ENTER to display the *y*-coordinate.

2.8.3 Graphing Polar Equations: The TI-82 graphs a polar function in the form $r = f(\theta)$. In the fourth line of the MODE menu, select Pol for polar graphs. You may now graph up to six different polar functions at a time.

For example, to graph $r = 4 \sin \theta$, press Y= for the polar graph editing screen. Then enter the expression $4 \sin \theta$ for r1 by pressing 4 sin X,T,θ; note that the X,T,θ key produces θ in polar mode. Choose a good viewing rectangle and an appropriate interval and increment for θ. In Figure 2.76, the viewing rectangle is roughly "square" and extends from -6 to 6 horizontally and from -4 to 4 vertically.

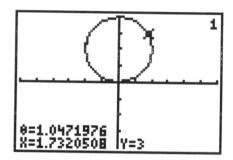

Figure 2.76: Polar graph of $r = 4 \sin \theta$

Figure 2.76 shows *rectangular* coordinates of the cursor's location on the graph. You may sometimes wish to trace along the curve and see *polar* coordinates of the cursor's location. The first line of the WINDOW FORMAT menu (Figure 2.18) has options for displaying the cursor's position in rectangular RectGC or polar PolarGC form.

2.9   Probability

2.9.1 Random Numbers: The command Rand generates a number between 0 and 1. You will find this command in the PRB (probability) sub-menu of the MATH menu. Press MATH ◀ 1 ENTER to generate a random number. Press ENTER to generate another number; keep pressing ENTER to generate more of them.

If you need a random number between, say, 0 and 10, then press 10 MATH ◀ 1 ENTER. To get a random number between 5 and 15, press 5 + 10 MATH ◀ 1 ENTER.

2.9.2 Permutations and Combinations: To calculate the number of *permutations* of 12 objects taken 7 at a time, $_{12}P_7$, press 12 MATH ◀ 2 7 ENTER. Thus $_{12}P_7 = 3,991,680$, as shown in Figure 2.77.

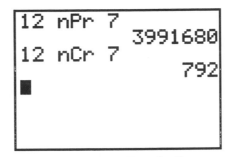

Figure 2.77: $_{12}P_7$ and $_{12}C_7$

For the number of *combinations* of 12 objects taken 7 at a time, $_{12}C_7$, press 12 MATH ◀ 3 7 ENTER. So $_{12}C_7 = 792$.

2.9.3 Probability of Winning: A state lottery is configured so that each player chooses six different numbers from 1 to 40. If these six numbers match the six numbers drawn by the State Lottery Commission, the player wins the top prize. There are $_{40}C_6$ ways for the six numbers to be drawn. If you purchase a single lottery ticket, your probability of winning is 1 in $_{40}C_6$. Press 1 ÷ 40 MATH ◀ 3 6 ENTER to calculate your chances, but don't be disappointed.

2.10   Programming

2.10.1 Entering a Program: The TI-82 is a programmable calculator that can store sequences of commands for later replay. Here's an example to show you how to enter a useful program that solves quadratic equations by the quadratic formula.

Press PRGM to access the programming menu. The TI-82 has space for many programs, each called by a name you give it. Create a new program now, so press PRGM ◀ 1.

For convenience, the cursor is a blinking **A**, indicating that the calculator is set to receive alphabetic characters. Enter a descriptive title of up to eight characters, letters or numerals (but the first character must be a letter). Name this program QUADRAT and press ENTER to go to the program editor.

In the program, each line begins with a colon **:** supplied automatically by the calculator. Any command you could enter directly in the TI-82's home screen can be entered as a line in a program. There are also special programming commands.

Input the program QUADRAT by pressing the keystrokes given in the listing below. You may interrupt program input at any stage by pressing 2nd QUIT. To return later for more editing, press PRGM ▶, move the cursor down to this program's name, and press ENTER.

Program Line	Keystrokes
: Disp "Enter A"	PRGM ▶ 3 2nd A-LOCK " E N T E R ␣ A " ENTER

    displays the words *Enter A* on the TI-82 screen

: Input A	PRGM ▶ 1 ALPHA A ENTER

    waits for you to input a value that will be assigned to the variable A

: Disp "Enter B"	PRGM ▶ 3 2nd A-LOCK " E N T E R ␣ B " ENTER
: Input B	PRGM ▶ 1 ALPHA B ENTER
: Disp "Enter C"	PRGM ▶ 3 2nd A-LOCK " E N T E R ␣ C " ENTER
: Input C	PRGM ▶ 1 ALPHA C ENTER
: B^2-4AC → D	ALPHA B x² - 4 ALPHA A ALPHA C STO▶ ALPHA D ENTER

    calculates the discriminant and stores its value as D

: If D>0	PRGM 1 ALPHA D 2nd TEST 3 0 ENTER

    tests to see if the discriminant is positive

: Then	PRGM 2 ENTER

    in case the discriminant is positive, continues on to the next line;
    if the discriminant is not positive, jumps to the command after **Else** below

: Disp "TWO REAL ROOTS"	PRGM ▶ 3 2nd A-LOCK " T W O ␣ R E A L ␣ R O O T S " ENTER
: (-B+√D)/(2A) → M	((-) ALPHA B + 2nd √ ALPHA D) ÷ (2 ALPHA A) STO▶ ALPHA M ENTER

    calculates one root and stores it as M

: Disp M	PRGM ▶ 3 ALPHA M ENTER

    displays one root

: (-B-√D)/(2A) → N	((-) ALPHA B - 2nd √ ALPHA D) ÷ (2 ALPHA A) STO▶ ALPHA N ENTER
: Disp N	PRGM ▶ 3 ALPHA N ENTER

: Else   PRGM 3 ENTER

    continues from here if the discriminant is not positive

: If D=0   PRGM 1 ALPHA D 2nd TEST 1 0 ENTER

    tests to see if the discriminant is zero

: Then   PRGM 2 ENTER

    in case the discriminant is zero, continues on to the next line;
    if the discriminant is not zero, jumps to the command after Else below

: Disp "DOUBLE   PRGM ▶ 3 2nd A-LOCK " D O U B L E ␣ R O O T " ENTER
ROOT"

    displays a message in case there is a double root

: -B/(2A) → M   (-) ALPHA B ÷ (2 ALPHA A) STO▶ ALPHA M ENTER

    the quadratic formula reduces to $\dfrac{-b}{2a}$ when $D = 0$

: Disp M   PRGM ▶ 3 ALPHA M ENTER

: Else   PRGM 3 ENTER

    continues from here if the discriminant is not zero

: Disp "COMPLEX   PRGM ▶ 3 2nd A-LOCK " C O M P L E X ␣ R O O T S "
ROOTS"   ENTER

    displays a message in case the roots are complex numbers

: Disp "REAL PART"   PRGM ▶ 3 2nd A-LOCK " R E A L ␣ P A R T " ENTER

: -B/(2A) → R   (-) ALPHA B ÷ (2 ALPHA A) STO▶ ALPHA R ENTER

    calculates the real part $\dfrac{-b}{2a}$ of the complex roots

: Disp R   PRGM ▶ 3 ALPHA R ENTER

: Disp "IMAGINARY   PRGM ▶ 3 2nd A-LOCK " I M A G I N A R Y ␣ P A R T "
PART"   ENTER

: √-D/(2A) → I   2nd √ (-) ALPHA D ÷ (2 ALPHA A) STO▶ ALPHA I ENTER

    calculates the imaginary part $\dfrac{\sqrt{-D}}{2a}$ of the complex roots;
    since $D < 0$, we must use $-D$ as the radicand

:Disp 1	PRGM ▶ 3 ALPHA I ENTER
:End	PRGM 7 ENTER
marks the end of an If-Then-Else group of commands	
:End	PRGM 7

When you have finished, press 2nd QUIT to leave the program editor.

You may remove a program from memory by pressing 2nd MEM 2 *[Delete...]* 6 *[Prgm...]*. Then move the cursor to the program's name and press ENTER to delete the entire program.

2.10.2 Executing a Program: To execute the program just enetered, press PRGM and then the number or letter that it was named. If you have forgotten its name, use the arrow keys to move through the program listing to find its description QUADRAT. Then press ENTER to select this program and enter again to execute it.

The program has been written to prompt you for values of coefficients a, b, and c in a quadratic equation $ax^2 + bx + c = 0$. Input a value, then press ENTER to continue the program.

If you need to interrupt a program during execution, press ON.

The instruction manual for your TI-82 gives detailed information about programming. Refer to it to learn more about programming and how to use other features of your calculator.

2.11   Differentiation

2.11.1 Limits: Suppose you need to find this limit: $\lim\limits_{x \to 0} \dfrac{\sin 4x}{x}$. Plot the graph of $f(x) = \dfrac{\sin 4x}{x}$ in a convenient viewing rectangle that contains the point where the function appears to intersect the line $x = 0$ (because you want the limit as $x \to 0$). Your graph should lend support to the conclusion that $\lim\limits_{x \to 0} \dfrac{\sin 4x}{x} = 4$.

To test the reasonableness of the conclusion that $\lim\limits_{x \to \infty} \dfrac{2x-1}{x+1} = 2$, evaluate the function $f(x) = \dfrac{2x-1}{x+1}$ for several large positive values of x (since you want the limit as $x \to \infty$). For example, evaluate $f(100)$, $f(1000)$, and $f(10,000)$. Another way to test the reasonableness of this result is to examine the graph of $f(x) = \dfrac{2x-1}{x+1}$ in a viewing rectangle that extends over large values of x. See, as in Figure 2.79 (where the viewing rectangle extends horizontally from 0 to 100), whether the graph is asymptotic to the horizontal line $y = 2$ (enter $\dfrac{2x-1}{x+1}$ for Y$_1$ and 2 for Y$_2$).

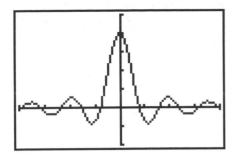

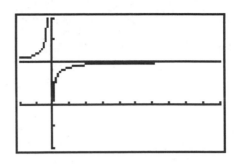

Figure 2.78: Checking $\lim\limits_{x\to 0} \dfrac{\sin 4x}{x} = 4$

Figure 2.79: Checking $\lim\limits_{x\to\infty} \dfrac{2x-1}{x+1} = 2$

1.11.2 Numerical Derivatives: The derivative of a function f at x can be defined as the limit of the slopes of secant lines, so $f'(x) = \lim\limits_{\Delta x \to 0} \dfrac{f(x+\Delta x)-f(x-\Delta x)}{2\Delta x}$. And for small values of Δx, the expression $\dfrac{f(x+\Delta x)-f(x-\Delta x)}{2\Delta x}$ gives a good approximation to the limit.

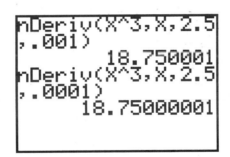

Figure 2.80: Using nDeriv(

The TI-82 has a function nDeriv(in the MATH menu to calculate the *symmetric difference*, $\dfrac{f(x+\Delta x)-f(x-\Delta x)}{2\Delta x}$. So to find a numerical approximation to $f'(2.5)$ when $f(x) = x^3$ and with $\Delta x = 0.001$, press MATH 8 X,т,θ ∧ 3, ALPHA X , 2.5 , .001) ENTER as shown in Figure 2.80. The format of this command is nDeriv(*expression, variable, value, Δx*). The same derivative is also approximated in Figure 2.80 using $\Delta x = 0.0001$. For most purposes, $\Delta x = 0.001$ gives a very good approximation to the derivative and is the TI-82's default. So if you do use $\Delta x = 0.001$, just enter nDeriv(*expression, variable, value*).

Technology Tip: It is sometimes helpful to plot both a function and its derivative together. In Figure 2.82, the function $f(x) = \dfrac{5x-2}{x^2+1}$ and its numerical derivative (actually, an approximation to the derivative given by the symmetric difference) are graphed. You can duplicate this graph by first entering $\dfrac{5x-2}{x^2+1}$ for Y_1 and then entering its numerical derivative for Y_2 by pressing MATH 8 2nd Y-VARS 1 1 , X,T,θ , X,T,θ) as you see in Figure 2.81.

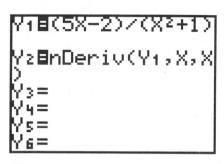

Figure 2.81: Entering $f(x)$ and $f'(x)$

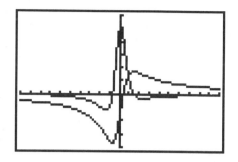

Figure 2.82: Graphs of $f(x)$ and $f'(x)$

Technology Tip: To approximate the *second* derivative $f''(x)$ of a function $y = f(x)$ or to plot the second derivative, first enter the expression for Y_1 and its derivative for Y_2 as above. Then enter the second derivative for Y_3 by pressing MATH 8 2nd Y-VARS 1 2 , X,T,θ , X,T,θ).

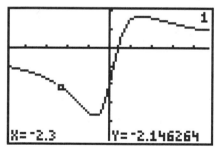

Figure 2.83: $f(x) = \dfrac{5x-2}{x^2+1}$ at $x = -2.3$

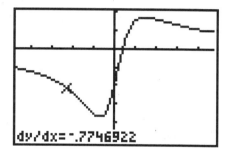

Figure 2.84: dy/dx

You may also approximate a derivative while you are examining the graph of a function. When you are in a graph window, press 2nd CALC 6, then use the arrow keys to trace along the curve to a point where you want the derivative (see figure 2.83 for the graph of $f(x) = \dfrac{5x - 2}{x^2 + 1}$ at $x = -2.3$) and press ENTER. The TI-82 uses $\Delta x = 0.001$ for this approximation.

1.11.3 Newton's Method: With your TI-82, you may iterate using Newton's method to find the zeros of a function. Recall that Newton's Method determines each successive approximation by the formula $x_{n+1} = x_n - \dfrac{f(x_n)}{f'(x_n)}$.

As an example of the technique, consider $f(x) = 2x^3 + x^2 - x + 1$. Enter this function as Y_1. A look at its graph suggests that it has a zero near $x = -1$, so start the iteration by going to the home screen and storing -1 as x (see figure 2.85). Then press these keystrokes: X,T,θ − 2nd Y-VARS 1 1 ÷ MATH 8 2nd Y-VARS 1 1 , X,T,θ , X,T,θ) STO♦ X,T,θ. Press ENTER repeatedly until two successive approximations differ by less than some predetermined value, say 0.0001. Note that each time you press ENTER, the TI-82 will use its *current* value of x, and that value is changing as you continue the iteration.

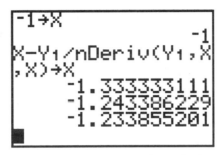

Figure 2.85: Newton's method

Technology Tip: Newton's Method is sensitive to your seed value for x, so look carefully at the function's graph to make a good first estimate. Also, remember that the method sometimes fails to converge!

You may want to save the Newton's Method formula as a short program. See your calculator's manual for further information on programming the TI-82.

2.12   Integration

2.12.1 Approximating Definite Integrals: The TI-82 has the function fnInt(in the MATH menu to approximate an integral. So to find a numerical approximation to $\displaystyle\int_0^1 \cos x^2\, dx$ with a tolerance of 0.001 (which controls the accuracy of the approximation), press MATH 9 COS X,T,θ x^2 , X,T,θ , 0 , 1 , .001) ENTER as shown in Figure 2.86. The format of this command is fnInt(*expression, variable, lower limit, upper limit, tolerance*). The same integral is also approximated in Figure 2.86 using a tolerance of 0.00001, the TI-82's default that is used when no other tolerance is specified.

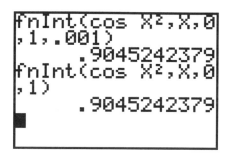

Figure 2.86 fnInt(

2.12.2 Areas: You may approximate the area under the graph of a function $y = f(x)$ between $x = A$ and $x = B$ with your TI-82. For example, here are keystrokes for finding the area under the graph of the function $y = \cos x^2$ between $x = 0$ and $x = 1$. This area is represented by the definite integral $\int_0^1 \cos x^2 dx$. So graph $f(x) = \cos x^2$ and press 2nd CALC 7. Use the arrow keys to trace along the curve to the lower limit and press ENTER; then trace again to the upper limit (see Figure 2.87) and press ENTER. The region under the graph between the lower limit and the upper limit is shaded and the area is displayed as in Figure 2.88. The TI-82 uses fnInt(with the default tolerance of 10^{-5} in this calculation.

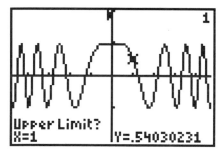

Figure 2.87: Setting the upper limit

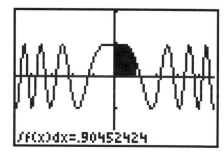

Figure 2.88: $\int f(x)dx$

Technology Tip: When approximating the area under $f(x) = \cos x^2$ between $x = 0$ and $x = 1$, you must trace along the curve to *exactly* where $x = 0$ and $x = 1$. Now to trace along the curve to $x = a$, the viewing rectangle must be chosen so that the function is evaluated at $x = a$. The window shown in Figure 2.87 was made first by pressing ZOOM 4 *[Zdecimal]*, then by changing its vertical dimensions to appropriate values. By contrast, find the area under $f(x) = \cos x^2$ between $x = 0$ and $x = 1$ in ZOOM 7 *[Ztrig]* window.

Technology Tip: Suppose that you want to find the area between two functions, $y = f(x)$ and $y = g(x)$, from $x = A$ to $x = B$. If $f(x) \geq g(x)$ for $A \leq x \leq B$, then enter the expression $f(x) - g(x)$ for Y_1 and proceed as before to find the required area.

Chapter 3

Texas Instruments TI-83

3.1 Getting started with the TI-83

3.1.1 Basics: Press the ON key to begin using your TI-83. If you need to adjust the display contrast, first press 2nd, then press and hold ▼ (the *down* arrow key) to lighten or ▲ (the *up* arrow key) to darken. As you press and hold ▼ or ▲, an integer between 0 (lightest) and 9 (darkest) appears in the upper right corner of the display. When you have finished with the calculator, turn it off to conserve battery power by pressing 2nd and then OFF.

Check your TI-83's settings by pressing MODE. If necessary, use the arrow keys to move the blinking cursor to a setting you want to change. Press ENTER to select a new setting. To start with, select the options along the left side of the MODE menu as illustrated in Figure 3.1: normal display, floating decimals, radian measure, function graphs, connected lines, sequential plotting, real numbers, and full screen display. Details on alternative options will be given later in this guide. For now, leave the MODE menu by pressing CLEAR.

Figure 3.1: MODE menu

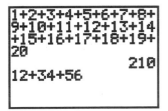

Figure 3.2: Home screen

3.1.2 Editing: One advantage of the TI-83 is that up to eight lines are visible at one time, so you can *see* a long calculation. For example, type this sum (Figure 3.2):

$$1 + 2 + 3 + 4 + 5 + 6 + 7 + 8 + 9 + 10 + 11 + 12 + 13 + 14 + 15 + 16 + 17 + 18 + 19 + 20$$

Then press ENTER to see the answer too.

Often we do not notice a mistake until we see how unreasonable an answer is. The TI-83 permits you to re-display an entire calculation, edit it easily, then execute the *corrected* calculation.

Suppose you had typed 12 + 34 + 56 as in Figure 3.2 but had *not yet* pressed ENTER, when you realize that 34 should have been 74. Simply press ◄ (the *left* arrow key) as many times as necessary to move the blinking cursor left to 3, then type 7 to write over it. On the other hand, if 34 should have been 384, move the cursor back to 4, press 2nd INS (the cursor changes to a blinking underline) and then type 8 (inserts at the cursor position and the other characters are pushed to the right). If the 34 should have been 3 only, move the cursor to 4, and press DEL to delete it.

Technology Tip: To move quickly to the *beginning* of an expression you are currently editing, press ▲ (the *up* arrow key); to jump to the *end* of that expression, press ▼ (the *down* arrow key).

Even if you had pressed ENTER, you may still edit the previous expression. Press 2nd and then ENTER to recall the last expression that was entered. Now you can change it. In fact, the TI-83 retains many prior entries in a "last entry" storage area. Press 2nd ENTRY repeatedly until the previous line you want replaces the current line.

Technology Tip: When you need to evaluate a formula for different values of a variable, use the editing feature to simplify the process. For example, suppose you want to find the balance in an investment account if there is now $5000 in the account and interest is compounded annually at the rate of 8.5%. The formula for the balance is $P = \left(1 + \frac{r}{n}\right)^{nt}$, where P = principal, r = rate of interest (expressed as a decimal), n = number of times interest is compounded each year, and t = number of years. In our example, this becomes $5000(1+.085)^t$. Here are the keystrokes for finding the balance after $t = 3$, 5, and 10 years.

Years	Keystrokes	Balance
3	5000 (1 + .085) ∧ 3 ENTER	$6386.45
5	2nd ENTRY ◄ 5 ENTER	$7518.28
10	2nd ENTRY ◄ 10 ENTER	$11,304.92

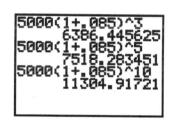

Figure 3.3: Editing expressions

Then to find the balance from the same initial investment but after 5 years when the annual interest rate is 7.5%, press the keys to change the last calculation above: 2nd ENTRY ◄ DEL ◄ 5 ◄ ◄ ◄ ◄ ◄ 7 ENTER.

3.1.3 Key Functions: Most keys on the TI-83 offer access to more than one function, just as the keys on a computer keyboard can produce more than one letter ("g" and "G") or even quite different characters ("5" and "%"). The primary function of a key is indicated on the key itself, and you access that function by a simple press on the key.

To access the *second* function indicated to the *left* above a key, first press 2nd (the cursor changes to a blinking ♠) and *then* press the key. For example to calculate $\sqrt{25}$, press 2nd $\sqrt{}$ 25 ENTER.

Technology Tip: The TI-83 automatically places a left parenthesis, (, after many functions and operators (including LOG, 2nd 10^x, LN, 2nd e^x, SIN, COS, TAN, and 2nd $\sqrt{}$). If no other calculations are being done, a matching right parenthesis does *not* need to be entered.

When you want to use a letter or other character printed to the *right* above a key, first press ALPHA (the cursor changes to a blinking **A**) and then the key. For example, to use the letter K in a formula, press ALPHA K. If you need several letters in a row, press 2nd A-LOCK, which is like the CAPS LOCK key on a computer keyboard, and then press all the letters you want. Remember to press ALPHA when you are finished and want to restore the keys to their primary functions.

3.1.4 Order of Operations:
The TI-83 performs calculations according to the standard algebraic rules. Working outwards from inner parentheses, calculations are performed from left to right. Powers and roots are evaluated first, followed by multiplications and divisions, and then additions and subtractions.

Enter these expressions to practice using your TI-83.

Expression	Keystrokes	Display
$7 - 5 \cdot 3$	7 – 5 × 3 ENTER	–8
$(7 - 5) \cdot 3$	(7 – 5) × 3 ENTER	6
$120 - 10^2$	120 – 10 x² ENTER	20
$(120 - 10)^2$	(120 – 10) x² ENTER	12100
$\dfrac{24}{2^3}$	24 ÷ 2 ∧ 3 ENTER	3
$\left(\dfrac{24}{2}\right)^3$	(24 ÷ 2) ∧ 3 ENTER	1728
$(7 - -5) \cdot -3$	(7 – (–) 5) × (–) 3 ENTER	–36

3.1.5 Algebraic Expressions and Memory:
Your calculator can evaluate expressions such as $\dfrac{N(N+1)}{2}$ after you have entered a value for N. Suppose you want $N = 200$. Press 200 STO◆ ALPHA N ENTER to store the value 200 in memory location N. Whenever you use N in an expression, the calculator will substitute the value 200 until you make a change by storing *another* number in N. Next enter the expression $\dfrac{N(N+1)}{2}$ by typing ALPHA N (ALPHA N + 1) ÷ 2 ENTER. For $N = 200$, you will find that $\dfrac{N(N+1)}{2} = 20100$.

The contents of any memory location may be revealed by typing just its letter name and then ENTER. And the TI-83 retains memorized values even when it is turned off, so long as its batteries are good.

3.1.6 Repeated Operations with ANS:
The result of your *last* calculation is always stored in memory location ANS and replaces any previous result. This makes it easy to use the answer from one computation in an-

other computation. For example, press 30 + 15 ENTER so that 45 is the last result displayed. Then press 2nd ANS ÷ 9 ENTER and get 5 because 45 ÷ 9 = 5.

With a function like division, you press the ÷ *after* you enter an argument. For such functions, whenever you would start a new calculation with the previous answer followed by pressing the function key, you may press just the function key. So instead of 2nd ANS ÷ 9 in the previous example, you could have pressed simply ÷ 9 to achieve the same result. This technique also works for these functions: + − × ∧ x^2 x^{-1}.

Here is a situation where this is especially useful. Suppose a person makes $5.85 per hour and you are asked to calculate earnings for a day, a week, and a year. Execute the given keystrokes to find the person's incomes during these periods (results are shown in Figure 3.4).

Pay Period	Keystrokes	Earnings
8-hour day	5.85 × 8 ENTER	$46.80
5-day week	× 5 ENTER	$234
52-week year	× 52 ENTER	$12,168

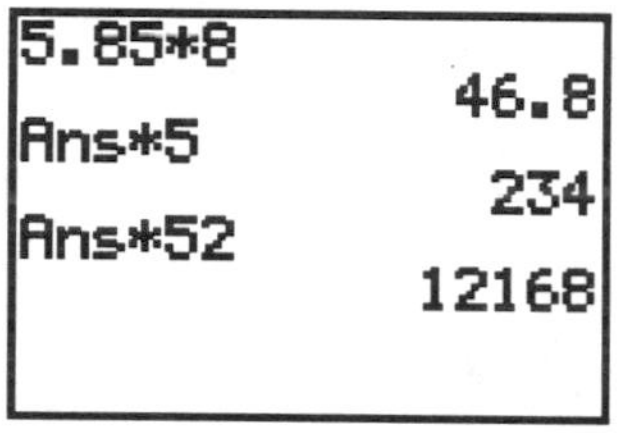

Figure 3.4: ANS variable

3.1.7 The MATH Menu:

3.1.7 The MATH Menu: Operators and functions associated with a scientific calculator are available either immediately from the keys of the TI-83 or by the 2nd keys. You have direct access to common arithmetic operations (x^2, 2nd $\sqrt{\ }$, 2nd x^{-1}, ∧), trigonometric functions (SIN, COS, TAN), and their inverses (2nd SIN^{-1}, 2nd COS^{-1}, 2nd TAN^{-1}), exponential and logarithmic functions (LOG, 2nd 10^X, LN, 2nd e^X), and a famous constant (2nd π).

A significant difference between the TI-83 graphing calculators and most scientific calculators is that TI-83 requires the argument of a function *after* the function, as you would see in a formula written in your textbook. For example, on the TI-83 you calculate $\sqrt{16}$ by pressing the keys 2nd $\sqrt{\ }$ 16 in that order.

Here are keystrokes for basic mathematical operations. Try them for practice on your TI-83.

Expression	Keystrokes	Display
$\sqrt{3^2 + 4^2}$	2nd $\sqrt{\ }$ (3 x^2 + 4 x^2) ENTER	5
$2\frac{1}{3}$	2 + 3 x^{-1} ENTER	2.333333333
log 200	LOG 200 ENTER	2.301029996
$2.34 \cdot 10^5$	2.34 × 2nd 10^x 5 ENTER	234000

Additional mathematical operations and functions are available from the MATH menu. Press MATH to see the various options (Figure 3.5). You will learn in your mathematics textbook how to apply many of them. As an example, calculate $\sqrt[3]{7}$ by pressing MATH then *either* 4[$\sqrt[3]{\ }$] or ▼ ▼ ▼ ENTER; finally press 7 to see 1.912931183. To leave the MATH menu and take no other action, press 2nd QUIT or just CLEAR.

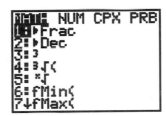

Figure 3.5: MATH menu

The *factorial* of a non-negative integer is the *product* of *all* the integers from 1 up to the given integer. The symbol for factorial is the exclamation point. So 4! (pronounced *four factorial*) is $1 \cdot 2 \cdot 3 \cdot 4 = 24$. You will learn more about applications of factorials in your textbook, but for now use the TI-83 to calculate 4! Press these keystrokes: 4 MATH ◄ 4[!] ENTER *or* 4 MATH ◄ ▼ ▼ ▼ ENTER ENTER.

Note that you can select a sub-menu from the MATH menu by pressing either ◄ or ►. It is easier to press ◄ once than to press ► three times to get to the PRB menu.

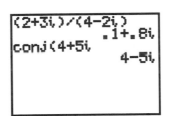

Figure 3.6: Complex number calculations

On the TI-83 it is possible to do calculations with complex numbers. To enter the imaginary number i, press 2nd i. For example, to divide $2 + 3i$ by $4 - 2i$, press (2 + 3 2nd i) ÷ (4 − 2 2nd i) ENTER. The result is $0.1 + 0.8i$ (Figure 3.6).

To find the complex conjugate of $4 + 5i$ press MATH ▶ ▶ ENTER 4 + 5 2nd i ENTER (Figure 3.6).

3.2   Functions and Graphs

3.2.1 *Evaluating Functions:* Suppose you receive a monthly salary of \$1975 plus a commission of 10% of sales. Let x = your sales in dollars; then your wages W in dollars are given by the equation $W = 1975 + .10x$. If your January sales were \$2230 and your February sales were \$1865, what was your income during those months?

Here's one method to use your TI-83 to perform this task. Press the Y= at the top of the calculator to display the function editing screen (Figure 3.7). You may enter as many as ten different functions for the TI-83 to use at one time. If there is already a function Y_1 press ▲ or ▼ as many times as necessary to move the cursor to Y_1 and then press CLEAR to delete whatever was there. Then enter the expression $1975 + .10x$ by pressing these keys: 1975 + .10 X,T,θ,n. (The X,T,θ,n key lets you enter the variable X easily without having to use the ALPHA key.) Now press 2nd QUIT to return to the main calculations screen.

Assign the value 2230 to the variable x by these keystrokes (see Figure 3.8): 2230 STO▸ X,T,θ,n. Then press ALPHA : to allow another expression to be entered on the same command line. Next press the following key-strokes to evaluate Y_1 and find January's wages: VARS ▶ 1*[Function]* 1*[Y₁]* ENTER.

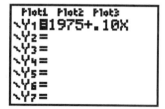

Figure 3.7: Y= screen

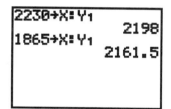

Figure 3.8: Evaluating a function

It is not necessary to repeat all these steps to find the February wages. Simply press 2nd ENTRY to recall the entire previous line, change 2230 to 1865, and press ENTER. Each time the TI-83 evaluates the function Y_1, it uses the *current* value of x.

Like your textbook, the TI-83 uses standard function notation. So to evaluate $Y_1(2230)$ when $Y_1(x) = 1975 + .10x$, press VARS ▶ 1*[Function]* 1*[Y₁]* (2230) ENTER (see Figure 3.9). Then to evaluate $Y_1(1865)$, press 2nd ENTRY to recall the last line and change 2230 to 1865.

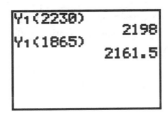

Figure 3.9: Function notation

You may also have the TI-83 make a table of values for the function. Press 2nd TBLSET to set up the table (Figure 3.10). Move the blinking cursor onto Ask beside Indpnt:, then press ENTER. This configuration permits you to input values for x one at a time. Now press 2nd TABLE, enter 2230 in the x column, and press ENTER (see Figure 3.11). Continue to enter additional values for x and the calculator automatically completes the table with corresponding values of Y_1. Press 2nd QUIT to leave the TABLE screen.

Figure 3.10: TBLSET screen

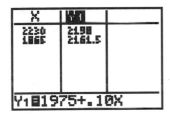

Figure 3.11: Table of values

Technology Tip: The TI-83 does not require multiplication to be expressed between variables, so *xxx* means x^3. It is often easier to press two or three x's together than to search for the square key or the powers key. Of course, expressed multiplication is also not required between a constant and a variable. Hence to enter $2x^3 + 3x^2 - 4x + 5$ in the TI-83, you might save keystrokes and press just these keys: 2 X,T,θ,n X,T,θ,n X,T,θ,n + 3 X,T,θ,n X,T,θ,n − 4 X,T,θ,n + 5

3.2.2 Functions in a Graph Window: Once you have entered a function in the Y= screen of the TI-83, just press GRAPH to see its graph. The ability to draw a graph contributes substantially to our ability to solve problems.

For example, here is how to graph $y = -x^3 + 4x$. First press Y= and delete anything that may be there by moving with the arrow keys to Y_1 or to any of the other lines and pressing CLEAR wherever necessary. Then, with the cursor on the top line Y_1, press (−) X,T,θ,n ∧ 3 + 4 X,T,θ,n to enter the function (as in Figure 3.12). Now press GRAPH and the TI-83 changes to a window with the graph of $y = -x^3 + 4x$ (Figure 3.13).

While the TI-83 is calculating coordinates for a plot, it displays a busy indicator at the top right of the graph window.

Technology Tip: If you would like to see a function in the Y= menu and its graph in a graph window, both at the same time, open the MODE menu, move the cursor down to the last line, and select Horiz screen. Your TI-83's screen is now divided horizontally (see Figure 3.12), with an upper graph window and a lower window that can display the home screen or an editing screen. The split screen is also useful when you need to do some calculations as you trace along a graph. For now, restore the TI-83 to Full screen.

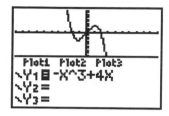

Figure 3.12: Split screen: Y= below

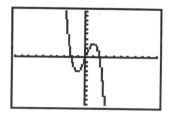

Figure 3.13: Graph of $y = -x^3 + 4x$

Technology Tip: The TI-83 also can also show a vertically split screen with a graph on the left side and a table of values on the right side (the G-T option on screen display line of the MODE menu). This feature allows you to see both the graph and a corresponding table of values. Note that unless you use 2nd TBLSET to generate the table (as in Section 3.2.1), the table that is shown may not correspond to the current function.

Your graph window may look like the one in Figure 3.13 or it may be different. Since the graph of $y = -x^3 + 4x$ extends infinitely far left and right and also infinitely far up and down, the TI-83 can display only a piece of the actual graph. This displayed rectangular part is called a *viewing rectangle*. You can easily change the viewing rectangle to enhance your investigation of a graph.

The viewing rectangle in Figure 3.13 shows the part of the graph that extends horizontally from −10 to 10 and vertically from −10 to 10. Press WINDOW to see information about your viewing rectangle. Figure 3.14 shows the WINDOW screen that corresponds to the viewing rectangle in Figure 3.13. This is the *standard* viewing rectangle for the TI-83.

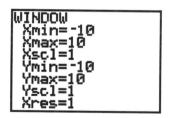

Figure 3.14: Standard WINDOW

The variables Xmin and Xmax are the minimum and maximum x-values of the viewing rectangle; Ymin and Ymax are the minimum and maximum y-values.

Xscl and Yscl set the spacing between tick marks on the axes.

Xres sets pixel resolution (1 through 8) for function graphs.

Technology Tip: Small Xres values improve graph resolution, but may cause the TI-83 to draw graphs more slowly.

Use the arrow keys ▲ and ▼ to move up and down from one line to another in this list; pressing the ENTER key will move down the list. Enter a new value to over-write a previous value and then press ENTER. Remember that a minimum *must* be less than the corresponding maximum or the TI-83 will issue an error message. Also, remember to use the (−) key, not − (which is subtraction), when you want to enter a negative value. Figures 3.13-14, 3.15-16, and 3.17-18 show different WINDOW screens and the corresponding viewing rectangle for each one.

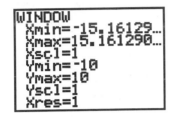

Figure 3.15: Square window

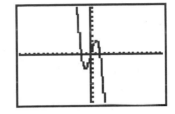

Figure 3.16: Graph of $y = -x^3 + 4x$

To initialize the viewing rectangle quickly to the *standard* viewing rectangle (Figure 3.14), press ZOOM 6[ZStandard]. To set the viewing rectangle quickly to a square (Figure 3.15), press ZOOM 5[ZSquare]. More information about square windows is presented later in Section 3.2.4.

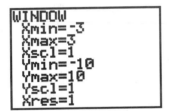

Figure 3.17: Custom window

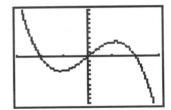

Figure 3.18: Graph of $y = -x^3 + 4x$

Sometimes you may wish to display grid points corresponding to tick marks on the axes. This and other graph format options may be changed by pressing 2nd FORMAT to display the FORMAT menu (Figure 3.19). Use arrow keys to move the blinking cursor to GridOn; press ENTER and then GRAPH to redraw the graph. Figure 3.20 shows the same graph as in Figure 3.18 but with the grid turned on. In general, you'll want the grid turned *off*, so do that now by pressing 2nd FORMAT, use the arrow keys to move the blinking cursor to GridOff, and press ENTER and CLEAR.

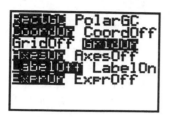

Figure 3.19: FORMAT menu

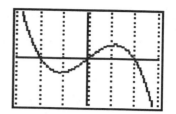

Figure 3.20: Grid turned on for $y = -x^3 + 4x$

Technology Tip: On the TI-83, the style of your graph can be changed by changing the icon to the left of Y_1 on the Y= screen. To change the icon press Y= ◄ ◄ and then ENTER repeatedly to scroll through the different styles available.

3.2.3 Graphing Step and Piecewise-Defined Functions: The greatest integer function, written $[[x]]$, gives the greatest *integer* less than or equal to a number x. On the TI-83, the greatest integer function is called int and is located under the NUM sub-menu of the MATH menu (Figure 3.5). From the home screen, calculate $[[6.78]] = 6$ by pressing MATH ▶ 5*[int(]* 6.78 ENTER.

To graph $y = [[x]]$, go into the Y= menu, move beside Y_1 and press CLEAR MATH ▶ 5*[int(]* X,T,θ,n GRAPH. Figure 3.21 shows this graph in a viewing rectangle from −5 to 5 in both directions.

The true graph of the greatest integer function is a step graph, like the one in Figure 3.22. For the graph of $y = [[x]]$, a segment should *not* be drawn between every pair of successive points. You can change from Connected line to Dot graph on the TI-83 by opening the MODE menu. Move the cursor down to the fifth line; select whichever graph type you require; press ENTER to put it into effect, and GRAPH to see the result.

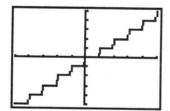

Figure 3.21: **Connected** graph of $y = [[x]]$

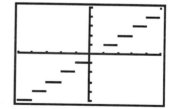

Figure 3.22: **Dot** graph of $y = [[x]]$

Make sure to change your TI-83 back to Connected line, since most of the functions that you will be graphing should be viewed this way.

The TI-83 can graph piecewise-defined functions by using the options in the TEST menu (Figure 3.23) that is displayed by pressing 2nd TEST. Each TEST function returns the value 1 if the statement is true, and the value 0 if the statement is false.

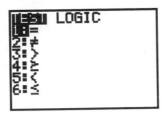

Figure 3.23: 2nd TEST menu

For example, to graph the function $f(x) = \begin{cases} x^2 + 2, & x < 0 \\ x - 1, & x \geq 0 \end{cases}$, enter the following keystrokes: Y= (X,T,θ,n x^2 +

1) (X,T,θ,n 2nd TEST 5[<] 0) + (X,T,θ,n − 1) (X,T,0,n 2nd TEST 4[>] 0) (Figure 3.24). Then press GRAPH to display the graph. Figure 3.25 shows this graph in a viewing rectangle from −5 to 5 in both directions.

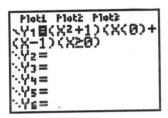

Figure 3.24: Piecewise-defined function

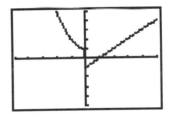

Figure 3.25: Piecewise-defined graph

3.2.4 Graphing a Circle: Here is a useful technique for graphs that are not functions but can be "split" into a top part and a bottom part, or into multiple parts. Suppose you wish to graph the circle of radius 6 whose equation is $x^2 + y^2 = 36$. First solve for y and get an equation for the top semicircle, $y = \sqrt{36 - x^2}$, and for the bottom semicircle, $y = -\sqrt{36 - x^2}$. Then graph the two semicircles simultaneously.

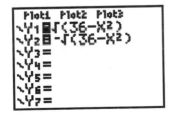

Figure 3.26: Two semicircles

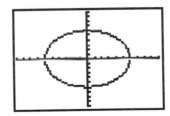

Figure 3.27: Circle's graph - standard WINDOW

Use the following keystrokes to draw this circle's graph. Enter $\sqrt{36 - x^2}$ as Y$_1$ and $-\sqrt{36 - x^2}$ as Y$_2$ (see Figure 3.26) by pressing Y= CLEAR 2nd $\sqrt{}$ 36 – x,T,θ,n x^2) ENTER CLEAR (–) 2nd $\sqrt{}$ 36 – x,T,θ,n x^2). Then press GRAPH to draw them both (Figure 3.27). Make sure that the WINDOW is set large enough to display a circle of radius 6.

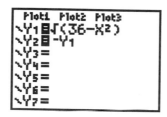

Figure 3.28: Using the VARS menu

Instead of entering $-\sqrt{36 - x^2}$ as Y$_2$, you could have entered –Y$_1$ as Y$_2$ and saved some keystrokes. On the TI-83, try this by going into the Y= screen and pressing ▼ to move the cursor down to Y$_2$. Then press CLEAR (–) VARS ▶ 1*[Function]* ENTER (Figure 3.28). The graph should be as before.

If your range were set to a viewing rectangle extending from –10 to 10 in both directions, your graph would look like Figure 3.27. Now this does *not* look a circle, because the units along the axes are not the same. You need what is called a "square" viewing rectangle. Press ZOOM 5*[ZSquare]* and see a graph that appears more circular.

Technology Tip: Another way to get a square graph is to change the range variables so that the value of Ymax – Ymin is approximately $\frac{2}{3}$ times Xmax – Xmin. For example, see the WINDOW in Figure 3.29 to get the corresponding graph in Figure 3.30. This method works because the dimensions of the TI-83's display are such that the ratio of vertical to horizontal is approximately $\frac{2}{3}$.

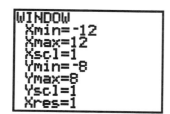

Figure 3.29: $\frac{\text{vertical}}{\text{horizontal}} = \frac{16}{24} = \frac{2}{3}$

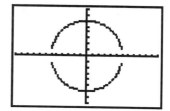

Figure 3.30: A "square" circle

The two semicircles in Figure 3.30 do not meet because of an idiosyncrasy in the way the TI-83 plots a graph.

3.2.5 TRACE: Graph the function $y = -x^3 + 4x$ from Section 3.2.2 using the standard viewing rectangle. (Remember to clear any other functions in the Y= screen.) Press any of the arrow keys ▲ ▼ ◄ ► and see the cursor move from the center of the viewing rectangle. The coordinates of the cursor's location are displayed at the bottom of the screen, as in Figure 3.31, in floating decimal format. This cursor is called a *free-moving cursor* because it can move from dot to dot *anywhere* in the graph window.

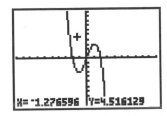

Figure 3.31: Free-moving cursor

Remove the free-moving cursor and its coordinates from the window by pressing GRAPH, CLEAR, or ENTER. Press an arrow key again and the free-moving cursor will reappear at the same point you left it.

Press TRACE to enable the left ◄ and right ► arrow keys to move the cursor along the function. The cursor is no longer free-moving, but is now constrained to the function. The coordinates that are displayed belong to points on the function's graph, so the y-coordinate is the calculated value of the function at the corresponding x-coordinate (Figure 3.32). The TI-83 displays the function that is being traced in the upper left of the screen while the TRACE feature is being used.

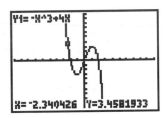

Figure 3.32: TRACE

Now plot a second function, $y = -.25x$, along with $y = -x^3 + 4x$. Press Y= and move the cursor to the Y_2 line and enter $-.25x$, then press GRAPH to see both functions.

Notice that in Figure 3.33 the equal signs next to Y_1 and Y_2 are *both* highlighted. This means that *both* functions will be graphed. In the Y= screen, move the cursor directly on top of the equal sign next to Y_1 and press ENTER. This equal sign should no longer be highlighted (Figure 3.35). Now press GRAPH and see that only Y_2 is plotted (Figure 3.36).

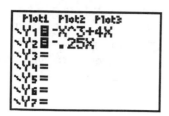

Figure 3.33: Two functions

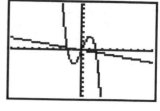

Figure 3.34: $y = -x^3 + 4x$ and $y = -.25x$

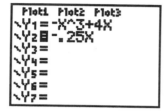

Figure 3.35: only Y_2 active

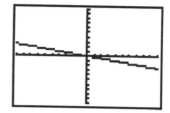

Figure 3.36: Graph of $y = -.25x$

Many different functions can be stored in the Y= list and any combination of them may be graphed simultaneously. You can make a function active or inactive for graphing by pressing ENTER on its equal sign to highlight (activate) or remove the highlight (deactivate). Now go back to the Y= screen and do what is needed in order to graph Y_1 but not Y_2.

Now activate both functions so that both graphs are plotted. Press TRACE and the cursor appears first on the graph of $y = -x^3 + 4x$ because it is higher up on the Y= list. You know that the cursor is on this function, Y_1, because this function is displayed in the upper left of the screen. Press the up ▲ or down ▼ arrow key to move the cursor vertically to the graph of $y = -.25x$. Now the function $Y_2 = -.25x$ is shown in the upper left of the screen. Next press the left and right arrow keys to trace along the graph of $y = -.25x$. When more than one function is plotted, you can move the trace cursor vertically from one graph to another with the ▲ and ▼ keys.

Technology Tip: By the way, trace the graph of $y = -.25x$ and press and hold either ◄ or ►. Eventually you will reach the left or right edge of the window. Keep pressing the arrow key and the TI-83 will allow you to continue the trace by panning the viewing rectangle. Check the WINDOW screen to see that the Xmin and Xmax are automatically updated.

The TI-83 has a display of 95 horizontal columns of pixels and 63 vertical rows, so when you trace a curve across a graph window, you are actually moving from Xmin to Xmax in 94 equal jumps, each called Δx. You would calculate the size of each jump to be $\Delta x = \dfrac{\text{Xmax} - \text{Xmin}}{94}$. Sometimes you may want the jumps to be friendly numbers like 0.1 or 0.25 so that, when you trace along the curve, the x-coordinates will be incremented by such a convenient amount. Just set your viewing rectangle for a particular increment Δx by making

TI-83 Graphics Calculator

Xmax = Xmin + 94 · Δx. For example, if you want Xmin = −5 and $\Delta x = 0.3$, set Xmax = −5 + 94 · 0.3 = 23.2. Likewise, set Ymax = Ymin + Δy if you want the vertical increment to be some special Δy.

To center your window around a particular point, say (h, k), and also have a certain Δx, set Xmin = $h - 47 \cdot \Delta x$ and make Xmax = $h + 47 \cdot \Delta x$. Likewise, make Ymin = $k - 31 \cdot \Delta y$ and make Ymax = $k + 31 \cdot \Delta y$. For example, to center a window around the origin $(0, 0)$, with both horizontal and vertical increments of 0.25, set the range so that Xmin = 0 − 47 · 0.25 = −11.75, Xmax = 0 + 47 · 0.25 = 11.75, Ymin = 0 − 31 · 0.25 = −7.75 and Ymax = 0 + 31 · 0.25 = 7.75.

See the benefit by first plotting $y = x^2 + 2x + 1$ in a standard graphing window. Trace near its y-intercept, which is $(0, 1)$, and move towards its x-intercept, which is $(-1, 0)$. Then press ZOOM 4[ZDecimal] and trace again near the intercepts.

3.2.6 ZOOM: Plot again the two graphs, for $y = -x^3 + 4x$ and $y = -.25x$. There appears to be an intersection near $x = 2$. The TI-83 provides several ways to enlarge the view around this point. You can change the viewing rectangle directly by pressing WINDOW and editing the values of Xmin, Xmax, Ymin, and Ymax. Figure 3.38 shows a new viewing rectangle for the range displayed in Figure 3.37. The cursor has been moved near the point of intersection; move your cursor closer to get the best approximation possible for the coordinates of the intersection.

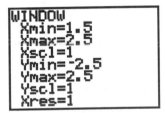

Figure 3.37: New WINDOW

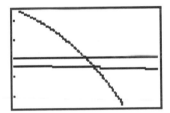

Figure 3.38: Closer view

A more efficient method for enlarging the view is to draw a new viewing rectangle with the cursor. Start again with a graph of the two functions $y = -x^3 + 4x$ and $y = -.25x$ in a standard viewing rectangle. (Press ZOOM 6[ZStandard] for the standard viewing window.)

Now imagine a small rectangular box around the intersection point, near $x = 2$. Press ZOOM 1[ZBox] (Figure 3.39) to draw a box to define this new viewing rectangle. Use the arrow keys to move the cursor, whose coordinates are displayed at the bottom of the window, to one corner of the new viewing rectangle you imagine.

Press ENTER to fix the corner where you moved the cursor; it changes shape and becomes a blinking square (Figure 3.40). Use the arrow keys again to move the cursor to the diagonally opposite corner of the new rectangle (Figure 3.41). If this box looks all right to you, press ENTER. The rectangular area you have enclosed will now enlarge to fill the graph window (Figure 3.42).

Figure 3.39: ZOOM menu

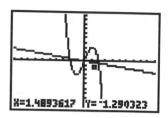

Figure 3.40: One corner selected

You may cancel the zoom any time *before* you press this last ENTER. Press ZOOM once more and start over. Press CLEAR or GRAPH to cancel the zoom, or press 2nd QUIT to cancel the zoom and return to the home screen.

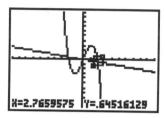

Figure 3.41: Box drawn

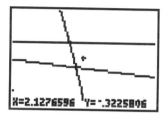

Figure 3.42: New viewing rectangle

You can also quickly magnify a graph around the cursor's location. Return once more to the standard window for the graph of the two functions $y = -x^3 + 4x$ and $y = -.25x$. Press ZOOM 2[Zoom In] and then press arrow keys to move the cursor as close as you can to the point of intersection near $x = 2$ (see Figure 3.43). Then press ENTER and the calculator draws a magnified graph, centered at the cursor's position (Figure 3.44). The range variables are changed to reflect this new viewing rectangle. Look in the WINDOW menu to verify this.

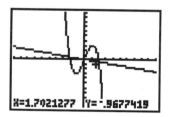

Figure 3.43: Before a zoom in

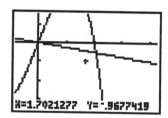

Figure 3.44: After a zoom in

As you see in the ZOOM menu (Figure 3.39), the TI-83 can zoom in (press ZOOM 2) or zoom out (press ZOOM 3). Zoom out to see a larger view of the graph, centered at the cursor position. You can change the horizontal and vertical scale of the magnification by pressing ZOOM ▶ 4[SetFactors] (see Figure 3.45) and editing XFact and YFact, the horizontal and vertical magnification factors (Figure 3.46).

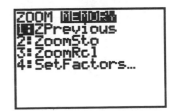

Figure 3.45: ZOOM MEMORY menu

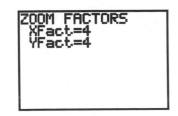

Figure 3.46: ZOOM MEMORY SetFactors

Technology Tip: An advantage of zooming in from square viewing window is that subsequent windows will also be square. Likewise, if you zoom in from a friendly viewing rectangle, the zoomed windows will also be friendly.

The default zoom factor is 4 in both direction. It is not necessary for Xfact and Yfact to be equal. sometimes, you may prefer to zoom in one direction only, so the other factor should be set to 1. As usual, press 2nd QUIT to leave the ZOOM menu.

Technology Tip: The TI-83 remembers the window it displayed before a zoom. So if you should zoom in too much and lose the curve, press ZOOM ▶ 1*[ZPrevious]* to go back to the window before. If you want to execute a series of zooms but then return to a particular window, press ZOOM ▶ 2*[ZoomSto]* to store the current window's dimensions. Later, press ZOOM ▶ 3*[ZoomRcl]* to recall the stored window.

3.2.7 Relative Minimums and Maximums: Graph $y = -x^3 + 4x$ once again in the standard viewing rectangle. This function appears to have a relative minimum near $x = -1$ and a relative maximum near $x = 1$. You may zoom and trace to approximate these extreme values.

First trace along the curve near the local minimum. Notice by how much the x-values and y-values change as you move from point to point. Trace along the curve until the y-coordinate is as *small* as you can get it, so that you are as close as possible to the local minimum, and zoom in (press ZOOM 2*[Zoom In]* ENTER or use a zoom box). Now trace again along the curve and, as you move from point to point, see that the coordinates change by smaller amounts than before. Keep zooming and tracing until you find the coordinates of the local minimum point as accurately as you need them, approximately $(-1.15, -3.08)$.

Follow a similar procedure to find the local maximum. Trace along the curve until the y-coordinate is as *great* as you can get it, so that you are as close as possible to the local maximum, and zoom in. The local maximum point on the graph of $y = -x^3 + 4x$ is approximately $(1.15, 3.08)$.

The TI-83 can automatically find the maximum and minimum points. Press 2nd CALC to display the CALCULATE menu (Figure 3.47). Choose 3*[minimum]* to calculate the minimum value of the function and 4*[maximum]* for the maximum. You will be prompted to trace the cursor along the graph first to a point *left* of the minimum/maximum (press ENTER to set this *left bound*). Then move to a point *right* of the minimum/maximum and set a *right bound* by pressing ENTER. Note the two arrows near the top of the display marking the left and right bounds (as in Figure 3.48).

Figure 3.47: CALCULATE menu

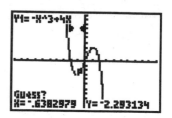

Figure 3.48: Finding a minimum

Next move the cursor along the graph between the two bounds and as close to the minimum/maximum as you can. This serves as a *guess* for the TI-83 to start its search. Good choices for the left bound, right bound, and guess can help the calculator work more efficiently and quickly. Press ENTER and the coordinates of the relative minimum/maximum point will be displayed (see Figure 3.49).

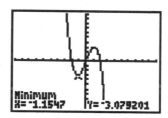

Figure 3.49: Relative minimum on $y = -x^3 + 4x$

Note that if you have more than one graph on the screen, the upper left corner of the TI-83 screen will show the equation of the function whose minimum/maximum is being calculated.

3.3   Solving Equations and Inequalities

3.3.1 Intercepts and Intersections: Tracing and zooming are also used to locate an x-intercept of a graph, where a curve crosses the x-axis. For example, the graph of $y = x^3 - 8x$ crosses the x-axis three times (Figure 3.50). After tracing over to the x-intercept point that is farthest to the left, zoom in (Figure 3.51). Continue this process until you have located all three intercepts with as much accuracy as you need. The three x-intercepts of $y = x^3 - 8x$ are approximately –2.828, 0, and 2.828.

Technology Tip: As you zoom in, you may also wish to change the spacing between tick marks on the x-axis so that the viewing rectangle shows scale marks near the intercept point. Then the accuracy of your approximation will be such that the error is less than the distance between two tick marks. Change the x-scale on the TI-83 from the WINDOW menu. Move the cursor down to Xscl and enter an appropriate value.

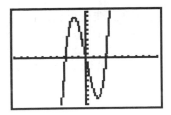

Figure 3.50: Graph of $y = x^3 - 8x$

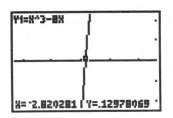

Figure 3.51: Near an x-intercept of $y = x^3 - 8x$

The x-intercept of a function's graph is a *zero* of the function, so press 2nd CALC to display the CALCULATE menu (Figure 3.47) and choose 2*[zero]* to find a zero of this function. Set a left bound, right bound, and guess as described in Section 3.2.7. The TI-83 shows the coordinates of the point and indicates that it is a zero (Figure 3.52)

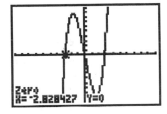

Figure 3.52: A zero of $y = x^3 - 8x$

TRACE and ZOOM are especially important for locating the intersection points of two graphs, say the graphs of $y = -x^3 + 4x$ and $y = -.25x$. Trace along one of the graphs until you arrive close to an intersection point. Then press ▲ or ▼ to jump to the other graph. Notice that the x-coordinate does not change, but the y-coordinate is likely to be different (Figures 3.53 and 3.54).

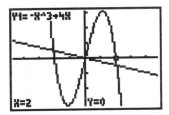

Figure 3.53: Trace on $y = -x^3 + 4x$

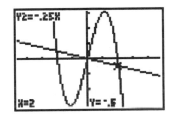

Figure 3.54: Trace on $y = -.25x$

When the two y-coordinates are as close as they can get, you have come as close as you now can to the point of intersection. So zoom in around the intersection point, then trace again until the two y-coordinates are as close as possible. Continue this process until you have located the point of intersection with as much accuracy as necessary.

You can also find the point of intersection of two graphs by pressing 2nd CALC 5*[intersect]*. Trace with the cursor first along one graph near the intersection and press ENTER; then trace with the cursor along the other graph and press ENTER. Marks + are placed on the graphs at these points. Finally, move the cursor near the point of intersection and press ENTER again. Coordinates of the intersection will be displayed at the bottom of the window (Figure 3.55).

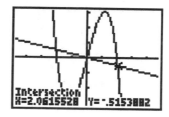

Figure 3.55: An intersection of $y = -x^3 + 4x$ and $y = -.25x$

3.3.2 *Solving Equations by Graphing:*

Suppose you need to solve the equation $24x^3 - 36x + 17 = 0$. First graph $y = 24x^3 - 36x + 17$ in a window large enough to exhibit *all* its x-intercepts, corresponding to all the equation's zeros (roots). Then use trace and zoom, or the TI-83's zero finder, to locate each one. In fact, this equation has just one solution, approximately $x = -1.414$.

Remember that when an equation has more than one x-intercept, it may be necessary to change the viewing rectangle a few times to locate all of them.

Technology Tip: To solve an equation like $24x^3 + 17 = 36x$, you may first transform it into standard form, $24x^3 - 36x + 17 = 0$, and proceed as above. However, you may also graph the *two* functions $y = 24x^3 + 17$ and $y = 36x$, then zoom and trace to locate their point of intersection.

3.3.3 *Solving Systems by Graphing:*

The solutions to a system of equations correspond to the points of intersection of their graphs (Figure 3.56). For example, to solve the system $y = x^3 + 3x^2 - 2x - 1$ and $y = x^2 - 3x - 4$, first graph them together. Then use zoom and trace or the intersect option in the CALCULATE menu, to locate their point of intersection, approximately $(-2.17, 7.25)$.

If you did not use the Intersect option, you must judge whether the two current y-coordinates are sufficiently close for $x = -2.17$ or whether you should continue to zoom and trace to improve the approximation. The solutions of the system of two equations $y = x^3 + 3x^2 - 2x - 1$ and $y = x^2 - 3x - 4$ correspond to the solutions of the single equation $x^3 + 3x^2 - 2x - 1 = x^2 - 3x - 4$, which simplifies to $x^3 + 2x^2 + x + 3 = 0$. So you may also graph $y = x^3 + 2x^2 + x + 3$ and find its x-intercepts to solve the system.

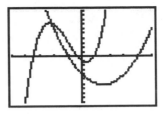

Figure 3.56: Graph of $y = x^3 + 3x^2 - 2x - 1$ and $y = x^2 - 3x - 4$

3.3.4 Solving Inequalities by Graphing: Consider the inequality $1 - \dfrac{3x}{2} \geq x - 4$. To solve it with your TI-83, graph the two functions $y = 1 - \dfrac{3x}{2}$ and $y = x - 4$ (Figure 3.57). First locate their point of intersection, at $x = 2$. The inequality is true when the graph of $y = 1 - \dfrac{3x}{2}$ lies *above* the graph of $y = x - 4$, and that occurs when $x < 2$. So the solution is the half-line $x \leq 2$, or $(-\infty, 2]$.

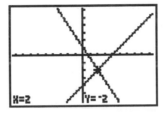

Figure 3.57: Solving $1 - \dfrac{3x}{2} \geq x - 4$

The TI-83 is capable of shading the region above or below a graph, or between two graphs. For example, to graph $y \geq x^2 - 1$, first graph the function $y = x^2 - 1$ as Y_1. Then press 2nd DRAW 7*[Shade(]* VARS ▶ 1*[Function]* 1*[Y₁]* , 10) ENTER (see Figure 3.58).

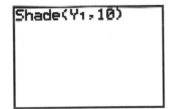

Figure 3.58: DRAW Shade

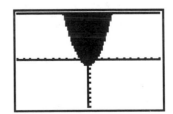

Figure 3.59: Graph of $y \geq x^2 - 1$

These keystrokes instruct the TI-83 to shade the region *above* $y = x^2 - 1$ and *below* $y = 10$ (chosen because this is the greatest y-value in the graph window) using the default shading option of vertical lines. The result is shown in Figure 3.59.

Now use shading to solve the previous inequality, $1 - \dfrac{3x}{2} \geq x - 4$. The function whose graph forms the lower boundary is named *first* in the SHADE command (see Figure 3.60). To enter this in your TI-83, press these keys: 2nd DRAW 7*[Shade(]* X,T,θ,n − 4 , 1 − 3 X,T,θ,n ÷ 2) ENTER (Figure 3.61). The shading extends left from $x = -2$, hence the solution to $1 - \dfrac{3x}{2} \geq x - 4$ is the half-line $x \leq 2$, or $(-\infty, \ 2]$.

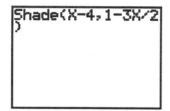

Figure 3.60: DRAW Shade command

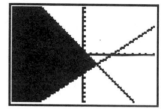

Figure 3.61: Graph of $1 - \dfrac{3x}{2} \geq x - 4$

More information about the DRAW menu is in the TI-83 manual.

3.4   Trigonometry

3.4.1 Degrees and Radians: The trigonometric functions can be applied to angles measured either in radians or degrees, but you should take care that the TI-83 is configured for whichever measure you need. Press MODE to see the current settings. Press ▼ twice and move down to the third line of the mode menu where angle measure is selected. Then press ◄ or ► to move between the displayed options. When the blinking cursor is on the measure you want, press ENTER to select it. Then press CLEAR or 2nd QUIT to leave the mode menu.

It's a good idea to check the angle measure setting before executing a calculation that depends on a particular measure. You may change a mode setting at any time and not interfere with pending calculations. Try the following keystrokes to see this in action.

Expression	Keystrokes	Display
sin 45°	MODE ▼ ▼ ▶ ENTER CLEAR	
	SIN 45 ENTER	.7071067812
sin π°	SIN 2nd π ENTER	.05480366515
sin π	MODE ▼ ▼ ENTER CLEAR	
	SIN 2nd π ENTER	0
sin 45	SIN 45 ENTER	.8509035245
$\sin\dfrac{\pi}{6}$	SIN (2nd π ÷ 6) ENTER	.5

The first line of keystrokes sets the TI-83 in degree mode and calculates the sine of 45 *degrees*. While the calculator is still in degree mode, the second line of keystrokes calculates the sine of π *degrees*, approximately 3.1415°. The third line changes to radian mode just before calculating the sine of π *radians*. The fourth line calculates the sine of 45 *radians* (the calculator remains in radian mode).

The TI-83 makes it possible to mix degrees and radians in a calculation. Execute these keystrokes to calculate $\tan 45° + \sin\frac{\pi}{6}$ as shown in Figure 3.62: TAN 45 2nd ANGLE 1) + SIN (2nd π ÷ 6) 2nd ANGLE 3) ENTER. Do you get 1.5 whether your calculator is in set *either* in degree mode *or* in radian mode?

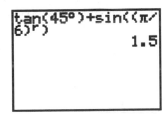

Figure 3.62: Angle measure

Technology Tip: The automatic left parenthesis that the TI-83 places after functions such as sine, cosine, and tangent (as noted in Section 3.1.3) *can* affect the outcome of calculations. In the previous example, the degree sign must be *inside* of the parentheses so that when the TI-83 is in radian mode, it calculates the tangent of 45 degrees, rather than converting the tangent of 45 radians into an equivalent number of degrees. Also, the parentheses around the fraction $\frac{\pi}{6}$ are required so that when the TI-83 is in radian mode, it converts $\frac{\pi}{6}$ into radians, rather than converting merely the 6 to radians. Experiment with the placement of parentheses to see how they affect the result of the computation.

3.4.2 Graphs of Trigonometric Functions: When you graph a trigonometric function, you need to pay careful attention to the choice of graph window. For example, graph $y = \dfrac{\sin 30x}{30}$ in the standard viewing rectangle. Trace along the curve to see where it is. Zoom in to a better window, or use the period and amplitude to establish better WINDOW values.

Technology Tip:. Since $\pi \approx 3.1$, set Xmin = 0 and Xmax = 6.2 to cover the interval from 0 to 2π.

Next graph $y = \tan x$ in the standard window first, then press ZOOM 7*[ZTrig]* to change to a special window for trigonometric functions in which the horizontal increment is $\frac{\pi}{24}$ or 7.5° and the vertical range is from −4 to 4. The TI-83 plots consecutive points and then connects them with a segment, so the graph is not exactly what you should expect. You may wish to change from Connected line to Dot graph (see Section 3.2.3) when you plot the tangent function.

3.5   Scatter Plots

3.5.1 Entering Data: The table shows the total prize money (in millions of dollars) awarded at the Indianapolis 500 race from 1981 to 1989. (*Source*: Indianapolis Motor Speedway Hall of Fame.)

Year	1981	1982	1983	1984	1985	1986	1987	1988	1989
Prize ($million)	$1.61	$2.07	$2.41	$2.80	$3.27	$4.00	$4.49	$5.03	$5.72

We'll now use the TI-83 to construct a scatter plot that represents these points and to find a linear model that approximates the given data.

The TI-83 holds data in *lists*. There are six list names in memory (L_1, L_2, L_3, L_4, L_5, L_6), but you can create as many list names as your TI-83 memory has space to store. Before entering data, clear all the data lists. The keystrokes to clear the six standard lists are: STAT 4*[ClrList]* 2nd L1 , 2nd L2 , 2nd L3 , 2nd L4 , 2nd L5 , 2nd L6 ENTER. This can also be done from within the list editor by highlighting each list title (L1, etc.) and pressing CLEAR ENTER.

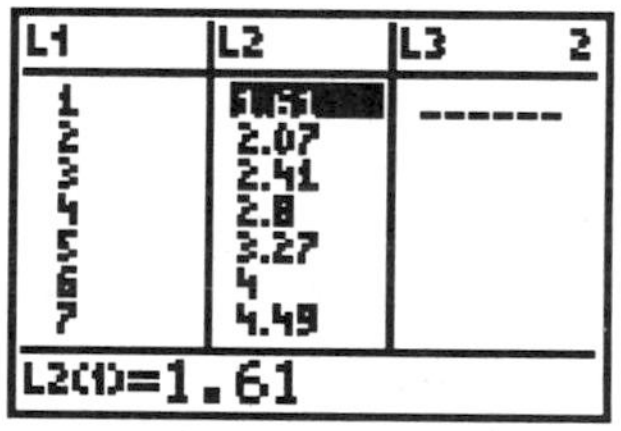

Figure 3.63: Entering data points

Now press STAT 1*[Edit]* to reach the list editor. Instead of entering the full year 198x, enter only x. Here are the keystrokes for the first three years: 1 ENTER 2 ENTER 3 ENTER and so on, then press ▶ to move to the first element of the next list and press 1.61 ENTER 2.07 ENTER 2.41 ENTER and so on (see Figure 3.63). Press 2nd QUIT when you have finished.

You may edit statistical data in the same way you edit expressions in the home screen. Move the cursor to any value you wish to change, then type the correction. To insert or delete data, move the cursor over the data point you wish to add or delete. Press 2nd INS and a new data point is created; press DEL and the data point is deleted.

3.5.2 Plotting Data: Once all the data points have been entered, press 2nd STAT PLOT to display the Plot1 screen. Press ENTER to turn Plot1 on, select the other options shown in Figure 3.64, and press GRAPH. (Make sure that you have cleared or turned off any functions in the Y= screen, or those functions will be graphed simultaneously.) Figure 3.65 shows this plot in a window from 0 to 10 horizontally and vertically. You may now press TRACE to move from data point to data point.

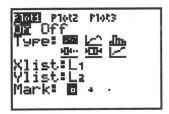

Figure 3.64: Plot1 menu

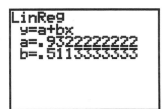

Figure 3.65: Scatter plot

To draw the scatter plot in a window adjusted automatically to include all the data you entered, press ZOOM 9 *[ZoomStat]*.

When you no longer want to see the scatter plot, press 2nd STAT PLOT 1, move the cursor to OFF, and press ENTER. The TI-83 still retains all the data you entered.

3.5.3 Regression Line: The TI-83 calculates slope and *y*-intercept for the line that best fits all the data. The TI-83 can calculate regression lines in two equivalent forms. After the data points have been entered, press STAT ▶ 4*[LinReg(ax+b)]* ENTER to calculate a linear regression model with the slope named a and the *y*-intercept named b (Figure 3.66). Pressing STAT ▶ 8*[LinReg(a+bx)]* ENTER produces a linear regression model with the roles of a and b reversed (Figure 3.67)

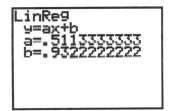

Figure 3.66: Linear regression: STAT ▶ 4

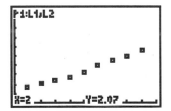

Figure 3.67: Linear regression: STAT ▶ 8

Technology Tip: The number r (between −1 and 1) is called the *correlation coefficient* and measures the goodness of fit of the linear regression with the data. The closer the absolute value of r is to 1, the better the fit; the closer the absolute value of r is to 0, the worse the fit. Press VARS 5*[Statistics]* ▶ ▶*[EQ]* 7*[r]* ENTER for r. The TI-83 will display both the correlation coefficient and the coefficient of determination (r^2) if the Diagnostic mode is on. This is done by pressing 2nd CATALOG and then using the ▼ (*down* arrow) to scroll to DiagnosticOn and then pressing ENTER ENTER. (Pressing CATALOG displays an alphabetical list of all functions and instructions on the TI-83.) For now, scroll to DiagnosticOff in the CATALOG and press ENTER ENTER to turn the diagnostics off.

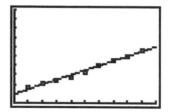

Figure 3.68: Linear regression line

Turn Plot1 on again, if it is not currently displayed. Graph the regression line $y = ax + b$ by pressing Y=, in-activating any existing functions, moving to a free line or clearing one, the pressing VARS 5*[Statistics]* ▶ ▶ 1*[RegEQ]* GRAPH. See how well this line fits with your data points (Figure 3.68).

3.5.4 Exponential Growth Model: The table shows the world population (in millions) from 1980 to 1992.

Year	1980	1985	1986	1987	1988	1989	1990	1991	1992
Population (millions)	4453	4850	4936	5024	5112	5202	5294	5384	5478

Follow the procedure described above to enter the data in order to find an exponential model that approximates the given data. Use 0 for 1980, 5 for 1985, and so on.

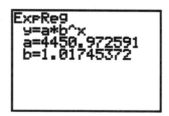

Figure 3.69: Exponential growth model

The TI-83 will not compute the exponential growth model $y = ae^{cx}$. The exponential regression that The TI-83 will compute is of the form $y = ab^x$. To get this exponential growth model press STAT ▶ 0[ExpReg] ENTER to find the values of **a** and **b** (Figure 3.69). In this case, the exponential growth model is $y = 4451(1.0174537^x)$. To convert this to the form $y = ae^{cx}$, the required equation is $c = \ln b$, and the exponential growth model in this case is $y = 4451e^{x\ln 1.0174537}$ or $y = 4451e^{0.017303t}$.

If you wish to plot and graph the data, follow the method for linear regression. Set an appropriate range for the data and then press 2nd STAT PLOT ENTER ENTER GRAPH. The data will now be plotted in the range. As in the linear regression model, press Y=, inactivating any existing functions, moving to a free line or clearing one, the pressing VARS 5[Statistics] ▶ ▶ 1[RegEQ] GRAPH to graph the exponential growth model. Note that the exponential regression model does not need to be converted to the form $y = ae^{cx}$ before graphing.

3.6 Matrices

3.6.1 Making a Matrix:
The TI-83 can work with 10 different matrices (A through J). Here's how to create this 3×4 matrix $\begin{bmatrix} 1 & -4 & 3 & 5 \\ -1 & 3 & -1 & -3 \\ 2 & 0 & -4 & 6 \end{bmatrix}$ in your calculator.

Press MATRX to see the matrix menu (Figure 3.70); then press ▶ ▶ or just ◀ to switch to the matrix EDIT menu. Whenever you enter the matrix EDIT menu, the cursor starts at the top matrix. Move to another matrix by repeatedly pressing ▼. For now, press ENTER to edit matrix [A].

The display will show the dimension of matrix [A] if the matrix exists; otherwise, it will display 1×1 (Figure 3.71). Change the dimensions of matrix [A] by pressing 3 ENTER 4 ENTER. Simply press ENTER or an arrow key to accept an existing dimension. The matrix shown in the window changes in size to reflect a changed dimension.

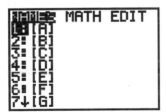

Figure 3.70: MATRX menu

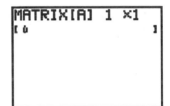

Figure 3.71: Editing a matrix

Use the arrow keys or press ENTER repeatedly to move the cursor to a matrix element you want to change. If you press ENTER, you will move right across a row and then back to the first column of the next row. At the right edge of the screen in Figure 3.72, there are dashes to indicate more columns than are shown. Go to them by pressing ▶ as many times as necessary. The ordered pair at the bottom left of the screen shows the cursor's current location within the matrix. The element in the second row and first column in Figure 3.72 is highlighted, so that the ordered pair at the bottom of the window is 2 , 1 and the screen shows that element's current value. Continue to enter all the elements of matrix [A]; press ENTER after inputing each value.

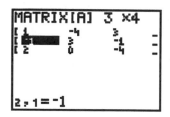

Figure 3.72: Editing a matrix

When you are finished, leave the editing screen by pressing 2nd QUIT to return to the home screen.

3.6.2 Matrix Math: From the home screen, you can perform many calculations with matrices. To see matrix [A], press MATRX 1 ENTER (Figure 3.73).

Perform the scalar multiplication 2[A] pressing 2 MATRX 1 ENTER. The resulting matrix is displayed on the screen. To replace matrix [B] by 2[A] press 2 MATRX 1 STO♦ MATRX 2 ENTER (Figure 3.74), or if you do this immediately after calculating 2[A], press only STO♦ MATRX 2 ENTER. The calculator will display the matrix. Press MATRX to verify that the dimensions of matrix [B] have been changed automatically to reflect this new value.

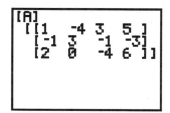

Figure 3.73: Matrix [A]

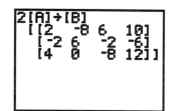

Figure 3.74: Matrix [B]

To add two matrices, say [A] and [B], create [B] (with the same dimensions as [A]) and then press MATRX 1 + MATRX 2 ENTER. Again, if you want to store the answer as a specific matrix, say [C], then press STO♦ MATRX 3. Subtraction is performed in similar manner.

Now set the dimensions of [C] to 2×3 and enter the matrix $\begin{bmatrix} 2 & 0 & 3 \\ 1 & -5 & -1 \end{bmatrix}$ as [C]. For matrix multiplication of [C] by [A], press MATRX 3 × MATRX 1 ENTER. If you tried to multiply [A] by [C], your TI-83 would signal an error because the dimensions of the two matrices do not permit multiplication in this way.

The *transpose* of a matrix is another matrix with the rows and columns interchanged. The symbol for the transpose of [A] is [A]$^\mathsf{T}$. To calculate [A]$^\mathsf{T}$, press MATRX 1 MATRX ▶ 2*[$^\mathsf{T}$]* ENTER.

3.6.3 Row Operations: Here are the keystrokes necessary to perform elementary row operations on a matrix. Your textbook provides a more careful explanation of the elementary row operations and their uses.

To interchange the second and third rows of the matrix [A] that was defined above, press MATRX ▶ ALPHA C*[rowSwap(]* MATRX 1 , 2 , 3) ENTER (see Figure 3.75). The format of this command is rowSwap(*matrix, row1, row2*).

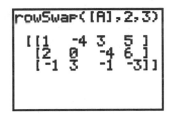

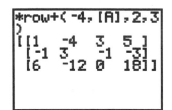

Figure 3.75: Swap rows 2 and 3   Figure 3.76: Add –4 times row 2 to row 3

To add row 2 and row 3 and *store* the results in row 3, press MATRX ▶ ALPHA D*[row+(]* MATRX 1 , 2 , 3) ENTER. The format of this command is row+(*matrix, row1, row2*).

To multiply row 2 by –4 and *store* the results in row 2, thereby replacing row 2 with new values, press MATRX ▶ ALPHA E*[*row(]* (–) 4 , MATRX 1 , 2) ENTER. The format of this command is *row(*value, matrix, row*).

To multiply row 2 by –4 and *add* the results to row 3, thereby replacing row 3 with new values, press MATRX ▶ ALPHA F*[*row+(]* (–) 4 , MATRX 1 , 2, 3) ENTER (see Figure 3.76). The format of this command is *row+(*scalar, matrix, row1, row2*).

Note that your TI-83 does *not* store a matrix obtained as the result of any row operation. So, when you need to perform several row operations in succession, it is a good idea to store the result of each one in a temporary place. You may wish to use matrix [J] to hold such intermediate results.

For example, use row operations to solve this system of linear equations: $\begin{cases} x - 2y + 3z = 9 \\ -x + 3y = -4 \\ 2x - 5y + 5z = 17 \end{cases}$.

First enter this *augmented matrix* as [A] in your TI-83: $\begin{bmatrix} 1 & -2 & 3 & 9 \\ -1 & 3 & 0 & -4 \\ 2 & -5 & 5 & 17 \end{bmatrix}$. Next store this matrix as [E] (press

MATRX 1 STO◆ MATRX 5 ENTER), so you may keep the original in case you need to recall it.

Here are the row operations and their associated keystrokes. At each step, the result is stored in [E] and replaces the previous matrix [E]. The completion of the row operations is shown in Figure 3.77.

Row Operations	Keystrokes
add row 1 to row 2	MATRX ▶ ALPHA D MATRX 5 , 1 , 2) STO◆ MATRX 5 ENTER
add −2 times row 1 to row 3	MATRX ▶ ALPHA F (−) 2 , MATRX 5 , 1 , 3) STO◆ MATRX 5 ENTER
add row 2 to row 3	MATRX ▶ ALPHA D MATRX 5 , 2, 3,) STO◆ MATRX 5 ENTER
multiply row 3 by $\frac{1}{2}$	MATRX ▶ ALPHA E 1 ÷ 2 , MATRX 5 , 3) STO◆ MATRX 5 ENTER

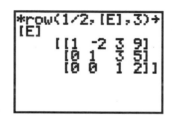

Figure 3.77: Final matrix after row operations

Thus $z = 2$, so $y = -1$ and $x = 1$.

Technology Tip: The TI-83 can produce a row-echelon form and the reduced row-echelon form of a matrix. The row-echelon form of matrix [A] is obtained by pressing MATRX ▶ ALPHA A*[ref(]* MATRX 1) ENTER and the reduced row-echelon form is obtained by pressing MATRX ▶ ALPHA B*[rref(]* MATRX 1) ENTER. Note that the row-echelon form of a matrix is not unique, so your calculator may not get exactly the same matrix as you do by using row operations. However, the matrix that the TI-83 produces will result in the same solution to the system.

3.6.4 Determinants and Inverses: Enter this 3×3 square matrix as [A]: $\begin{bmatrix} 1 & -2 & 3 \\ -1 & 3 & 0 \\ 2 & -5 & 5 \end{bmatrix}$. To calculate its de-

terminant $\begin{vmatrix} 1 & -2 & 3 \\ -1 & 3 & 0 \\ 2 & -5 & 5 \end{vmatrix}$, go to the home screen and press MATRX ▶ 1*[det(]* MATRX 1) ENTER. You

should find that the determinant is 2 as shown in Figure 3.78.

Since the determinant of the matrix is not zero, it has an inverse matrix. Press MATRX 1 x^{-1} ENTER to calculate the inverse. The result is shown in Figure 3.79.

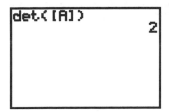

Figure 3.78: Determinant of [A]

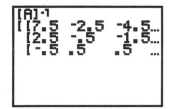

Figure 3.79: Inverse of [A]

Now let's solve a system of linear equations by matrix inversion. Once again, consider $\begin{cases} x - 2y + 3z = 9 \\ -x + 3y = -4 \\ 2x - 5y + 5z = 17 \end{cases}$.

The coefficient matrix for this system is the matrix $\begin{bmatrix} 1 & -2 & 3 \\ -1 & 3 & 0 \\ 2 & -5 & 5 \end{bmatrix}$ which was entered as matrix [A] in the previous example. Now enter the matrix $\begin{bmatrix} 9 \\ -4 \\ 17 \end{bmatrix}$ as [B]. Then press MATRX 1 x^{-1} × MATRX 2 ENTER to get the answer as shown in Figure 3.80

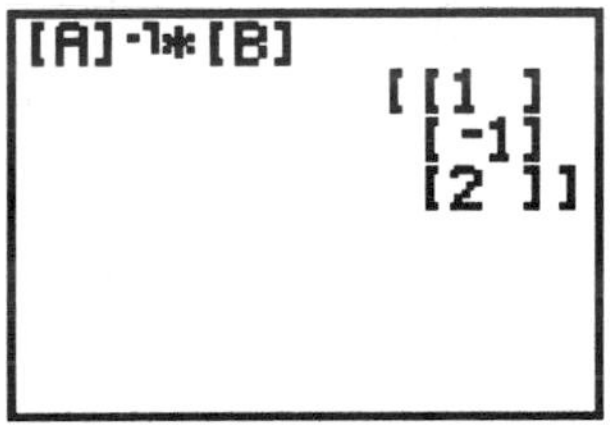

Figure 3.80: Solution matrix

The solution is still $x = 1$, $y = -1$, and $z = 2$.

3.7   Sequences

*3.7.1 Iteration with the **ANS** key:* The ANS key enables you to perform *iteration*, the process of evaluating a function repeatedly. As an example, calculate $\dfrac{n-1}{3}$ for $n = 27$. Then calculate $\dfrac{n-1}{3}$ for $n =$ the answer to the previous calculation. Continue to use each answer as n in the *next* calculation. here are keystrokes to accomplish this iteration on the TI-83 calculator. (See the results in Figure 3.81.) Notice that when you use ANS in place of n in a formula, it is sufficient to press ENTER to continue an iteration.

Iteration	Keystrokes	Display
1	27 ENTER	27
2	(2nd ANS – 1) ÷ 3 ENTER	8.666666667
3	ENTER	2.555555556
4	ENTER	.5185185185
5	ENTER	–.1604938272

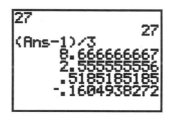

Figure 3.81: Iteration

Press ENTER several more times and see what happens with this iteration. You may wish to try it again with a different starting value.

3.7.2 Arithmetic and Geometric Sequences: Use iteration with the ANS variable to determine the n-th term of a sequence. For example, find the 18th term of an *arithmetic* sequence whose first term is 7 and whose common difference is 4. Enter the first term 7, then start the progression with the recursion formula, 2nd ANS + 4 ENTER. This yields the 2nd term, so press ENTER sixteen more times to find the 18th term. For a *geometric* sequence whose common ratio is 4, start the progression with 2nd ANS × 4 ENTER.

You can also define the sequence recursively with the TI-83 by selecting Seq in the MODE menu (see Figure 3.1). Once again, let's find the 18th term of an *arithmetic* sequence whose first term is 7 and whose common difference is 4. Press MODE ▼ ▼ ▼ ► ► ► ENTER 2nd QUIT. Then press Y= to edit any of the TI-83's three sequences, u_n, v_n, or w_n. Make sure that nMin is set to 1, because the first term is u_1 where $n = 1$. Make $u_n = u_{n-1} + 4$ and $u_1 = 7$ by pressing 2nd u (X,T,θ,n – 1) + 4 ENTER 7 ENTER (Figure 3.82). Now, when you press the variable key, X,T,θ,n, you get an n because the calculator is in sequence mode. Press 2nd QUIT to return to the home screen. To find the 18th term of this sequence, calculate u_{18} by pressing 2nd u (18) ENTER (Figure 3.83).

TI-83 Graphics Calculator

Figure 3.82: Sequential Y= menu

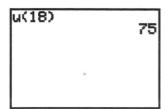

Figure 3.83: Sequence mode

Of course, you could also use the *explicit* formula for the n-th term of an arithmetic sequence $t_n = a + (n-1)d$. First enter values for the variables a, d, and n, then evaluate the formula by pressing ALPHA A + (ALPHA N – 1) ALPHA D ENTER. For a geometric sequence whose n-th term is given by $t_n = a \cdot r^{n-1}$, enter values for the variables a, d, and r, then evaluate the formula by pressing ALPHA A ALPHA R $\wedge$ (ALPHA N – 1) ENTER.

To use the explicit formula in Seq MODE, make $u_n = 7 + (n-1) \cdot 4$ by pressing Y= 7 + (X,T,θ,n – 1) × 4 ENTER 2nd QUIT. Once more, calculate u_{18} by pressing 2nd u (18) ENTER.

3.7.3 Finding Sums of Sequences: You can find the sum of a sequence by combining the sum(feature on the LIST MATH menu with the seq(feature on the LIST OPS menu. The format of the sum(command is sum(*list, start, end*), where the optional arguments *start* and *end* determine which elements of *list* are summed. The format of the seq(command is seq(*expression, variable, begin, end, increment*), where the optional argument *increment* indicates the difference between successive points at which *expression* is evaluated.

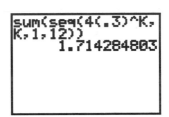

Figure 3.84: $\displaystyle\sum_{n=1}^{12} 4(0.3)^n$

For example, suppose you want to find the sum $\displaystyle\sum_{n=1}^{12} 4(0.3)^n$. Press 2nd LIST ◀ 5*[sum(]* 2nd LIST ▶ 5*[seq(]* 4 (.3) $\wedge$ ALPHA K , ALPHA K , 1, 12)) ENTER (Figure 3.84). Note that the sum(command does not need a starting or ending point, since every term in the sequence is being summed. Also, any letter can be used for the variable in the sum, i.e., the K could just have easily been an A or an N.

Now calculate the sum starting at $n = 0$ by using 2nd ENTRY to edit the range. You should obtain a sum of approximately 5.712848.

3.8   Parametric and Polar Graphs

3.8.1 Graphing Parametric Equations: The TI-83 plots parametric equations as easily as it plots functions. Up to six pairs of parametric equations can be plotted. In the MODE menu (Figure 3.1) go to the fourth line from the top and change the setting to Par. Be sure, if the independent parameter is an angle measure, that the angle measure in the MODE menu has been set to whichever you need, Radian or Degree.

You can now enter the parametric functions. For example, here are the keystrokes need to graph the parametric equations $x = \cos^3 t$ and $y = \sin^3 t$. First check that angle measure is in radians. Then press Y= (COS X,T,θ,n)) ∧ 3 ENTER (SIN X,T,θ,n)) ∧ 3 ENTER (Figure 3.85). Note that when you press the variable key X,T,θ,n, you now get a T because the calculator is in parametric mode.

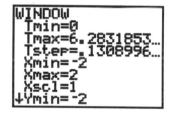

Figure 3.85: $x = \cos^3 t$ and $y = \sin^3 t$ Figure 3.86: Parametric WINDOW menu

Press WINDOW to set the graphing window and to initialize the values of T. In the standard window, the values of T go from 0 to 2π in steps of $\frac{\pi}{24} \approx 0.1309$, with the view from -10 to 10 in both directions. In order to provide a better viewing rectangle press ENTER three times to move the cursor down, then set the rectangle to go from -2 to 2 horizontally and vertically (Figure 3.86). Now press GRAPH to draw the graph (Figure 3.87).

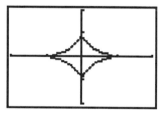

Figure 3.87: Parametric graph of $x = \cos^3 t$ and $y = \sin^3 t$

You may ZOOM and TRACE along parametric graphs just as you did with function graphs. However, unlike with function graphs, the cursor will not move to values outside of the T range, so the left arrow ◄ will not work when T = 0, and the right arrow ► will not work when T = 2π. As you trace along this graph, notice that the cursor moves in the *counterclockwise* direction as T increases.

3.8.2 Rectangular-Polar Coordinate Conversion: The 2nd ANGLE menu provides function for converting between rectangular and polar coordinate systems. These functions use the current angle measure setting, so it is a good idea to check the default angle measure before any conversion. Of course, you may override the current angle measure setting, as explained in Section 3.4.1. For the following examples, the TI-83 is set to radian measure.

Given the rectangular coordinates $(x, y) = (4, -3)$, convert to polar coordinates (r, θ) by pressing 2nd ANGLE 5[R ►Pr(] 4 , (–) 3) ENTER. The value of r is displayed; now press 2nd ANGLE 6[R ►Pθ(] 4, (–) 3) ENTER to display the value of θ (Figure 3.88). The polar coordinates are approximately (5, –0.6435).

Suppose $(r, \theta) = (3, \pi)$. Convert to rectangular coordinates (x, y) by pressing 2nd ANGLE 7[P ►Rx(] 3 , 2nd π) ENTER. The x-coordinate is displayed; press 2nd ANGLE 8[P ►Ry] 3 , 2nd π) ENTER to display the y-coordinate (Figure 3.89). The rectangular coordinates are (–3, 0).

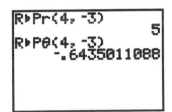

Figure 3.88: Rectangular to polar coordinates

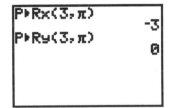

Figure 3.89: Polar to rectangular coordinates

3.8.3 Graphing Polar Equations: The TI-83 graphs polar functions in the form $r = f(\theta)$. In the fourth line of the MODE menu, select Pol for polar graphs. You may now graph up to six polar functions at a time. Be sure that the angle measure has been set to whichever you need, Radian or Degree. Here we will use radian measure.

For example, to graph $r = 4\sin\theta$, press Y= for the polar graph editing screen. Then enter the expression $4\sin\theta$ by pressing 4 SIN X,T,θ,n) ENTER. Now, when you press the variable key X,T,θ,n, you get a θ because the calculator is in polar mode..

Choose a good viewing rectangle and an appropriate interval and increment for θ. In Figure 3.90, the viewing rectangle is roughly "square" and extends from –6 to 6 horizontally and from –4 to 4 vertically. That is, Xmin, Xmax, Ymin, and Ymax are –6, 6, –4, and 4, respectively. (Refer back to the Technology Tip in Section 3.2.4.)

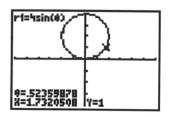

Figure 3.90: Polar graph of $r = 4\sin\theta$

Figure 3.90 shows *rectangular* coordinates of the cursor's location on the graph. You may sometimes wish to trace along the curve and see *polar* coordinates of the cursor's location. The first line of the FORMAT menu (Figure 3.19) has options for displaying the cursor's position in rectangular (RectGC) or polar (PolarGC) form.

3.9   Probability

3.9.1 Random Numbers: The command rand generates a number between 0 and 1. You will find this command in the PRB sub-menu of the MATH menu. Press MATH ◄ 1*[rand]* ENTER to generate a random number. Press ENTER to generate another number; keep pressing ENTER to generate more of them.

If you need a random number between, say, 0 and 10, then press 10 MATH ◄ 1*[rand]* ENTER. To get a random number between 5 and 15, press 5 + 10 MATH ◄ 1*[rand]* ENTER.

3.9.2 Permutations and Combinations: To calculate the number of permutations of 12 objects taken 7 at a time, $_{12}P_7$, press 12 MATH ◄ 2*[nPr]* 7 ENTER (Figure 3.91). Thus $_{12}P_7 = 3,991,680$.

For the number of combinations of 12 objects taken 7 at a time, $_{12}C_7$, press 12 MATH ◄ 3*[nCr]* 7 ENTER (Figure 3.91). Thus $_{12}C_7 = 792$.

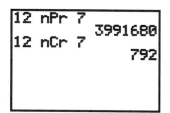

Figure 3.91: $_{12}P_7$ and $_{12}C_7$

3.9.3 Probability of Winning: A state lottery is configured so that each player chooses six different numbers from 1 to 40. If these six numbers match the six numbers drawn by the State Lottery Commission, the player wins the top prize. There are $_{40}C_6$ ways for the six numbers to be drawn. If you purchase a single lottery ticket, your probability of winning is 1 in $_{40}C_6$. Press 1 ÷ 40 MATH ◄ 3*[nCr]* 6 ENTER to calculate your chances, but don't be disappointed.

3.10   Programming

3.10.1 Entering a Program: The TI-83 is a programmable calculator that can store sequences of commands for later replay. Here's an example to show you how to enter a useful program that solves quadratic equations by the quadratic formula.

Press PRGM to access the programming menu. The TI-83 has space for many programs, each named by a name you give it. To create a new program now, start by pressing PRGM ◄ 1*[Create New]*.

For convenience, the cursor is a blinking **A**, indicating that the calculator is set to receive alphabetic characters. Enter a descriptive title of up to eight characters, letter, or numerals (but the first character must be a letter or θ). Name this program QUADRAT and press ENTER to go to the program editor.

In the program, each line begins with a colon (:) supplied automatically by the calculator. Any command you could enter directly in the TI-83's home screen can be entered as a line in a program. There are also special programming commands.

Input the program QUADRAT by pressing the keystrokes given in the listing below. You may interrupt program input at any stage by pressing 2nd QUIT. To return late for more editing, press PRGM ►, move the cursor down to this program's name, and press ENTER.

Each time you press ENTER while writing a program, the TI-83 *automatically* inserts the : character at the beginning of the next line.

The instruction manual for your TI-83 gives detailed information about programming. Refer to it to learn more about programming and how to use other features of your calculator.

Note that this program makes use of the TI-83's ability to compute complex numbers. Make sure that the type of numbers in the MODE menu (Figure 3.1) is set to a+b*i*.

Enter the program QUADRAT by pressing the given keystrokes.

Program Line	Keystrokes
: Disp "ENTER A"	PRGM ► 3 2nd A-LOCK " E N T E R ⌴ A " ENTER

  displays the words ENTER A on the TI-83 screen

| : Input A | PRGM ► 1 ALPHA A ENTER |

  waits for you to input a value that will be assigned to the variable A

| : Disp "ENTER B" | PRGM ► 3 2nd A-LOCK " E N T E R ⌴ B " ENTER |
| : Input B | PRGM ► 1 ALPHA B ENTER |

: Disp "ENTER C"	PRGM ► 3 2nd A-LOCK " E N T E R ⌴ C " ENTER
: Input C	PRGM ► 1 ALPHA C ENTER
: $B^2-4AC{\rightarrow}D$	ALPHA B x^2 – 4 ALPHA A ALPHA C STO• ALPHA D ENTER

calculates the discriminant and stores its value as D

: $(-B+\sqrt{}(D))/(2A){\rightarrow}M$	(((–) ALPHA B + 2nd $\sqrt{}$ ALPHA D)) ÷ (2 ALPHA A) STO• ALPHA M ENTER

calculates one root and stores it as M

: $(-B-\sqrt{}(D))/(2A){\rightarrow}N$	(((–) ALPHA B – 2nd $\sqrt{}$ ALPHA D)) ÷ (2 ALPHA A) STO• ALPHA N ENTER

: If D<0	PRGM 1 ALPHA D 2nd TEST 5 0 ENTER

tests to see if the discriminant is negative;

: Goto 1	PRGM 0 1 ENTER

if the discriminant is negative, jumps to the line Lbl 1 below; if the discriminant is not negative, continues on to the next line

: If D=0	PRGM 1 ALPHA D 2nd TEST 1 0 ENTER

tests to see if the discriminant is zero;

: Goto 2	PRGM 0 2 ENTER

if the discriminant is zero, jumps to the line Lbl 2 below; if the discriminant is not zero, continues on to the next line

: Disp "TWO REAL ROOTS", M	PRGM ► 3 2nd A-LOCK " T W O ⌴ R E A L ⌴ R O O T S " ALPHA , ALPHA M ENTER
: Pause	PRGM 8 ENTER

displays the message "TWO REAL ROOTS" and one root then pauses

:Disp N	PRGM ► 3 ALPHA N ENTER

displays the other root

: Stop	PRGM ALPHA F ENTER

stops program execution

: Lbl 1	PRGM 9 1 ENTER

jumping point for the Goto command above

: Disp "COMPLEX ROOTS", M	PRGM ▶ 3 2nd A-LOCK " C O M P L E X ⌴ R O O T S " ALPHA , ALPHA M ENTER
: Pause	PRGM 8 ENTER

displays the message "COMPLEX ROOTS" and one root, then pauses

:Disp N	PRGM ▶ 3 ALPHA N ENTER

displays the other root

: Stop	PRGM ALPHA F ENTER
: Lbl 2	PRGM 9 2 ENTER
:Disp "DOUBLE ROOT", M	PRGM ▶ 3 2nd A-LOCK " D O U B L E ⌴ R O O T " ALPHA , ALPHA M ENTER

displays a message in case there is a double root, and the solution (root)

When you have finished, press 2nd QUIT to leave the program editor and move on.

If you want to remove a program from memory, press 2nd MEM 2*[Delete]* 7*[Prgm]*. Then use the down arrow ▼ to move the indicator next to the name of the program you want to delete, and when the indicator is next to its name, press ENTER to remove it from the calculator's memory.

3.10.2 Executing a Program: To execute the program you have entered, press PRGM and the number corresponding to the program; and press ENTER to execute it. If you have forgotten its name, use the arrow keys to move through the program listing to find its description QUADRAT. Then press ENTER to execute it.

The program has been written to prompt you for values of the coefficients a, b, and c in a quadratic equation $ax^2 + bx + c = 0$. Input a value, then press ENTER to continue the program.

If you need to interrupt a program during execution, press ON 1.

The instruction manual for your TI-83 gives detailed information about programming. Refer to it to learn more about programming and how to use other features of your calculator.

3.11   Differentiation

3.11.1 Limits: Suppose you need to find this limit: $\lim_{x \to 0} \dfrac{\sin 4x}{x}$. With the calculator in FUNCTION graphing mode, plot the graph of $f(x) = \dfrac{\sin 4x}{x}$ in a convenient viewing rectangle that contains the point where the function appears to intersect the line $x = 0$ (because you want the limit as $x \to 0$). Your graph should support the conclusion that $\lim\limits_{x \to 0} \dfrac{\sin 4x}{x} = 4$ (Figure 3.92).

To test whether the conclusion that $\lim\limits_{x \to \infty} \dfrac{2x - 1}{x + 1} = 2$ is reasonable, evaluate $f(x) = \dfrac{2x - 1}{x + 1}$ for several large positive values of x (since you want the limit as $x \to \infty$). For example, evaluate $f(100)$, $f(1000)$, and $f(10,000)$. Another way to test the conclusion is to examine the graph of $f(x) = \dfrac{2x - 1}{x + 1}$ in a viewing rectangle that extends over large values of x. See, as in Figure 3.93 (where the viewing rectangle extends horizontally from 0 to 90), whether the graph is asymptotic to the horizontal line $y = 2$. Enter $\dfrac{2x - 1}{x + 1}$ for Y_1 and 2 for Y_2.

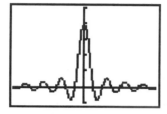

Figure 3.92: Checking $\lim\limits_{x \to 0} \dfrac{\sin 4x}{x} = 4$

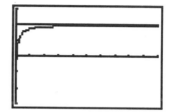

Figure 3.93: Checking $\lim\limits_{x \to \infty} \dfrac{2x - 1}{x + 1} = 2$

3.11.2 Numerical Derivatives: The derivative of a function f at x can be defined as the limit of the slopes of secant lines, so $f'(x) = \lim\limits_{\Delta x \to 0} \dfrac{f(x + \Delta x) - f(x - \Delta x)}{2\Delta x}$. And for small values of Δx, the expression $\dfrac{f(x + \Delta x) - f(x - \Delta x)}{2\Delta x}$ gives a good approximation to the limit.

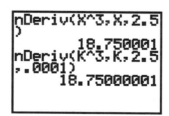

Figure 3.94: Using nDeriv(

TI-83 Graphics Calculator

The TI-83 has a function, nDeriv(which is available in the MATH menu that will calculate the *symmetric difference*, $\dfrac{f(x+\Delta x)-f(x-\Delta x)}{2\Delta x}$. So, to find a numerical approximation to $f'(2.5)$ when $f(x)=x^3$ and with $\Delta x = 0.001$, press MATH 8[nDeriv(] X,T,θ,n ∧ 3 , X,T,θ,n , 2.5 , .001) ENTER as shown in Figure 3.94. The format of this command is nDeriv(*expression, variable, value,* Δx), where the optional argument Δx controls the accuracy of the approximation. If no value for Δx is provided, the TI-83 automatically uses $\Delta x = 0.001$. The same derivative is also approximated in Figure 3.94 using $\Delta x = 0.0001$. For most purposes, $\Delta x = 0.001$ gives a very good approximation to the derivative. Note that any letter can be used for the variable.

Technology Tip: It is sometimes helpful to plot both a function and its derivative together. In Figure 3.96, the function $f(x)=\dfrac{5x-2}{x^2+1}$ and its numerical derivative (actually, an approximation to the derivative given by the symmetric difference) are graphed on viewing window that extends from −6 to 6 vertically and horizontally. You can duplicate this graph by first entering $\dfrac{5x-2}{x^2+1}$ for Y_1 and then entering its numerical derivative for Y_2 by pressing MATH 8[nDeriv(] VARS ▶ 1[Function] 1[Y_1] , X,T,θ,n , X,T,θ,n) (Figure 3.95).

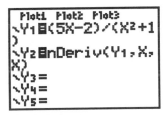

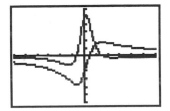

Figure 3.95: Entering $f(x)$ and $f'(x)$ Figure 3.96: Graphs of $f(x)$ and $f'(x)$

Technology Tip: To approximate the *second* derivative $f''(x)$ of a function $y=f(x)$ or to plot the second derivative, first enter the expression for Y_1 and its derivative for Y_2 as above. Then enter the second derivative for Y_3 by pressing MATH 8[nDeriv(] VARS ▶ 1[Function] 2[Y_2] , X,T,θ,n , X,T,θ,n).

You may also approximate a derivative while you are examining the graph of a function. When you are in a graph window, press 2nd CALC 6[dy/dx], then use the arrow keys to trace along the curve to a point where you want the derivative or enter a value and press ENTER. The TI-83 uses $\Delta x = 0.001$ for this approximation. For example, graph the function $f(x)=\dfrac{5x-2}{x^2+1}$ in the standard viewing rectangle. Then press 2nd CALC 6[dy/dx]. The coordinates of the point in the center of the of the range will appear. To find the numerical derivative at $x = -2.3$, press (−) 2.3 ENTER. Figure 3.97 shows the derivative at that point to be about − 0.7746922.

If more than one function is graphed you can use ▲ and ▼ to scroll between the functions.

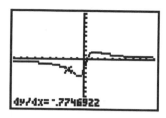

Figure 3.97: Derivative of $f(x) = \dfrac{5x-2}{x^2+1}$ at $x = -2.3$

3.11.3 Newton's Method: With the TI-83, you may iterate using Newton's method to find the zeros of a function. Recall that Newton's Method determines each successive approximation by the formula $x_{n+1} = x_n - \dfrac{f(x_n)}{f'(x_n)}$.

As an example of the technique, consider $f(x) = 2x^3 + x^2 - x + 1$. Enter this function as Y_1 and graph it in the standard viewing window. A look at its graph suggests that it has a zero near $x = -1$, so start the iteration by going to the home screen and storing -1 as x. Then press these keystrokes: X,T,θ,n – VARS ▶ 1*[Function]* 1*[Y₁]* ÷ MATH 8*[nDeriv(]* VARS ▶ 1*[Function]* 1*[Y₁]* , X,T,θ,n , X,T,θ,n) STO✦ X,T,θ,n ENTER ENTER (Figure 3.98) to calculate the first two iterations of Newton's method. Press ENTER repeatedly until two successive approximations differ by less than some predetermined value, say 0.0001. Note that each time you press ENTER, the TI-83 will use the *current* value of x, and that value is changing as you continue the iteration.

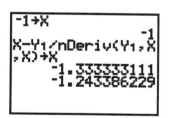

Figure 3.98: Newton's method

Technology Tip: Newton's Method is sensitive to your initial value for x, so look carefully at the function's graph to make a good first estimate. Also, remember that the method sometimes fails to converge!

You may want to write a short program for Newton's Method. See your calculator's manual for further information.

3.12   Integration

3.12.1 Approximating Definite Integrals: The TI-83 has a function, fnInt(, which is available in the MATH menu that will approximate a definite integral. For example, to find a numerical approximation to $\int_0^1 \cos x^2 dx$ press MATH 9*[fnInt(]* COS X,T,θ,n x^2) , X,T,θ,n , 0 , 1 , .001) ENTER (Figure 3.99). The TI-83 uses a method known as the Gauss-Kronrod method to perform the calculation. The format of this command is fnInt(*expression, variable, lower limit, upper limit, tolerance*), where the tolerance controls the accuracy of the approximation. The same integral is also approximated in Figure 3.99 using a tolerance of 0.00001, the TI-83's default that is used when no other tolerance is specified.

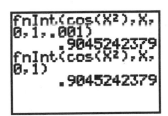

Figure 3.99: Using fnInt(

3.12.2 Areas: You may approximate the area under the graph of a function $y = f(x)$ between $x = A$ and $x = B$ with your TI-83. To do this you use the 2nd CALC menu when you have a graph displayed. For example, here are the keystrokes for finding the area under the graph of the function $y = \cos x^2$ between $x = 0$ and $x = 1$. The area is represented by the definite integral $\int_0^1 \cos x^2 dx$. First clear any existing graphs and then press Y= COS X,T,θ,n x^2) followed by GRAPH to draw the graph. The range in Figure 3.100 extends from −5 to 5 horizontally and from −2 to 2 vertically. Now press 2nd CALC 7*[∫f(x)dx]*. The TI-83 will prompt you for the lower and upper limits which are entered by pressing 0 ENTER 1 ENTER. The region between the graph and the x-axis from the lower limit to the upper limit is shaded and the approximate value of the integral is displayed (Figure 3.101).

Technology Tip: If the function takes on negative values between the lower and upper limits, the value that the TI-83 displays it the value of the integral, *not* the area of the shaded region.

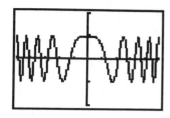

Figure 3.100: Graph of $y = \cos x^2$

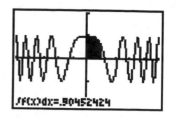

$\int f(x)dx = .90452424$

Figure 3.101: Graph and area

Technology Tip: Suppose that you want to find the area between two functions, $y = f(x)$ and $y = g(x)$ from $x = A$ and $x = B$. If $f(x) \ge g(x)$ for $A \le x \le B$, then graph the expression $f(x) - g(x)$ and use the method above to find the required area.

Chapter 4

Texas Instruments TI-85

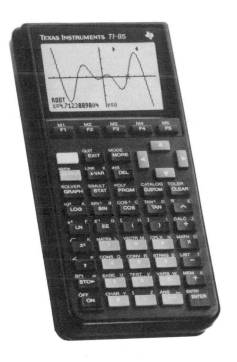

4.1   Getting started with the TI-85

4.1.1 Basics: Press the ON key to begin using your TI-85 calculator. If you need to adjust the display contrast, first press 2nd, then press and hold ▲ (the *up* arrow key) to increase the contrast or ▼ (the *down* arrow key) to decrease the contrast. As you press and hold ▲ or ▼, an integer between 0 (lightest) and 9 (darkest) appears in the upper right corner of the display. When you have finished with the calculator, turn it off to conserve battery power by pressing 2nd and then OFF.

Check the TI-85's settings by pressing 2nd MODE. If necessary, use the arrow keys to move the blinking cursor to a setting you want to change. Press ENTER to select a new setting. To start with, select the options along the left side of the MODE menu as illustrated in Figure 4.1: normal display, floating decimals, radian measure, rectangular coordinates, function graphs, decimal number system, rectangular vectors, and differentiation type. Details on alternative options will be given later in this guide. For now, leave the MODE menu by pressing EXIT or 2nd QUIT or CLEAR.

Figure 4.1: MODE menu

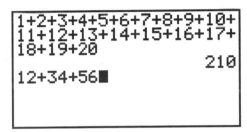

Figure 4.2: Home screen

4.1.2 Editing: One advantage of the TI-85 is that up to 8 lines are visible at one time, so you can *see* a long calculation. For example, type this sum (see Figure 4.2):

$$1 + 2 + 3 + 4 + 5 + 6 + 7 + 8 + 9 + 10 + 11 + 12 + 13 + 14 + 15 + 16 + 17 + 18 + 19 + 20$$

Then press ENTER to see the answer, too.

Often we do not notice a mistake until we see how unreasonable an answer is. The TI-85 permits you to re-display an entire calculation, edit it easily, then execute the *corrected* calculation.

Suppose you had typed 12 + 34 + 56 as in Figure 4.2 but had *not* yet pressed ENTER, when you realize that 34 should have been 74. Simply press ◄ (the *left* arrow key) as many times as necessary to move the blinking cursor left to 3, then type 7 to write over it. On the other hand, if 34 should have been 384, move the cursor back to 4, press 2nd INS (the cursor changes to a blinking underline) and then type 8 (inserts at the cursor position and other characters are pushed to the right). If the 34 should have been 3 only, move the cursor to 4 and press DEL to delete it.

While you are editing an expression, pressing the *up* (or *down*) arrow key causes the cursor to jump quickly to the *left* (or *right*) end of the expression.

Even if you had pressed ENTER, you may still edit the previous expression. Press 2nd and then ENTRY to *recall* the last expression that was entered. Now you can change it.

Technology Tip: When you need to evaluate a formula for different values of a variable, use the editing feature to simplify the process. For example, suppose you want to find the balance in an investment account if there is now $5000 in the account and interest is compounded annually at the rate of 8.5%. The formula for the balance is $P\left(1+\frac{r}{n}\right)^{nt}$, where P = principal, r = rate of interest (expressed as a decimal), n = number of times interest is compounded each year, and t = number of years. In our example, this becomes $5000(1+.085)^t$. Here are the keystrokes for finding the balance after $t = 3, 5,$ and 10 years.

Years	Keystrokes	Balance
3	5000 (1 + .085) ^ 3 ENTER	$6386.45
5	2nd ENTRY ◀ 5 ENTER	$7518.28
10	2nd ENTRY ◀ 10 ENTER	$11,304.92

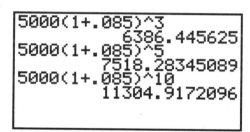

Figure 4.3: Editing expressions

Then to find the balance from the same initial investment but after 5 years when the annual interest rate is 7.5%, press these keys to change the last calculation above: 2nd ENTRY ◀ DEL ◀ 5 ◀ ◀ ◀ ◀ ◀ 7 ENTER.

4.1.3 Key Functions: Most keys on the TI-85 offer access to more than one function, just as the keys on a computer keyboard can produce more than one letter ("g" and "G") or even quite different characters ("5" and "%"). The primary function of a key is indicated on the key itself, and you access that function by a simple press on the key.

To access the *second* function indicated to the *left* above a key, first press 2nd (the cursor changes to a blinking ↑) and *then* press the key. For example, to calculate $\sqrt{25}$, press 2nd √ 25 ENTER.

When you want to use a capital letter or other character printed to the *right* above a key, first press ALPHA (the cursor changes to a blinking **A**) and then the key. For example, to use the letter K in a formula, press ALPHA K. If you need several letters in a row, press ALPHA twice in succession, which is like pressing CAPS

LOCK on a computer keyboard, and then press all the letters you want. Remember to press **ALPHA** when you are finished and want to restore keys to their primary functions. To type lowercase letters, press **2nd alpha** (the cursor changes to a blinking **a**). To lock in lowercase letters, press **2nd alpha 2nd alpha** or **2nd alpha ALPHA**. To unlock from lowercase, press **ALPHA ALPHA** (you'll see the cursor change from blinking **a** to blinking **A** and then to the standard blinking rectangle).

4.1.4 *Order of Operations:*

The TI-85 performs calculations according to the standard algebraic rules. Working outwards from inner parentheses, calculations are performed from left to right. Powers and roots are evaluated first, followed by multiplications and divisions, and then additions and subtractions.

Note that the TI-85 distinguishes between *subtraction* and the *negative sign*. If you wish to enter a negative number, it is necessary to use the (-) key. For example, you would evaluate $-5-(4\cdot-3)$ by pressing **(-) 5 − (4 × (-) 3) ENTER** to get 7.

Enter these expressions to practice using your TI-85.

Expression	Keystrokes	Display
$7-5\cdot3$	7 − 5 × 3 ENTER	-8
$(7-5)\cdot3$	(7 − 5) × 3 ENTER	6
$120-10^2$	120 − 10 x² ENTER	20
$(120-10)^2$	(120 − 10) x² ENTER	12100
$\dfrac{24}{2^3}$	24 ÷ 2 ^ 3 ENTER	3
$\left(\dfrac{24}{2}\right)^3$	(24 ÷ 2) ^ 3 ENTER	1728
$(7--5)\cdot-3$	(7 − (-) 5) × (-) 3 ENTER	-36

4.1.5 *Algebraic Expressions and Memory:*

Your calculator can evaluate expressions such as $\dfrac{N(N+1)}{2}$ *after* you have entered a value for N. Suppose you want $N = 200$. Press **200 STO ▶ N ENTER** to store the value 200 in memory location N. (The **STO ▶** key prepares the TI-85 for an alphabetical entry, so it is *not* necessary to press **ALPHA** also.) Whenever you use N in an expression, the calculator will substitute the value 200 until you make a change by storing *another* number in N. Next enter the expression $\dfrac{N(N+1)}{2}$ by typing **ALPHA N (ALPHA N + 1) ÷ 2 ENTER**. For $N = 200$, you will find that $\dfrac{N(N+1)}{2} = 20100$.

The contents of any memory location may be revealed by typing just its letter name and then **ENTER**. And the TI-85 retains memorized values even when it is turned off, so long as its batteries are good.

A variable name in the TI-85 can be a single letter, or a string of up to eight characters that begins with a letter followed by other letters, numerals, and various symbols. Variable names are case sensitive, which means that that length and Length and LENGTH may represent *different* quantities.

4.1.6 Repeated Operations with ANS: The result of your *last* calculation is always stored in memory location ANS and replaces any previous result. This makes it easy to use the answer from one computation in another computation. For example, press 30 + 15 ENTER so that 45 is the last result displayed. Then press 2nd ANS ÷ 9 ENTER and get 5 because $\frac{45}{9} = 5$.

With a function like division, you press the ÷ key *after* you enter an argument. For such functions, whenever you would start a new calculation with the previous answer followed by pressing the function key, you may press just the function key. So instead of 2nd ANS ÷ 9 in the previous example, you could have pressed simply ÷ 9 to achieve the same result. This technique also works for these functions: + − × x^2 ^ x^{-1}.

Here is a situation where this is especially useful. Suppose a person makes \$5.85 per hour and you are asked to calculate earnings for a day, a week, and a year. Execute the given keystrokes to find the person's incomes during these periods (results are shown in Figure 4.4):

Pay period	Keystrokes	Earnings
8-hour day	5.85 × 8 ENTER	\$46.80
5-day week	× 5 ENTER	\$234
52-week year	× 52 ENTER	\$12,168

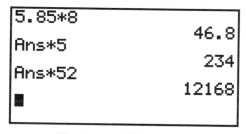

Figure 4.4: ANS variable

4.1.7 The MATH Menu: Operators and functions associated with a scientific calculator are available either immediately from the keys of the TI-85 or by 2nd keys. You have direct key access to common arithmetic operations (x^2, 2nd $\sqrt{\ }$, 2nd x^{-1}, and ^), trigonometric functions (SIN, COS, TAN) and their inverses (2nd SIN^{-1}, 2nd COS^{-1}, 2nd TAN^{-1}), exponential and logarithmic functions (LOG, 2nd 10^x, LN, 2nd e^x), and a famous constant (2nd π).

A significant difference between the TI-85 and many scientific calculators is that the TI-85 requires the argument of a function *after* the function, as you would see a formula written in your textbook. For example, on the TI-85 you calculate $\sqrt{16}$ by pressing the keys 2nd √ 16 in that order.

Here are keystrokes for basic mathematical operations. Try them for practice on your TI-85.

Expression	Keystrokes	Display
$\sqrt{3^2+4^2}$	2nd √ (3 x² + 4 x²) ENTER	5
$2\frac{1}{3}$	2 + 3 2nd x⁻¹ ENTER	2.33333333333
$\log 200$	LOG 200 ENTER	2.30102999566
$2.34 \cdot 10^5$	2.34 × 2nd 10ˣ 5 ENTER	234000
	or 2.34 × 10 ^ 5 ENTER	

Additional mathematical operations and functions are available from the MATH menu (Figure 4.5). Press 2nd MATH to see the various options that are listed across the bottom of the screen. These options are activated by pressing corresponding menu keys, F1 through F5.

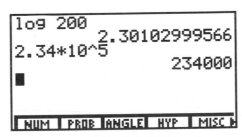

Figure 4.5: Basic MATH menu

For example, F1 brings up the NUM menu of numerical functions. You will learn in your mathematics textbook how to apply many of them. Note that the basic MATH menu items have moved up a line; these options are now available by pressing 2nd M1 through 2nd M5. As an example, determine |–5| by pressing 2nd MATH F1 and then F5 (-) 5 ENTER (see Figure 4.6).

Next calculate $\sqrt[3]{7}$ by pressing 2nd MATH F5 (when the MATH NUM menu is displayed, as in Figure 4.6; press just 2nd M5) to access the MISC menu of miscellaneous mathematical functions. The arrow at the right end of this menu indicates there are more items that you can access. You may press the MORE key repeatedly to move down the row of options and back again. To calculate $\sqrt[3]{7}$, press 2nd MATH F5 MORE 3 F4 [ˣ√] 7 ENTER to see 1.9129 (Figure 4.7). To leave the MATH menu or any other menu and take no further action, press EXIT a couple of times.

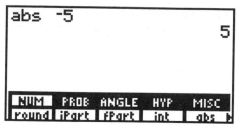

Figure 4.6: MATH NUM menu

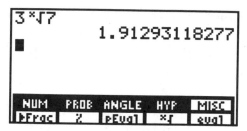

Figure 4.7: MATH MISC menu

The *factorial* of a non-negative integer is the *product* of *all* the integers from 1 up to the given integer. The symbol for factorial is the exclamation point. So 4! (pronounced *four factorial*) is $1 \cdot 2 \cdot 3 \cdot 4 = 24$. You will learn more about applications of factorials in your textbook, but for now use the TI-85 to calculate 4! Press these keystrokes: 2nd MATH F2 *[PROB]* 4 F1 *[!]* ENTER.

The complex number $a + bi$ is represented by the TI-85 as an ordered pair (a, b). Perform arithmetic with complex numbers by using this ordered pair notation. So to divide $2 + 3i$ by $4 - 2i$, press (2 , 3) ÷ (4 , (-) 2) ENTER to get $(.1, .8)$ for $.1 + .8i$.

4.2   Functions and Graphs

4.2.1 *Evaluating Functions:* Suppose you receive a monthly salary of $1975 plus a commission of 10% of sales. Let x = your sales in dollars; then your wages W in dollars are given by the equation $W = 1975 + .10x$. If your January sales were $2230 and your February sales were $1865, what was your income during those months?

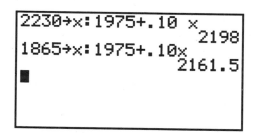

Figure 4.8: Evaluating a function

Here's how to use your TI-85 to perform this task. First press EXIT and CLEAR as necessary to get a blank home screen. Then set $x = 2230$ by pressing 2230 STO ► x-VAR. (The x-VAR key makes it easy to produce a lower case x for a variable name without having to use the 2nd alpha key.) Then press ALPHA to leave alphabetic entry and 2nd : to allow another expression to be entered on the same command line. Finally, enter the expression $1975 + .10x$ by pressing these keys: 1975 + .10 x-VAR. Now press ENTER to calculate the answer (Figure 4.8).

It is not necessary to repeat all these steps to find the February wages. Simply press 2nd ENTRY to recall the entire previous line and change 2230 to 1865.

Technology Tip: The TI-85 does not require multiplication to be expressed between variables, so *xxx* means x^3. It is often easier to press two or three *x*'s together than to search for the square key or the power key. Of course, expressed multiplication is also not required between a constant and a variable. Hence to enter $2x^3 + 3x^2 - 4x + 5$ in the TI-85, you might save some keystrokes and press just these keys: 2 x-VAR x-VAR x-VAR + 3 x-VAR x-VAR - 4 x-VAR + 5.

4.2.2 Functions in a Graph Window: On the TI-85, once you have entered a function, you can easily generate its graph. The ability to draw a graph contributes substantially to our ability to solve problems.

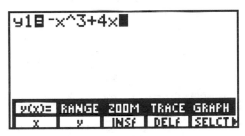

Figure 4.9: y(x)= screen

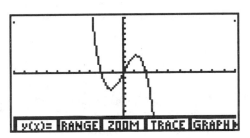

Figure 4.10: Graph of $y = -x^3 + 4x$

Here is how to graph $y = -x^3 + 4x$. First press the GRAPH key and then F1 to select y(x)=. This give you access to the function editing screen (Figure 4.9). Press F4 *[DELf]* as many times as necessary to delete any functions that may be there already. Then, with the cursor on the top line to the right of y1=, press (-) F1 ^ 3 + 4 F1 to enter the function. As you see, the TI-85 uses lower-case letters for its graphing variables, just like your mathematics textbook. Note that pressing F1 in this menu is the same as pressing either x-VAR or 2nd alpha X. Now press 2nd M5 *[GRAPH]* and the TI-85 changes to a window with the graph of $y = -x^3 + 4x$ (Figure 4.10).

While the TI-85 is calculating coordinates for a plot, it displays a busy indicator at the top right of the graph window.

Your graph window may look like the one in Figure 4.10 or it may be different. Since the graph of $y = -x^3 + 4x$ extends infinitely far left and right and also infinitely far up and down, the TI-85 can display only a piece of the actual graph. This displayed rectangular part is called a *viewing rectangle*. You can easily change the viewing rectangle to enhance your investigation of a graph.

The viewing rectangle in Figure 4.10 shows the part of the graph that extends horizontally from -10 to 10 and vertically from -10 to 10. Press F2 *[RANGE]* to see information about your viewing rectangle. Figure 4.11

shows the RANGE screen that corresponds to the viewing rectangle in Figure 4.10. This is the *standard* viewing rectangle for the TI-85.

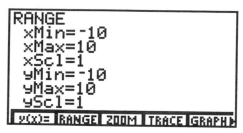

Figure 4.11: Standard RANGE

The variables xMin and xMax are the minimum and maximum *x*-values of the viewing rectangle; yMin and yMax are its minimum and maximum *y*-values.

xScl and yScl set the spacing between tick marks on the axes.

Use the arrow keys ▲ and ▼ to move up and down from one line to another in this list; pressing the ENTER key will move down the list. Press CLEAR to delete the current value and then enter a new value. You may also edit the entry as you would edit an expression. Remember that a minimum *must* be less than the corresponding maximum or the TI-85 will issue an error message. Also, remember to use the (-) key, not − (which is subtraction), when you want to enter a negative value. Figures 4.10-11, 4.12-13, and 4.14-15 show different RANGE screens and the corresponding viewing rectangle for each one.

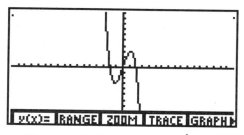

Figure 4.12: Square window

Figure 4.13: Graph of $y = -x^3 + 4x$

To set the range quickly to standard values (see Figure 4.11), press F3 F4 *[ZOOM ZSTD]*

To set the viewing rectangle quickly to a "square" window (Figure 4.12), in which the horizontal and vertical axes have the same scale, press F3 MORE F2 *[ZOOM ZSQR]* in the GRAPH menu. More information about square windows is presented later in Section 4.2.4.

Sometimes you may wish to display grid points corresponding to tick marks on the axes. This and other graph format options may be changed by pressing GRAPH MORE F3 (Figure 4.16). Use arrow keys to move the blinking cursor to GridOn, then press ENTER and EXIT. Figure 4.17 shows the same graph as in

Figure 4.15 but with the grid turned on. In general, you'll want the grid turned *off*, so do that now by pressing GRAPH MORE F3 again, use the arrow keys to move the blinking cursor to GridOff, and press ENTER EXIT.

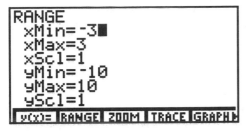

Figure 4.14: Custom window

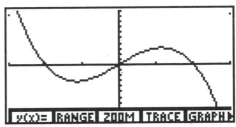

Figure 4.15: Graph of $y = -x^3 + 4x$

Figure 4.16: GRAPH FORMT menu

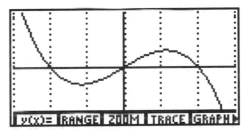

Figure 4.17: Grid on

4.2.3 Piecewise-Defined Functions: The greatest integer function, written $[[x]]$, gives the greatest *integer* less than or equal to a number x. On the TI-85, the greatest integer function is called int and is located under the NUM sub-menu of the MATH menu (see Figure 4.5). So calculate $[[6.78]] = 6$ by pressing 2nd MATH F1 F4 6.78 ENTER.

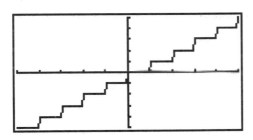

Figure 4.18: DrawLine graph of $y = [[x]]$

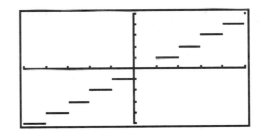

Figure 4.19: DrawDot graph of $y = [[x]]$

To graph $y = [[x]]$, go in the y(x)= menu, move beside y1 and press CLEAR 2nd MATH F1 F4 EXIT F1 2nd M5. Figure 4.18 shows this graph in a viewing rectangle from -5 to 5 in both directions. The bottom menu line has been cleared by pressing CLEAR once; you may restore it by pressing GRAPH again.

The true graph of the greatest integer function is a step graph, like the one in Figure 4.19. Calculators like the TI-85 graph a function by plotting points, then connecting successive points with segments. For the graph of $y = [[x]]$, a segment should *not* be drawn between every pair of successive points. You can change from DrawLine to DrawDot format on the TI-85 by opening the GRAPH FORMT menu (Figure 4.16).

You should also change to DrawDot format when plotting a piecewise-defined function. For example, to plot the graph of $f(x) = \begin{cases} x^2+1, & x<0 \\ x-1, & x\geq 0 \end{cases}$, enter the expression $(x^2+1)(x<0)+(x-1)(x\geq0)$ somewhere in your y(x)= list by pressing (x-VAR x² + 1) (x-VAR 2nd TEST F2 0) + (x-VAR − 1) (x-VAR 2nd TEST F5 0). Then change the format to DrawDot and draw the graph.

4.2.4 *Graphing a Circle:* Here is a useful technique for graphs that are not functions, but that can be "split" into a top part and a bottom part, or into multiple parts. Suppose you wish to graph the circle whose equation is $x^2 + y^2 = 36$. First solve for y and get an equation for the top semicircle, $y = \sqrt{36-x^2}$, and for the bottom semicircle, $y = -\sqrt{36-x^2}$. Then graph the two semicircles simultaneously.

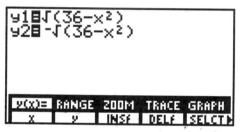

Figure 4.20: Two semicircles

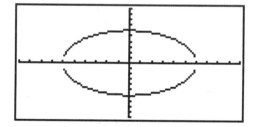

Figure 4.21: Circle's graph - standard view

The keystrokes to draw this circle's graph follow. Enter $\sqrt{36-x^2}$ as y1 and $-\sqrt{36-x^2}$ as y2 (see Figure 4.20) by pressing GRAPH F1 CLEAR 2nd √ (36 - F1 x²) ENTER (-) 2nd √ (36 - F1 x²). Then press 2nd M5 to draw them both.

If your range were set to the standard viewing rectangle, your graph would look like Figure 4.21. Now this does *not* look like a circle, because the units along the axes are not the same. This is where the square viewing rectangle is important. Press F3 MORE F2 and see a graph that appears more circular.

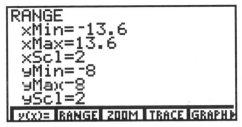

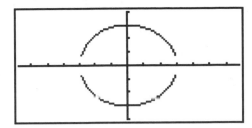

Figure 4.22: $\frac{\text{vertical}}{\text{horizontal}} = \frac{16}{27.2} = \frac{10}{17}$

Figure 4.23: A "square" circle

Technology Tip: Another way to get a square graph is to change the range variables so that the value of yMax - yMin is approximately $\frac{10}{17}$ times xMax - xMin. For example, see the RANGE in Figure 4.22 and the corresponding graph in Figure 4.23. The method works because the dimensions of the TI-85's display are such that the ratio of vertical to horizontal is approximately $\frac{10}{17}$.

The two semicircles in Figure 4.23 do not meet because of an idiosyncrasy in the way the TI-85 plots a graph.

Back when you entered $\sqrt{36 - x^2}$ as y1 and $-\sqrt{36 - x^2}$ as y2, you could have entered -y1 for y2 and saved some keystrokes. Try this by going back to the y(x)= menu and pressing the arrow key to move the cursor down to y2. Then press CLEAR (-) 2nd VARS MORE F3 ENTER. The graph should be just as it was before.

4.2.5 TRACE: Graph $y = -x^3 + 4x$ in the standard viewing rectangle. Press any of the arrow keys ▲ ▼ ◄ ► and see the cursor move from the center of the viewing rectangle. The coordinates of the cursor's location are displayed at the bottom of the screen, as in Figure 4.24, in floating decimal format. This cursor is called a *free-moving cursor* because it can move from dot to dot *anywhere* in the graph window.

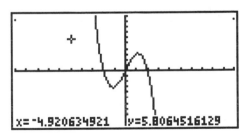

Figure 4.24: Free-moving cursor

Remove the free-moving cursor and its coordinates from the window by pressing ENTER, CLEAR, or GRAPH (this also restores the GRAPH menu line). An advantage of pressing ENTER or CLEAR to remove the free-moving cursor is that, if you press an arrow key once again, the cursor will reappear at the same point you left it.

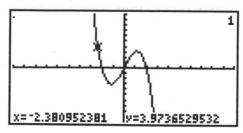

Figure 4.25: Trace on $y = -x^3 + 4x$

Press F4 *[TRACE]* to enable the left ◄ and right ► arrow keys to move the cursor along the function. The cursor is no longer free-moving, but is now constrained to the function. The coordinates that are displayed belong to points on the function's graph, so the *y*-coordinate is the calculated value of the function at the corresponding *x*-coordinate.

Now plot a second function, $y = -.25x$, along with $y = -x^3 + 4x$. Press GRAPH F1 for the y(x)= menu and enter $-.25x$ for y2, then press 2nd M5 to see their graphs (Figure 4.27).

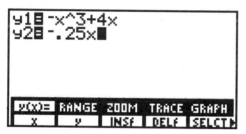

Figure 4.26: Two functions

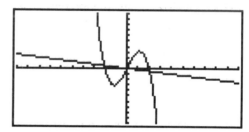

Figure 4.27: $y = -x^3 + 4x$ and $y = -.25x$

Note in Figure 4.26 that the equal signs next to y1 and y2 are *both* highlighted. This means *both* functions will be graphed. In the y(x)= screen, move the cursor to y1 and press F5 *[SELCT]* to turn function selection *off*. The equal sign beside y1 should no longer be highlighted (see Figure 4.28). The SELCT command operates as a toggle switch; executing it once more sets function selection *on*. Now press 2nd M5 *[GRAPH]* and see that only y2 is plotted.

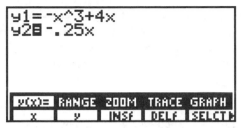

Figure 4.28: y(x)= screen with only y2 active

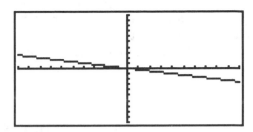

Figure 4.29: Graph of $y = -.25x$

Many different functions may be stored in the y(x)= list and any combination of them may be graphed simultaneously. You can make a function active or inactive for graphing by pressing SELCT to highlight (activate) or remove the highlight (deactivate). Go back to the y(x)= screen and do what is needed in order to graph y1 but not y2.

Now activate y2 again so that both graphs are plotted. Press GRAPH F4 *[TRACE]* and the cursor appears first on the graph of $y = -x^3 + 4x$ because it is higher up in the y(x)= list. You know that the cursor is on this function, y1, because of the numeral 1 displayed in the upper right corner of the window (see Figure 4.25). Press the up ▲ or down ▼ arrow key to move the cursor vertically to the graph of $y = -.25x$. Now the numeral 2 is displayed in the top right corner of the window. When more than one function is plotted, you can move the trace cursor vertically from one graph to another in this way. Next press the right and left arrow keys to trace along the graph of $y = -.25x$.

Technology Tip: By the way, trace along the graph of $y = -.25x$ and press and hold either ◀ or ▶. Eventually you will reach the *left* or *right* edge of the window. Keep pressing the arrow key and the TI-85 will allow you to continue the trace by panning the viewing rectangle. Check the RANGE screen to see that xMin and xMax are automatically updated.

If you trace along the graph of $y = -x^3 + 4x$, the cursor will eventually move *above* or *below* the viewing rectangle. The cursor's coordinates on the graph will still be displayed, though the cursor itself can no longer be seen.

When you are tracing along a graph, press ENTER and the window will quickly pan over so that the cursor's position on the function is centered in a new viewing rectangle. This feature is especially helpful when you trace near or beyond the edge of the current viewing rectangle.

The TI-85's display has 127 horizontal columns of pixels and 63 vertical rows. So when you trace a curve across a graph window, you are actually moving from xMin to xMax in 126 equal jumps, each called Δx. You would calculate the size of each jump to be $\Delta x = \dfrac{xMax - xMin}{126}$. Sometimes you may want the jumps to be friendly numbers like .1 or .25 so that, when you trace along the curve, the x-coordinates will be incremented by such a convenient amount. Just set your viewing rectangle for a particular increment Δx by making xMax = xMin + 126·Δx. For example, if you want xMin = -15 and $\Delta x = .25$, set xMax = -15 + 126·.25 = 16.5. Likewise, set yMax = yMin + 62·Δy if you want the vertical increment to be some special Δy.

To center your window around a particular point, say (h, k), and also have a certain Δx, set xMin = h - 63·Δx and xMax = h + 63·Δx. Likewise, make yMin = k - 31·Δy and yMax = k + 31·Δy. For example, to center a window around the origin, (0, 0), with both horizontal and vertical increments of .25, set the range so that xMin = 0 - 63·.25 = -15.75, xMax = 0 + 63·.25 = 15.75, yMin = 0 - 31·.25 = -7.75, and yMax = 0 + 31·.25 = 7.75.

See the benefit by first plotting $y = x^2 + 2x +$ in a standard graphing window. Trace near its y-intercept, which is (0, 1), and move towards its x-intercept, which is (-1, 0). Then change to a viewing rectangle that extends from -6.3 to 6.3 horizontally and from -3.1 to 3.1 vertically (center at the origin, Δx and Δy both .1), and trace again from the y-intercept. The TI-85 makes it easy to get this particular viewing rectangle: press GRAPH F3 MORE F4 *[ZOOM ZDECM]*.

4.2.6 ZOOM: Plot again the two graphs, for $y = -x^3 + 4x$ and for $y = -.25x$. There appears to be an intersection near $x = 2$. The TI-85 provides several ways to enlarge the view around this point. You can change the viewing rectangle directly by pressing RANGE and editing the values of xMin, xMax, yMin, and yMax. Figure 4.31 shows a new viewing rectangle for the range displayed in Figure 4.30. Trace has been turned on and the coordinates are displayed for a point on $y = -x^3 + 4x$ that is close to the intersection.

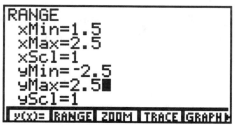
Figure 4.30: New RANGE

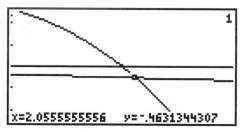
Figure 4.31: Closer view

A more efficient method for enlarging the view is to draw a new viewing rectangle with the cursor. Start again with a graph of the two functions $y = -x^3 + 4x$ and $y = -.25x$ in a standard viewing rectangle (press GRAPH F3 F4 for the standard window, from -10 to 10 along both axes).

First of all, imagine a small rectangular box around the intersection point, near $x = 2$. Press GRAPH F3 F1 *[ZOOM BOX]* to enable drawing a box (Figure 4.32) to define a new viewing rectangle. Use the arrow keys to move the cursor, whose coordinates are displayed at the bottom of the window, to one corner of the new viewing rectangle you are imagining (Figure 4.33).

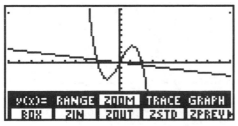

Figure 4.32: ZOOM menu

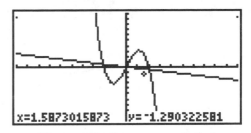

Figure 4.33: One corner selected

Press ENTER to fix the corner where you have moved the cursor; it changes shape and becomes a blinking square. Use the arrow keys again to move the cursor to the diagonally opposite corner of the new rectangle (Figure 4.34). If this box looks all right to you, press ENTER. The rectangular area you have enclosed will now enlarge to fill the graph window (Figure 4.35).

You may cancel the zoom any time *before* you press this last ENTER. Just press EXIT or GRAPH to interrupt the zoom and return to the current graph window. Even if you did execute the zoom, you may still return to the previous viewing rectangle by pressing F5 *[ZPREV]* in the ZOOM menu.

You can also gain a quick magnification of the graph around the cursor's location. Return once more to the standard range for the graph of the two functions $y = -x^3 + 4x$ and $y = -.25x$. Start the zoom by pressing GRAPH F3 F2 *[ZOOM ZIN]*; next use the arrow keys to move the cursor as close as you can to the point of intersection near $x = 2$ (see Figure 4.36). Then press ENTER and the calculator draws a magnified graph, centered at the cursor's position (Figure 4.37). The range variables are changed to reflect this new viewing rectangle. Look in the RANGE menu to check.

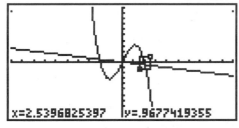

Figure 4.34: Box drawn

Figure 4.35: New viewing rectangle

As you see in the ZOOM menu (Figure 4.32), the TI-85 can zoom in *[ZIN]* and zoom out *[ZOUT]*. You would zoom out to see a larger view of the graph, centered at the cursor position. You can change the horizontal and vertical scale of the magnification by pressing GRAPH F3 MORE MORE F1 *[ZOOM ZFACT]* (see Figure 4.38) and editing xFact and yFact, the horizontal and vertical magnification factors.

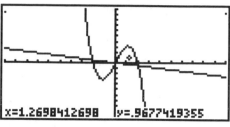

Figure 4.36: Before a zoom in

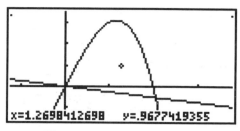

Figure 4.37: After a zoom in

The default zoom factor is 4 in both directions. It is not necessary for **xFact** and **yFact** to be equal. Sometimes, you may prefer to zoom in one direction only, so the other factor should be set to 1. Press **GRAPH** or **EXIT** to leave the ZOOM FACTORS menu.

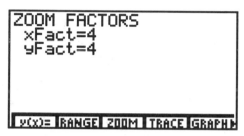

Figure 4.38: Set zoom factors

Technology Tip: If you should zoom in too much and lose the curve, zoom back to the standard viewing rectangle and start over.

4.2.7 *Relative Minimums and Maximums:* Graph $y = -x^3 + 4x$ once again by itself in the standard viewing rectangle (Figure 4.10). This function appears to have a relative minimum near $x = -1$ and a relative maximum near $x = 1$. You may zoom and trace to approximate these extreme values.

First trace along the curve near the local minimum. Notice by how much the x-values and y-values change as you move from point to point. Trace along the curve until the y-coordinate is as *small* as you can get it, so that you are as close as possible to the local minimum, and zoom in (use either **ZIN** or a zoom box). Now trace again along the curve and, as you move from point to point, see that the coordinates change by smaller amounts than before. Keep zooming and tracing until you find the coordinates of the local minimum point as accurately as you need them, approximately (-1.15, -3.08).

Follow a similar procedure to find the local maximum. Trace along the curve until the y-coordinate is as *great* as you can get it, so that you are as close as possible to the local maximum, and zoom in. The local maximum point on the graph of $y = -x^3 + 4x$ is approximately (1.15, 3.08).

TI-85 Advanced Scientific Calculator

Technology Tip: Trace along the function as near as possible to the minimum or maximum point and press ENTER to center the window at the cursor's location. Then you will not need to move the cursor again after you press ZIN.

4.3 Solving Equations and Inequalities

4.3.1 Intercepts and Intersections: Tracing and zooming are also used to locate an *x*-intercept of a graph, where a curve crosses the *x*-axis. For example, the graph of $y = x^3 - 8x$ crosses the *x*-axis three times (see Figure 4.39). After tracing over to the *x*-intercept point that is furthest to the left, zoom in (Figure 4.40). Continue this process until you have located all three intercepts with as much accuracy as you need. The three *x*-intercepts of $y = x^3 - 8x$ are approximately -2.828, 0, and 2.828.

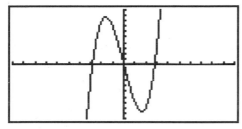   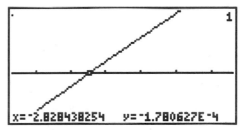

Figure 4.39: Graph of $y = x^3 - 8x$ Figure 4.40: An *x*-intercept of $y = x^3 - 8x$

Technology Tip: As you zoom in, you may also wish to change the spacing between tick marks on the *x*-axis so that the viewing rectangle shows scale marks near the intercept point. Then the accuracy of your approximation will be such that the error is less than the distance between two tick marks. Change the *x*-scale on the TI-85 from the GRAPH F2 *[RANGE]* menu. Move the cursor down to xScl and enter an appropriate value.

TRACE and ZOOM are especially important for locating the intersection points of two graphs, say the graphs of $y = -x^3 + 4x$ and $y = -.25x$. Trace along one of the graphs until you arrive close to an intersection point. Then press ▲ or ▼ to jump to the other graph. Notice that the *x*-coordinate does not change, but the *y*-coordinate is likely to be different (see Figures 4.41 and 4.42).

When the two *y*-coordinates are as close as they can get, you have come as close as you now can to the point of intersection. So zoom in around the intersection point, then trace again until the two *y*-coordinates are as close as possible. Continue this process until you have located the point of intersection with as much accuracy as necessary.

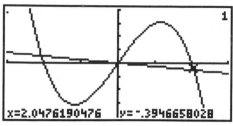 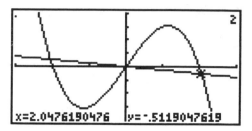

Figure 4.41: Trace on $y = -x^3 + 4x$ Figure 4.42: Trace on $y = -.25x$

4.3.2 Solving Equations by Graphing: Suppose you need to solve the equation $24x^3 - 36x + 17 = 0$. First graph $y = 24x^3 - 36x + 17$ in a window large enough to exhibit *all* its x-intercepts, corresponding to all its roots. Then use trace and zoom to locate each one. In fact, this equation has just one solution, approximately $x = -1.414$.

Remember that when an equation has more than one x-intercept, it may be necessary to change the viewing rectangle a few times to locate all of them.

Technology Tip: To solve an equation like $24x^3 + 17 = 36x$, you may first transform it into standard form, $24x^3 - 36x + 17 = 0$, and proceed as above to search for its x-intercepts. However, you may also graph the *two* functions $y = 24x^3 + 17$ and $y = 36x$, then zoom and trace to locate their point of intersection.

4.3.3 Solving Systems by Graphing: The solutions to a system of equations correspond to the points of intersection of their graphs (Figure 4.43). For example, to solve the system $y = x^2 - 3x - 4$ and $y = x^3 + 3x^2 - 2x - 1$, first graph them together. Then zoom and trace to locate their point of intersection, approximately (-2.17, 7.25).

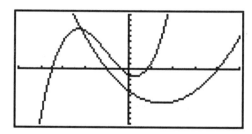

Figure 4.43: Solving a system of equations

You must judge whether the two current y-coordinates are sufficiently close for x = -2.17 or whether you should continue to zoom and trace to improve the approximation.

The solutions of the system of two equations $y = x^3 + 3x^2 - 2x - 1$ and $y = x^2 - 3x - 4$ correspond to the solutions of the single equation $x^3 + 3x^2 - 2x - 1 = x^2 - 3x - 4$, which simplifies to $x^3 + 2x^2 + x + 3 = 0$. So you may also graph $y = x^3 + 2x^2 + x + 3$ and find its x-intercepts to solve the system.

4.3.4 Solving Inequalities by Graphing: Consider the inequality $1 - \dfrac{3x}{2} \geq x - 4$. To solve it with your TI-85, graph the two functions $y = 1 - \dfrac{3x}{2}$ and $y = x - 4$ (Figure 4.44). First locate their point of intersection, at $x = 2$. The inequality is true when the graph of $y = 1 - \dfrac{3x}{2}$ lies *above* the graph of $y = x - 4$, and that occurs for $x < 2$. So the solution is the half-line $x \leq 2$, or $(-\infty, 2]$.

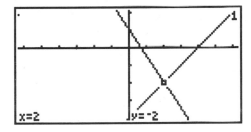

Figure 4.44: Solving $1 - \dfrac{3x}{2} \geq x - 4$

The TI-85 is capable of shading the region above or below a graph or between two graphs. For example, to graph $y \geq x^2 - 1$, first enter the function $y = x^2 - 1$ as y1 in the **GRAPH** y(x)= screen. Then press **GRAPH MORE F2** *[DRAW]* **F1** *[Shade]* **2nd VARS MORE F3** *[EQU]*, move the cursor to y1, and press **ENTER** , 100) (see Figure 4.45) and again **ENTER**. These keystrokes instruct the TI-85 to shade the region *above* $y = x^2 - 1$ and *below* $y = 100$ (chosen because this is a sufficiently large y-value). The result is shown in Figure 4.46.

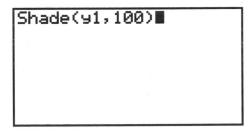

Figure 4.45: **DRAW** Shade

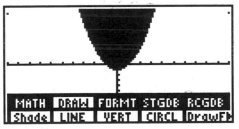

Figure 4.46: Graph of $y \geq x^2 - 1$

To clear the shading, when you are already in the DRAW menu, press MORE F5 *[CLDRW]*.

Now use shading to solve the previous inequality, $1 - \dfrac{3x}{2} \geq x - 4$. The function whose graph forms the lower boundary is named *first* in the SHADE command. To enter this in your TI-85 (see Figure 4.47), press these keys: GRAPH MORE F2 F1 x-VAR - 4 , 1 - 3 x-VAR ÷ 2) ENTER. The shading (see Figure 4.48) extends left from $x = 2$, hence the solution to $1 - \dfrac{3x}{2} \geq x - 4$ is the half-line $x \leq 2$, or $(-\infty, 2]$.

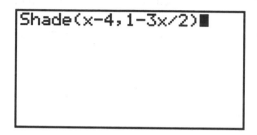

Figure 4.47: DRAW Shade command

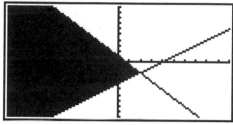

Figure 4.48: Graph of $1 - \dfrac{3x}{2} \geq x - 4$

More information about the DRAW menu is in the TI-85 manual.

4.4 Trigonometry

4.4.1 Degrees and Radians: The trigonometric functions can be applied to angles measured either in radians or degrees, but you should take care that the TI-85 is configured for whichever measure you need. Press 2nd MODE to see the current settings. Press ▼ twice and move down to the third line of the mode menu where angle measure is selected. Then press ◀ or ▶ to move between the displayed options. When the blinking cursor is on the measure you want, press ENTER to select it. Then press EXIT or 2nd QUIT to leave the mode menu.

It's a good idea to check the angle measure setting before executing a calculation that depends on a particular measure. You may change a mode setting at any time and not interfere with pending calculations. Try the following keystrokes to see this in action.

Expression	Keystrokes	Display
$\sin 45°$	2nd MODE ▼ ▼ ▶ ENTER	
	EXIT SIN 45 ENTER	.707106781187
$\sin \pi°$	SIN 2nd π ENTER	.054803665149

$\sin \pi$	SIN 2nd π 2nd MODE ▼ ▼	
	ENTER EXIT ENTER	0
$\sin 45$	SIN 45 ENTER	.850903524534
$\sin \frac{\pi}{6}$	SIN (2nd π ÷ 6) ENTER	.5

The first line of keystrokes sets the TI-85 in degree mode and calculates the sine of 45 *degrees*. While the calculator is still in degree mode, the second line of keystrokes calculates the sine of π *degrees*, 3.1415°. The third line changes to radian mode just before calculating the sine of π *radians*. The fourth line calculates the sine of 45 *radians* (the calculator is already in radian mode).

Figure 4.49: Angle measure

The TI-85 makes it possible to mix degrees and radians in a calculation. Execute these keystrokes to calculate $\tan 45° + \sin \frac{\pi}{6}$ as shown in Figure 4.49: TAN 45 2nd MATH F3 *[ANGLE]* F1 *[°]* + SIN (2nd π ÷ 6) F2 *[ᵣ]* ENTER. Do you get 1.5 whether your calculator is set *either* in degree mode *or* in radian mode?

4.4.2 Graphs of Trigonometric Functions: When you graph a trigonometric function, you need to pay careful attention to the choice of graph window. For example, graph $y = \dfrac{\sin 30x}{30}$ in the standard viewing rectangle. Trace along the curve to see where it is. Zoom in to a better window, or use the period and amplitude to establish better RANGE values.

Technology Tip: Since $\pi \approx 3.1$, set xMin = 0 and xMax = 6.3 to cover the interval from 0 to 2π in steps of 0.05.

Next graph $y - \tan x$ in the standard window. The TI-85 plots consecutive points and then connects them with a segment, so the graph is not exactly what you should expect. You may wish to change from DrawLine to DrawDot graph (see Section 4.2.3) when you plot the tangent function.

4.5   Scatter Plots

4.5.1 *Entering Data:* This table shows total prize money (in millions of dollars) awarded at the Indianapolis 500 race from 1981 to 1989. (*Source:* Indianapolis Motor Speedway Hall of Fame.)

Year	1981	1982	1983	1984	1985	1986	1987	1988	1989
Prize ($ million)	$1.61	$2.07	$2.41	$2.80	$3.27	$4.00	$4.49	$5.03	$5.72

We'll now use the TI-85 to construct a scatter plot that represents these points and to find a linear model that approximates the given data.

Press STAT F2 *[EDIT]* and enter Year for the name of xlist and Prize for the name of ylist (as shown in Figure 4.50).

Figure 4.50: STAT EDIT menu

Figure 4.51: Entering data points

Now press ENTER to prepare to input data from the table. Instead of entering the full year 198x, save keystrokes by entering only 8x. Here are the keystrokes for the first three years: 81 ENTER 1.61 ENTER 82 ENTER 2.07 ENTER 83 ENTER 2.41 ENTER and so on (see Figure 4.51). Continue to enter all the given data. Press EXIT when you have finished.

You may edit statistical data in the same way you edit expressions in the home screen. Move the cursor to the x or y value for any data point you wish to change, then type the correction. To insert or delete statistical data, move the cursor to the x or y value for any data point you wish to add or delete. Press F1 *[INSi]* and a new data point is created; press F2 *[DELi]* and the data point is deleted. To clear *all* data points, press F5 *[CLRxy]*.

4.5.2 *Plotting Data:* Once all the data points have been entered, press STAT F3 *[DRAW]* F2 *[SCAT]* to draw a scatter plot. Your viewing rectangle is important, so you may wish to change the RANGE first to improve the view of the data. If you change the RANGE *after* drawing the scatter plot, you will have to enter keystrokes to create the plot again. Figure 4.52 shows the scatter plot in a viewing rectangle extending from 80 to 90 for x and from 1 to 6 for y.

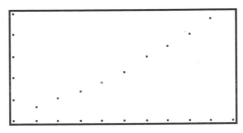

Figure 4.52: Scatter plot

4.5.3 Regression Line:

4.5.3 Regression Line: The TI-85 calculates the slope and *y*-intercept for the line that best fits all the data. After the data points have been entered, press STAT F1 *[CALC]*. You need to enter Year for the name of xlist and Prize for the name of ylist; note that these names are now assigned to function keys for ease of entry. Finally, press F2 *[LINR]* to calculate a linear regression model.

As you see in Figure 4.53, the TI-85 names the *y*-intercept a and calls the slope b. The number corr (between -1 and 1) is called the *correlation coefficient* and measures the goodness of fit of the linear regression equation with the data. The closer $|corr|$ is to 1, the better the fit; the closer $|corr|$ is to 0, the worse the fit. There are $n = 9$ data points.

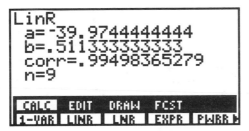

Figure 4.53: Linear regression model

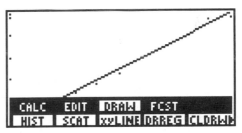

Figure 4.54: Linear regression line

Draw the regression line $y = a + bx$ through the data points by pressing 2nd M3 *[DRAW]* F4 *[DRREG]* (see Figure 4.54).

When you have no further need for some data, press 2nd MEM F2 *[DELETE]* F4 *[LIST]* and move the cursor to the name of a list you wish to delete. Press ENTER to remove that list from your calculator's memory. Then press EXIT to return to the home screen.

4.5.4 Exponential Growth Model:

4.5.4 Exponential Growth Model: After data points have been entered, press STAT F1 *[CALC]*, enter the names of the two lists, then press F4 *[EXPR]* to calculate an exponential growth model $y = a \cdot b^x$ for the data.

4.6 Matrices

4.6.1 Making a Matrix: The TI-85 can display and use many different matrices, each with up to 255 rows and up to 255 columns! Here's how to create this 3×4 matrix $\begin{bmatrix} 1 & -4 & 3 & 5 \\ -1 & 3 & -1 & -3 \\ 2 & 0 & -4 & 6 \end{bmatrix}$ in your calculator.

Press 2nd MATRX F2 *[EDIT]* to see the matrix edit menu (Figure 4.55). You must first name the matrix; let's name this matrix *A* (the TI-85 is already set for alphabetic entry) and press ENTER to continue.

You may now change the dimensions of matrix *A* to 3×4 by pressing 3 ENTER 4 ENTER. Simply press ENTER or the *down* arrow key to accept an existing dimension. Next enter 1 in the first row and first column of the matrix, then press ENTER to move horizontally across this row to the second column. Continue to enter the top row of elements. Press ENTER after the last element of the first row has been entered to move to the second row. You may use the up and down arrow keys to move vertically through the columns of the matrix.

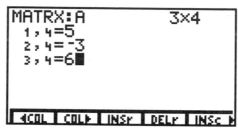

Figure 4.55: MATRX EDIT menu

Leave the matrix *A* editing screen by pressing EXIT or 2nd QUIT and return to the home screen.

4.6.2 Matrix Math: From the home screen you can perform many calculations with matrices. First, let's see matrix *A* itself by pressing ALPHA A ENTER (Figure 4.56).

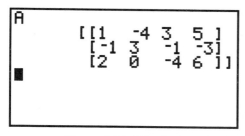

Figure 4.56: Matrix *A*

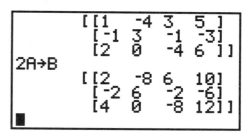

Figure 4.57: Matrix *B*

Calculate the scalar multiplication 2·A by pressing 2 ALPHA A ENTER. To create matrix B as 2·A, press 2 ALPHA A STO▶ B ENTER; or, if you do this immediately after calculating 2·A, press only STO▶ B ENTER (see Figure 4.57). Return to the matrix edit screen to verify that the dimensions of matrix B have been set automatically to reflect these new values.

Add the two matrices A and B by pressing ALPHA A + ALPHA B ENTER. Subtraction is similar.

Now create a matrix C with dimensions of 2×3 and enter this as C: $\begin{bmatrix} 2 & 0 & 3 \\ 1 & -5 & -1 \end{bmatrix}$. For matrix multiplication

of C by A, press ALPHA C × ALPHA A ENTER. If, on the other hand, you tried to multiply A by C, your TI-85 would signal an error because the dimensions of the two matrices do not permit multiplication in this order.

You may use exponential notation to abbreviate multiplying a matrix M by itself, but take care that M is a *square matrix* or such multiplication is not possible. For example, to calculate M·M·M, press ALPHA M ^ 3 ENTER.

The *transpose* of a matrix A is another matrix with the rows and columns interchanged. The symbol for the transpose of A is A^T. The transpose operator is found in the matrix math menu. So to calculate A^T, press ALPHA A 2nd MATRX F3 [MATH] F2 [T] ENTER.

4.6.3 Row Operations: Here are the keystrokes necessary to perform elementary row operations on a matrix. Your textbook provides more careful explanation of the elementary row operations and their uses.

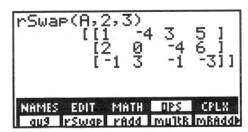

Figure 4.58: Swap rows 2 and 3

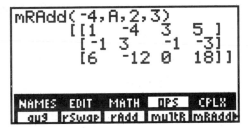

Figure 4.59: Add -4 times row 2 to row 3

To interchange the second and third rows of the matrix A that was defined above, press these keys: 2nd MATRX F4 [OPS] MORE F2 [rSwap] ALPHA A , 2 , 3) ENTER (see Figure 4.58). The format of this command is rSwap(*matrix, row1, row2*).

To add row 2 and row 3 and store the results in row 3, press 2nd MATRX F4 MORE F3 ALPHA A , 2 , 3) ENTER. The format of this command is rAdd(*matrix, row1, row2*).

To multiply row 2 by -4 and *store* the results in row 2, thereby replacing row 2 with new values, press 2nd MATRX F4 MORE F4 (-) 4 , ALPHA A , 2) ENTER. The format of this command is multR(*scalar, matrix, row*).

To multiply row 2 by -4 and *add* the results to row 3, thereby replacing row 3 with new values, press 2nd MATRX F4 MORE F5 (-) 4 , ALPHA A , 2 , 3) ENTER (see Figure 4.59). The format of this command is mRAdd(*scalar, matrix, row1, row2*).

Technology Tip: It is important to remember that your TI-85 does *not automatically* store a matrix obtained as the result of any row operations. So when you need to perform several row operations in succession, it is a good idea to store the result of each one in a temporary place.

For example, use elementary row operations to solve this system of linear equations: $\begin{cases} x - 2y + 3z = 9 \\ -x + 3y = -4 \\ 2x - 5y + 5z = 17 \end{cases}$.

First enter this *augmented matrix* as A in your TI-85: $\begin{bmatrix} 1 & -2 & 3 & 9 \\ -1 & 3 & 0 & -4 \\ 2 & -5 & 5 & 17 \end{bmatrix}$. Next store this matrix in C (press ALPHA A STO ► C ENTER) so you may keep the original in case you need to recall it.

Here are the row operations and their associated keystrokes. At each step, the result is stored in C and replaces the previous matrix C. The solution is shown in Figure 4.60.

Row Operation	Keystrokes
rAdd(C, 1, 2)	2nd MATRX F4 MORE F3 ALPHA C , 1 , 2) STO ► C ENTER
mRAdd(-2, C, 1, 3)	F5 (-) 2 , ALPHA C , 1 , 3) STO ► C ENTER
rAdd(C, 2, 3)	F3 ALPHA C , 2 , 3) STO ► C ENTER
multR(½, C, 3)	F4 1 ÷ 2 , ALPHA C , 3) STO ► C ENTER

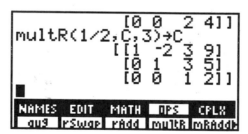

Figure 4.60: Final matrix after row operations

Thus $z = 2$, so $y = -1$ and $x = 1$.

4.6.4 Determinants and Inverses: Enter this 3×3 square matrix as A: $\begin{bmatrix} 1 & -2 & 3 \\ -1 & 3 & 0 \\ 2 & -5 & 5 \end{bmatrix}$. To calculate its determi-

nant, press 2nd MATRX F3 F1 ALPHA A ENTER. You should find that $|A| = 2$, as shown in Figure 4.61.

Since the determinant of matrix A is not zero, this matrix has an inverse, A^{-1}. Press ALPHA A 2nd x^{-1} ENTER to calculate the inverse of matrix A, also shown in Figure 4.61.

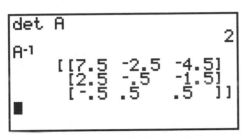

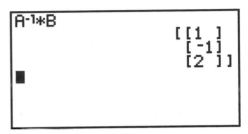

Figure 4.61: $|A|$ and A^{-1} Figure 4.62: Solution matrix

Now let's solve a system of linear equations by matrix inversion. Once more, consider the system of three

equations $\begin{cases} x - 2y + 3z = 9 \\ -x + 3y = -4 \\ 2x - 5y + 5z = 17 \end{cases}$. The coefficient matrix for this system is the matrix $\begin{bmatrix} 1 & -2 & 3 \\ -1 & 3 & 0 \\ 2 & -5 & 5 \end{bmatrix}$ that was

entered in the previous example as matrix A. If necessary, enter it again in your TI-85. Next enter the matrix $\begin{bmatrix} 9 \\ -4 \\ 17 \end{bmatrix}$ as B. Then enter $A^{-1} \cdot B$ by pressing ALPHA A 2nd x^{-1} × ALPHA B ENTER to calculate the solution

matrix (Figure 4.62). The solutions are still $x = 1$, $y = -1$, and $z = 2$.

4.7   Sequences

4.7.1 Iteration with the ANS Key: The ANS feature permits you to perform *iteration*, the process of evaluating a function repeatedly, on the TI-85 calculator.

As an example, calculate $\dfrac{n-1}{3}$ for $n = 27$. Then calculate $\dfrac{n-1}{3}$ for $n =$ the answer to the previous calculation. Continue to use each answer as n in the *next* calculation. Here are keystrokes to accomplish this iteration on the TI-85 (see the results in Figure 4.63). Notice that when you use ANS in place of n in a formula, it is sufficient to press ENTER to continue an iteration.

Iteration	Keystrokes	Display
1	27 ENTER	27
2	(2nd ANS - 1) ÷ 3 ENTER	8.66666666667
3	ENTER	2.55555555556
4	ENTER	.518518518519
5	ENTER	-.16049382716

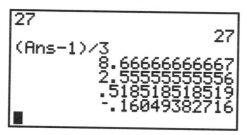

Figure 4.63: Iteration

Press ENTER several more times and see what happens with this iteration. You may wish to try it again with a different starting value.

4.7.2 Arithmetic and Geometric Sequences: Use iteration with the ANS variable to determine the n-th term of a sequence. For example, find the 18th term of an *arithmetic* sequence whose first term is 7 and whose common difference is 4. Enter the first term 7, then start the progression with the recursion formula, 2nd ANS + 4 ENTER. This yields the 2nd term, so press ENTER sixteen more times to find the 18th term. For a *geometric* sequence whose common ratio is 4, start the progression with 2nd ANS × 4 ENTER.

Of course, you could also use the *explicit* formula for the n-th term of an arithmetic sequence, $t_n = a + (n-1)d$. First enter values for the variables a, d, and n, then evaluate the formula by pressing 2nd alpha a + (2nd alpha n - 1) 2nd alpha d ENTER. For a geometric sequence whose n-th term is given by $t_n = a \cdot r^{n-1}$, enter values for the variables a, r, and n, then evaluate the formula by pressing 2nd alpha a 2nd alpha r ^ (2nd alpha n - 1) ENTER.

4.7.3 Sums of Sequences: Calculate the sum $\sum_{n=1}^{12} 4(0.3)^n$ on the TI-85 by pressing 2nd MATH F5 *[MISC]* F1 *[sum]* F3 *[seq]* 4 × .3 ^ ALPHA N , ALPHA N , 1 , 12 , 1) ENTER. You should get 1.71428480324. The format of this command is sum seq(*expression, variable, begin, end, increment*).

4.8   Parametric and Polar Graphs

4.8.1 Graphing Parametric Equations: The TI-85 plots parametric equations as easily as it plots functions. Just use the MODE menu (Figure 4.1), go to the fifth line from the top, and change the setting from Func for function graphs to Param for parametric graphs. Be sure, if the independent parameter is an angle measure, that MODE is set to whichever you need, Radian or Degree.

For example, here are the keystrokes needed to graph the parametric equations $x = \cos^3 t$ and $y = \sin^3 t$. First check that angles are currently being measured in radians. Change to parametric mode and press GRAPH F1 to examine the new parametric equation menu E(t)= (Figure 4.64). Enter the two parametric equations by pressing (COS F1) ^ 3 ENTER (SIN F1) ^ 3 ENTER.

Figure 4.64: Parametric E(t)= menu

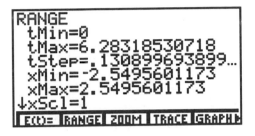

Figure 4.65: Parametric RANGE menu

Also look at the new parametric RANGE menu (Figure 4.65). In the standard viewing rectangle, the values of t go from 0 to 2π in steps of $\frac{\pi}{24} = .1309$. Press GRAPH to see the parametric graph (Figure 4.66).

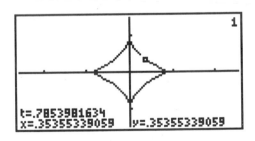

Figure 4.66: Parametric graph of $x = \cos^3 t$ and $y = \sin^3 t$

You may ZOOM and TRACE along parametric graphs just as you did with function graphs. As you trace along this graph, notice that the cursor moves in the *counterclockwise* direction as t increases.

4.8.2 Rectangular-Polar Coordinate Conversion: The CPLX menu (Figure 4.67) provides functions for converting between rectangular and polar coordinate systems.

Given rectangular coordinates $(x, y) = (4, -3)$, convert *from* these rectangular coordinates *to* polar coordinates (r, θ) by pressing 2nd CPLX MORE (4 , (-) 3) F2 ENTER. We see that $r = 5$ and $\theta = -.6435$. The measure of angle θ is displayed in radians, because that is the current default angle measure chosen in the MODE menu.

Suppose $(r, \theta) = (3, \pi)$. To convert *from* these polar coordinates *to* rectangular coordinates (x, y), press 2nd CPLX MORE (3 2nd ∠ π) F1 ENTER. Then $x = -3$ and $y = 0$.

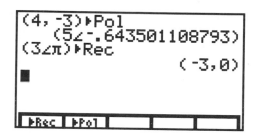

Figure 4.67: Coordinate conversions

4.8.3 Graphing Polar Equations: The TI-85 graphs a polar function in the form $r = f(\theta)$. In the fifth line of the MODE menu, select POL for polar graphs.

For example, to graph $r = 4 \sin \theta$, press GRAPH F1 for the r(θ)= menu. Then enter the expression $4 \sin \theta$ for r1. Choose a good viewing rectangle and an appropriate interval and increment for θ. In Figure 4.68, the viewing rectangle is roughly "square" and extends from -6.5 to 6.5 horizontally and from -4 to 4 vertically.

Figure 4.67 shows *rectangular* coordinates of the cursor's location on the graph. You may sometimes wish to trace along the curve and see *polar* coordinates of the cursor's location. The first line of the GRAPH FORMT menu (Figure 4.16) has options for displaying the cursor's position in rectangular RectGC or polar PolarGC form.

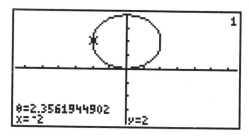

Figure 4.68: Polar graph of $r = 4 \sin \theta$

4.9   Probability

4.9.1 Random Numbers: The command rand generates a number between 0 and 1. You will find this command in the PROB (probability) sub-menu of the MATH menu. Press 2nd MATH F2 F4 ENTER to generate a random number. Press ENTER to generate another random number; keep pressing ENTER to generate more of them.

If you need a random number between, say, 0 and 10, then press 10 2nd MATH F2 F4 ENTER. To get a random number between 5 and 15, press 5 + 10 2nd MATH F2 F4 ENTER.

4.9.2 Permutations and Combinations: To calculate the number of *permutations* of 12 objects taken 7 at a time, $_{12}P_7$, press 2nd MATH F2 12 F2 7 ENTER. Then $_{12}P_7 = 3,991,680$, as shown in Figure 4.69.

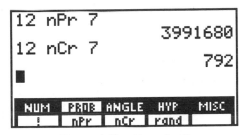

Figure 4.69: $_{12}P_7$ and $_{12}C_7$

For the number of *combinations* of 12 objects taken 7 at a time, $_{12}C_7$, press 2nd MATH F2 12 F3 7 ENTER. So $_{12}C_7 = 792$.

4.9.3 Probability of Winning: A state lottery is configured so that each player chooses six different numbers from 1 to 40. If these six numbers match the six numbers drawn by the State Lottery Commission, the player wins the top prize. There are $_{40}C_6$ ways for the six numbers to be drawn. If you purchase a single lottery ticket, your probability of winning is 1 in $_{40}C_6$. Press 1 ÷ 2nd MATH F2 40 F3 6 ENTER to calculate your chances, but don't be disappointed.

4.10   Programming

4.10.1 Entering a Program: The TI-85 is a programmable calculator that can store sequences of commands for later replay. Here's an example to show you how to enter a useful program that solves quadratic equations by the quadratic formula.

Press PRGM to access the programming menu. The TI-85 has space for many programs, each identified by a name that is up to eight characters long. The names of all your programs are listed alphabetically in the PRGM NAMES menu.

To create a new program, press PRGM F2 *[EDIT]* and enter its name. The cursor is now a blinking **A**, indicating the calculator is set to receive upper case alphabetic characters. Call this program QUADRAT and press ENTER when you have finished.

Within the program itself, each line begins with a colon **:** supplied automatically by the calculator after you press ENTER. Any command you could enter directly in the TI-85's home screen can be entered as a line in a program. There are also special programming commands.

Note that the TI-85 calculator checks for program errors as it *runs* a program, not while you enter or edit it.

Enter the program QUADRAT by pressing the given keystrokes.

Program Line	*Keystrokes*
: Disp "Enter A"	F3 F3 MORE F5 ALPHA E 2nd alpha ALPHA N T E R ⌴ ALPHA A F5 ENTER

displays the words *Enter A* on the TI-85 screen

| : Input A | MORE F1 ALPHA A ENTER |

waits for you to input a value that will be assigned to the variable A

: Disp "Enter B"	F3 MORE F5 ALPHA E 2nd alpha ALPHA N T E R ⌴ ALPHA B F5 ENTER
: Input B	MORE F1 ALPHA B ENTER
: Disp "Enter C"	F3 MORE F5 ALPHA E 2nd alpha ALPHA N T E R ⌴ ALPHA C F5 ENTER
: Input C	MORE F1 ALPHA C ENTER
: ClLCD	MORE F3 ENTER

clears the calculator's display

| : B²-4AC → D | ALPHA B x² - 4 ALPHA A × ALPHA C STO▸ D ENTER |

calculates the discriminant and stores its value as D

| : If D>0 | EXIT F4 F1 ALPHA D 2nd TEST F3 0 ENTER |

tests to see if the discriminant is positive

| : Then | EXIT F2 ENTER |

in case the discriminant is positive, continues on to the next line;
if the discriminant is not positive, jumps to the command after **Else** below

| : Disp "Two real roots" | EXIT F3 F3 MORE F5 ALPHA T 2nd alpha ALPHA W O ⌴ R E A L ⌴ R O O T S F5 ENTER |

: (-B+√D)/(2A) → M   ((-) ALPHA B + 2nd √ ALPHA D) ÷ (2 ALPHA A)
                          STO► M ENTER

    calculates one root and stores it as M

: Disp M   MORE F3 ALPHA M ENTER

    displays one root

: (-B-√D)/(2A) → N   ((-) ALPHA B – 2nd √ ALPHA D) ÷ (2 ALPHA A)
                          STO► N ENTER

: Disp N   F3 ALPHA N ENTER

: Else   EXIT F4 F3 ENTER

    continues from here if the discriminant is not positive

: If D==0   F1 ALPHA D 2nd TEST F1 0 ENTER

    tests to see if the discriminant is zero

: Then   EXIT F2 ENTER

    in case the discriminant is zero, continues on to the next line;
    if the discriminant is not zero, jumps to the command after **Else** below

: Disp "Double root"   EXIT F3 F3 MORE F5 ALPHA D 2nd alpha ALPHA O U B L E
                          ⌴ R O O T F5 ENTER

    displays a message in case there is a double root

: -B/(2A) → M   (-) ALPHA B ÷ (2 ALPHA A) STO► M ENTER

    the quadratic formula reduces to $\dfrac{-b}{2a}$ when $D = 0$

: Disp M   MORE F3 ALPHA M ENTER

: Else   EXIT F4 F3 ENTER

    continues from here if the discriminant is not zero

: Disp "Complex roots"   EXIT F3 F3 MORE F5 ALPHA C 2nd alpha ALPHA O M
                          P L E X ⌴ R O O T S F5 ENTER

    displays a message in case the roots are complex numbers

: Disp "Real part"   MORE F3 MORE F5 ALPHA R 2nd alpha ALPHA E A L ⌴
                          P A R T F5 ENTER

: -B/(2A) → R   (-) ALPHA B ÷ (2 ALPHA A) STO▸ R ENTER

calculates the real part $\dfrac{-b}{2a}$ of the complex roots

: Disp R   MORE F3 ALPHA R ENTER

: Disp "Imaginary part"   F3 MORE F5 ALPHA I 2nd alpha ALPHA M A G I N A R Y ␣ P A R T F5 ENTER

: √-D/(2A) → I   2nd √ (-) ALPHA D ÷ (2 ALPHA A) STO▸ I ENTER

calculates the imaginary part $\dfrac{\sqrt{-D}}{2a}$ of the complex roots;

since $D < 0$, wc must use $-D$ as the radicand

: Disp I   MORE F3 ALPHA I ENTER

: End   EXIT F4 F5 ENTER

marks the end of an If-Then-Else group of commands

: End   F5

When you have finished, press 2nd QUIT to leave the program editor.

4.10.2 Executing a Program: To execute the program just entered, press PRGM NAMES and look for QUADRAT. The names of programs are listed alphabetically; press MORE to advance through the listing. Press the function key above QUADRAT to select this program, then press ENTER to execute it.

The program has been written to prompt you for values of the coefficients a, b, and c in a quadratic equation $ax^2 + bx + c = 0$. Input a value, then press ENTER to continue the program.

If you need to interrupt a program during execution, press ON.

The instruction manual for your TI-85 gives detailed information about programming. Refer to it to learn more about programming and how to use other features of your calculator.

4.11 Differentiation

4.11.1 Limits: Suppose you need to find this limit: $\lim\limits_{x \to 0} \dfrac{\sin 4x}{x}$. Plot the graph of $f(x) = \dfrac{\sin 4x}{x}$ in a convenient viewing rectangle that contains the point where the function appears to intersect the line $x = 0$ (because
you want the limit as $x \to 0$). Your graph should lend support to the conclusion that $\lim\limits_{x \to 0} \dfrac{\sin 4x}{x} = 4$.

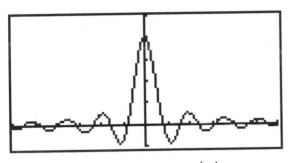 

Figure 4.70: Checking $\lim\limits_{x \to 0} \dfrac{\sin 4x}{x} = 4$ \qquad Figure 4.71: Checking $\lim\limits_{x \to \infty} \dfrac{2x-1}{x+1} = 2$

To test the reasonableness of the conclusion that $\lim\limits_{x \to \infty} \dfrac{2x-1}{x+1} = 2$, evaluate the function $f(x) = \dfrac{2x-1}{x+1}$ for
several large positive values of x (since you want the limit as $x \to \infty$). For example, evaluate $f(100)$,
$f(1000)$, and $f(10,000)$. Another way to test the reasonableness of this result is to examine the graph of
$f(x) = \dfrac{2x-1}{x+1}$ in a viewing rectangle that extends over large values of x. See, as in Figure 4.71 (where the
viewing rectangle extends horizontally from 0 to 100), whether the graph is asymptotic to the horizontal line
$y = 2$ (enter $\dfrac{2x-1}{x+1}$ for y1 and 2 for y2).

4.11.2 Numerical Derivatives: The derivative of a function f at x can be defined as the limit of the slopes of
secant lines, so $f'(x) = \lim\limits_{\Delta x \to 0} \dfrac{f(x + \Delta x) - f(x - \Delta x)}{2\Delta x}$. And for small values of Δx, the expression
$\dfrac{f(x + \Delta x) - f(x - \Delta x)}{2\Delta x}$ gives a good approximation to the limit.

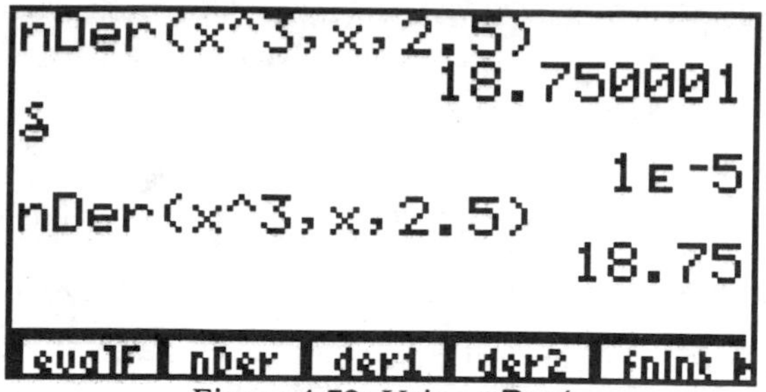

Figure 4.72: Using nDer(

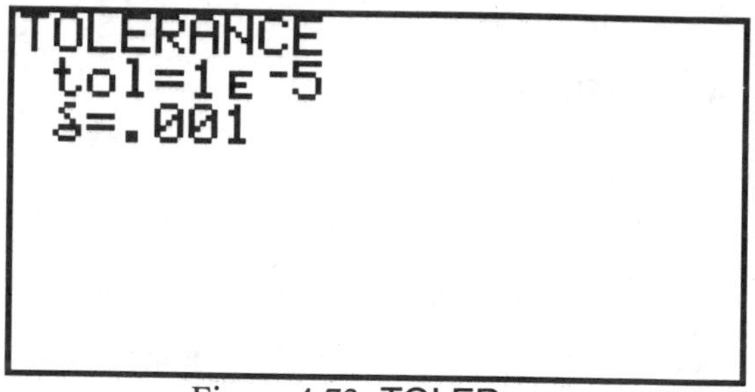

Figure 4.73: TOLER menu

The TI-82 has a function nDer(in the CALC menu to calculate the *symmetric difference*, $\dfrac{f(x+\Delta x)-f(x-\Delta x)}{2\Delta x}$. This calculator uses a value for step size Δx that has been stored in the variable δ. For most purposes, $\delta = 0.001$ gives a very good approximation to the derivative and is the TI-85's default (see Figure 4.73). So to find a numerical approximation to $f'(2.5)$ when $f(x) = x^3$ and with $\Delta x = 0.001$, press 2nd CALC F2 x- VAR $\wedge$ 3, x-VAR , 2.5) ENTER as shown in the top line of Figure 4.72. The format of this command is nDer(*expression, variable, value*). If you are evaluating for the *current* stored value of the variable, write just nDer(*expression, variable*). Figure 4.72 shows the same derivative approximated again after pressing 2nd TOLER and changing δ to 0.00001.

Note that the function nDer calculates an approximate numerical derivative whose accuracy depends on the choice of δ. For a calculation that employs the rules of differentiation to yield a value as accurate as the TI-85 can be, use the special functions der1 for a first derivative and der2 for a second derivative (both are in the CALC menu). These commands use the same format as nDER, so you would write der1(*expression, variable, value*).

Technology Tip: In general, you may use der1 and der2 with expressions involving the basic calculator operations and functions (for example, $\sin x$ and x^2) or built up from them with simple arithmetic (for example, $\sin x^2$ or $\sin x + x^2$). Refer to the TI-85 manual for more details about restrictions on expressions that can be used with der1 and der2.

You may *not* use nDer or der1 or der2 in the expression for a der1 or der2 command. However, you *can* use der1 and der2 in the expression for an nDer command. So to approximate the *third* derivative of $\sin x^2$ at $x = 2$, you may press 2nd CALC F2 F4 SIN x-VAR x^2 , x-VAR) , x-VAR , 2).

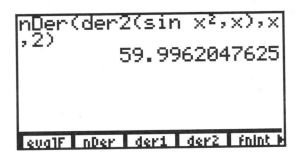

Figure 4.74: $f'''(2)$ for $f(x) = \sin x^2$

Technology Tip: It is sometimes helpful to plot both a function and its derivative together. In Figure 4.76, the function $f(x) = \dfrac{5x-2}{x^2+1}$ and its numerical derivative are graphed. You can duplicate this graph by first

entering $\dfrac{5x-2}{x^2+1}$ for y1 and then entering its numerical derivative for y2 by pressing 2nd CALC F3 2nd M2

1 , 2nd M1), as you see in Figure 4.75.

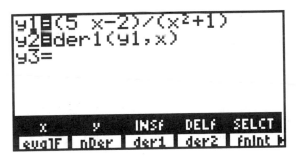
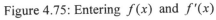

Figure 4.75: Entering $f(x)$ and $f'(x)$

Figure 4.76: Graphs of $f(x)$ and $f'(x)$

You may also approximate a derivative while you are examining the graph of a function. When you are in a graph window, press GRAPH MORE F1 *[MATH]* F4, then use the arrow keys to trace along the curve to a point where you want the derivative (see Figure 4.77 for the graph of $f(x) = \dfrac{5x-2}{x^2+1}$ at $x = -2.3$) and press ENTER. The TI-85 uses the current value of the variable δ for this approximation.

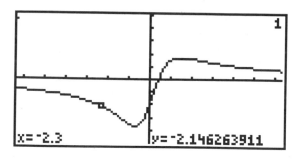 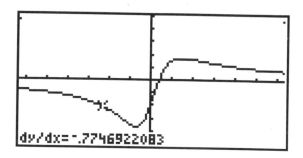

Figure 4.77: $f(x) = \dfrac{5x-2}{x^2+1}$ at $x = -2.3$

Figure 4.78: dy/dx

Technology Tip: With the TI-85, you can graph a function and the tangent line at one of its points. So graph the function $f(x) = \dfrac{5x-2}{x^2+1}$ once again. Then press GRAPH MORE F1 MORE MORE F3, trace with the cursor to a point of interest, and press ENTER. You may continue to draw more tangent lines, or clear them by pressing GRAPH MORE F2 MORE F5.

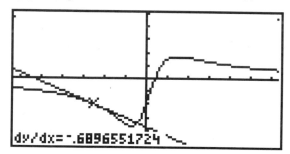

Figure 4.79: Tangent line

Figure 4.80: Point of inflection

In a similar way, the TI-85 locates a point of inflection. Graph the function, press GRAPH MORE F1 MORE F3, trace with the cursor near a point of inflection, and press ENTER.

You may choose the method of differentiation used by the TI-85 for graphing tangent lines and for locating points of inflection. In the MODE menu (Figure 4.1), move the cursor down to the last line and select either dxDer1 (exact differentiation) to use der1 or dxNDer (numeric differentiation) to use nDer. Recall that der1 is more accurate but also more restrictive, since it applies only to certain functions.

4.11.3 Newton's Method: With your TI-85, you may iterate using Newton's method to find the zeros of a function. Recall that Newton's Method determines each successive approximation by the formula
$$x_{n+1} = x_n - \frac{f(x_n)}{f'(x_n)}.$$

As an example of the technique, consider $f(x) = 2x^3 + x^2 - x + 1$. Enter this function as y1. A look at its graph suggests that it has a zero near $x = -1$, so start the iteration by going to the home screen and storing -1 as x (see figure 4.81). Then press these keystrokes: x-VAR − 2nd VARS MORE F3, select y1, keep on pressing ÷ 2nd CALC F3 2nd VARS MORE F3, again select y1, and conclude by pressing , x-VAR ,) STO ▶ x-VAR. Press ENTER repeatedly until two successive approximations differ by less than some predetermined value, say 0.0001. Note that each time you press ENTER, the TI-85 will use its *current* value of x, and that value is changing as you continue the iteration.

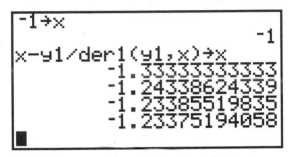

Figure 4.81: Newton's method

Technology Tip: Newton's Method is sensitive to your seed value for x, so look carefully at the function's graph to make a good first estimate. Also, remember that the method sometimes fails to converge!

You may want to save the Newton's Method formula as a short program. See your calculator's manual for further information on programming the TI-85.

4.12   Integration

4.12.1 Approximating Definite Integrals: The TI-85 has the function fnInt in the CALC menu to approximate an integral. So to find a numerical approximation to $\int_0^1 \cos x^2 \, dx$ press 2nd CALC F5 COS x-VAR x² , x-VAR , 0 , 1) ENTER as shown in Figure 4.82. The format of this command is fnInt(*expression, variable, lower limit, upper limit*). In this example, the tolerance tol (which controls the accuracy of the approximation) is 0.00001, the TI-85's default. You may reset the tolerance by pressing 2nd TOLER (refer back to Figure 4.73) or by assigning a value to the variable tol.

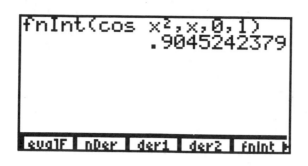

Figure 4.82: fnInt(

4.12.2 Areas: You may approximate the area under the graph of a function $y = f(x)$ between $x = A$ and $x = B$ with your TI-85. For example, here are keystrokes for finding the area under the graph of the function $y = \cos x^2$ between $x = 0$ and $x = 1$. This area is represented by the definite integral $\int_0^1 \cos x^2 dx$. So graph $f(x)$ $= \cos x^2$ and press GRAPH MORE F1 F5. Use the arrow keys to trace along the curve to the lower limit and press ENTER; then trace again to the upper limit (see Figure 4.83) and press ENTER. The region under the graph between the lower limit and the upper limit is marked and its area is displayed as in Figure 4.84. The TI-85 uses fnInt with the default tolerance tol $= 10^{-5}$ in this calculation.

Technology Tip: When approximating the area under $f(x) = \cos x^2$ between $x = 0$ and $x = 1$, both limit points must be visible in the current viewing rectangle. Also, you must trace along the curve to *exactly* where $x = 0$ and $x = 1$. Now to trace along the curve to $x = a$, the viewing rectangle must be chosen so that the function is evaluated at $x = a$. The window shown in Figure 4.83 was made first by pressing ZOOM MORE F4 *[ZDECM]*, then by changing its vertical dimensions to appropriate values. By contrast, find the area under $f(x) = \cos x^2$ between $x = 0$ and $x = 1$ in ZOOM MORE F3 *[ZTRIG]* window.

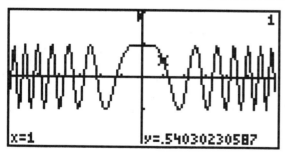

Figure 4.83: Setting the upper limit

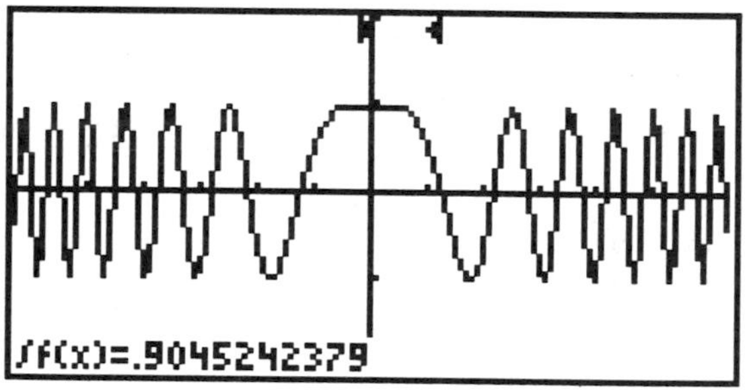

Figure 4.84: $\int f(x)dx$

Technology Tip: Suppose that you want to find the area between two functions, $y = f(x)$ and $y = g(x)$, from $x = A$ to $x = B$. If $f(x) \geq g(x)$ for $A \leq x \leq B$, then enter the expression $f(x) - g(x)$ for y1 and proceed as before to find the required area.

Chapter 5

Texas Instruments TI-92

5.1   Getting started with the TI-92

In this book, the key with the green diamond symbol inside a green border will be indicated by ♦, the key with the white arrow pointing up inside a white border (the *shift* key) will be indicated by ♠, and the key with the white arrow (the *backspace* key) pointing to the left will be indicated by ←. Although the cursor pad allows for movements in eight directions, we will mainly use the four directions of up, down, right, and left. These directions will be indicated by ↑, ↓, →, and ←, respectively

There are eight blue keys on the left side of the calculator labeled F1 through F8. These *function keys* have different effects depending on the screen that is currently showing. The effect or menu of the function keys corresponding to a screen are shown across the top of the display.

5.1.1 Basics: Press the ON key to begin using your TI-92. If you need to adjust the display contrast, first press ♦, then press − (the minus key) to lighten or + (the plus key) to darken. To lighten or darken the screen more, press ♦ then + or − again. When you have finished with the calculator, turn it off to conserve battery power by pressing 2nd and then OFF. Note that the TI-92 has three ENTER keys and two 2nd keys which can be used interchangeably.

Check your TI-92's settings by pressing MODE. If necessary, use the cursor pad to move the blinking cursor to a setting you want to change You can also use F1 to go to page 1 or F2 to go to page 2 of the MODE menu. To change a setting, use ↓ to get to the setting that you want to change, then press → to see the options available. Use ↑ or ↓ to highlight the setting that you want and press ENTER to select the setting. To start with, select the options shown in Figures 5.1 and 5.2: function graphs, main folder, floating decimals with 10 digits displayed, radian measure, normal exponential format, real numbers, rectangular vectors, pretty print, full screen display, Home screen showing, and approximate calculation mode. Note that some of the lines on page 2 of the MODE menu are not readable. These lines pertain to options that are not set as above. Details on alternative options will be given later in this guide. For now, leave the MODE menu by pressing ♦ HOME or 2nd QUIT. Some of the current settings are shown on the status line of the Home screen.

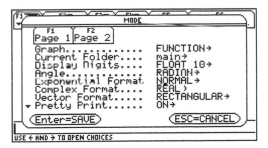

Figure 5.1: MODE menu, page 1

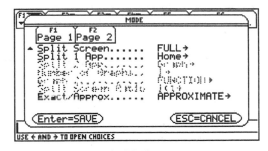

Figure 5.2: MODE menu, page 2

Technology Tip: There are many different ways to get to the most commonly used screens on your TI-92. One method is by using the APPS menu (Figure 5.3) which is accessed by pressing the blue APPS key on the right side of the calculator. Thus, to get to the Home screen you can press 2nd QUIT, ◆ HOME, or APPS ENTER.

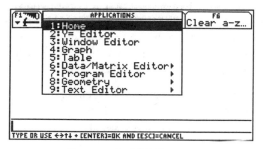

Figure 5.3: APPS menu

5.1.2 *Editing:* One advantage of the TI-92 is that you can use the cursor pad to scroll in order to see a long calculation. For example, type this sum (Figure 5.4):

$$1 + 2 + 3 + 4 + 5 + 6 + 7 + 8 + 9 + 10 + 11 + 12 + 13 + 14 + 15 + 16 + 17 + 18 + 19 + 20$$

Then press ENTER to see the answer. The sum is too long for both the entry line and the history area. The direction(s) in which the line extends off the screen is indicated by an ellipsis at the end of the entry line and arrows (← or →) in the history area. You can scroll through the entire calculation by using the cursor pad (↑ or ↓) to put the cursor on the appropriate line and then using → or ← to move the cursor to the part of the calculation that you wish to see.

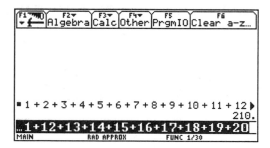

Figure 5.4: Home screen

Often we do not notice a mistake until we see how unreasonable an answer is. The TI-92 permits you to re-display an entire calculation, edit it easily, then execute the *corrected* calculation.

Suppose you had typed 12 + 34 + 56 as in Figure 5.5 but had *not yet* pressed ENTER, when you realize that 34 should have been 74. Simply press the ← direction on the cursor pad as many times as necessary to move the blinking cursor line until it is to the immediate right of the 3, press ← to delete the 3, and then type 7. On

TI-92 Graphics Calculator

the other hand, if 34 should have been 384, move the cursor until it is between the 3 and the 4 and then type 8. If the 34 should have been 3 only, move the cursor to right of the 4, and press ← to delete the 4.

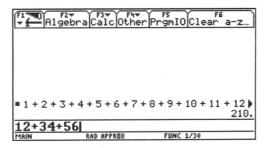

Figure 5.5: Editing a calculation

Technology Tip: The TI-92 has two different inputing modes: *insert* and *overtype*. The default mode is the insert mode, in which the cursor is a blinking vertical line and new text will be inserted at the cursor's position and other characters are pushed to the right. In the overtype mode, the cursor is a blinking square and the characters that you type replace the existing characters. To change from one mode to another, press 2nd INS. The TI-92 remains in whatever the last input mode was, even after being turned off.

Even if you had pressed ENTER, you may still edit the previous expression. Immediately after you press ENTER your entry remains on the entry line. Pressing the ← direction on the cursor pad moves the cursor to the beginning of the line, while pressing the → direction on the cursor pad puts the cursor at the end of the line. Now the expression can be edited as above. To edit a previous expression that is no longer on the entry line, press 2nd and then ENTRY to recall the prior expression. Now you can change it. In fact, the TI-92 retains as many entries as the current history area holds in a "last entry" storage area, including entries that have scrolled off of the screen. Press 2nd ENTRY repeatedly until the previous line you want is on the entry line. (The number of entries that the history area can hold may be changed, see your user's manual for more information.)

To clear the entry line, press CLEAR while the cursor is on that line. To clear previous entry/answer pairs from the history area, use the cursor pad to either the entry or the answer and press CLEAR (both the entry and the answer will be deleted from the display). To clear the entire history area, press F1*[Tools] 8[Clear Home]*, although this will not clear the entry line.

Technology Tip: When you need to evaluate a formula for different values of a variable, use the editing feature to simplify the process. For example, suppose you want to find the balance in an investment account if there is now \$5000 in the account and interest is compounded annually at the rate of 8.5%. The formula for the balance is $P = \left(1 + \frac{r}{n}\right)^{nt}$, where P = principal, r = rate of interest (expressed as a decimal), n = number of times interest is compounded each year, and t = number of years. In our example, this becomes $5000(1+.085)^t$. Here are the keystrokes for finding the balance after $t = 3$, 5, and 10 years.

Years	Keystrokes	Balance
3	5000 (1 + .085) ∧ 3 ENTER	$6386.45
5	→ ← 5 ENTER	$7518.28
10	→ ← 10 ENTER	$11,304.92

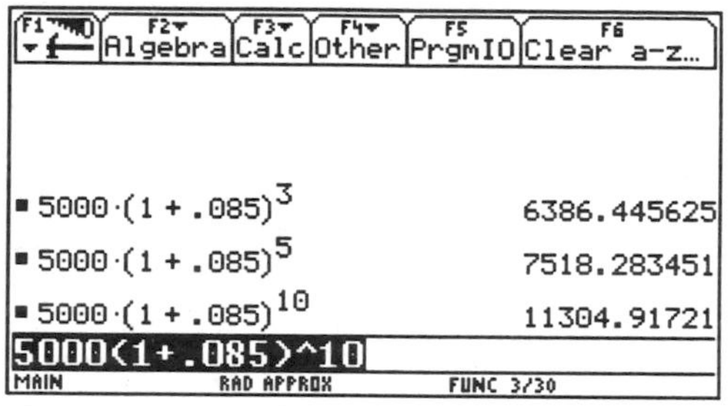

Figure 5.6: Editing expressions

Then, to find the balance from the same initial investment but after 5 years when the annual interest rate is 7.5%, press the following keys to change the last calculation above: → ← ← 5 ← ← ← ← ← 7 ENTER. You could also use the CLEAR key to erase everything to the right of the current location of the cursor. Then, changing the calculation from 10 years at the annual interest rate of 8.5% to 5 years at the annual interest rate of 7.5% is then done by pressing → ← ← CLEAR 5 ← ← ← ← ← 7 ENTER.

5.1.3 Key Functions: Most keys on the TI-92 offer access to more than one function, just as the keys on a computer keyboard can produce more than one letter ("g" and "G") or even quite different characters ("5" and "%"). The primary function of a key is indicated on the key itself, and you access that function by a simple press on the key.

To access the *second* function indicated in yellow or to the *left* above a key, first press 2nd ("2nd" appears on the status line) and *then* press the key. For example to calculate $\sqrt{25}$, press 2nd $\sqrt{\ }$ 25) ENTER.

Technology Tip: The TI-92 automatically places a left parenthesis, (, after many functions and operators (including LN, 2nd e^X, SIN, COS, TAN, and 2nd $\sqrt{\ }$). If a right parenthesis is not entered, the TI-92 will respond with an error message indicating that the right parenthesis is missing.

When you want to use a function printed in green or to the *right* above a key, first press ◆ ("◆"appears on the status line) and then press the key. For example, if you are in EXACT calculation mode and want to find the approximate value of $\sqrt{45}$ press 2nd $\sqrt{\ }$ 45) ◆ ≈. The QWERTY keyboard on the TI-92 is similar to a typewriter and can produce both upper and lower case letters. To switch from one case to another, press 2nd CAPS. For a single upper case letter, use the ♠ key. There are also additional symbols available from the keyboard by using the 2nd and ◆ keys. Some of the most commonly used symbols are marked on the keyboard, but most are not. See your TI-92 user's manual for more information.

TI-92 Graphics Calculator

5.1.4 Order of Operations: The TI-92 performs calculations according to the standard algebraic rules. Working outwards from inner parentheses, calculations are performed from left to right. Powers and roots are evaluated first, followed by multiplications and divisions, and then additions and subtractions.

Enter these expressions to practice using your TI-92?

Expression	Keystrokes	Display
$7 - 5 \cdot 3$	7 – 5 × 3 ENTER	–8
$(7 - 5) \cdot 3$	(7 – 5) × 3 ENTER	6
$120 - 10^2$	120 – 10 ∧ 2 ENTER	20
$(120 - 10)^2$	(120 – 10) ∧ 2 ENTER	12100
$\dfrac{24}{2^3}$	24 ÷ 2 ∧ 3 ENTER	3
$\left(\dfrac{24}{2}\right)^3$	(24 ÷ 2) ∧ 3 ENTER	1728
$(7 - -5) \cdot -3$	(7 – (–) 5) × (–) 3 ENTER	–36

5.1.5 Algebraic Expressions and Memory: Your calculator can evaluate expressions such as $\dfrac{N(N+1)}{2}$ after you have entered a value for N. Suppose you want $N = 200$. Press 200 STO▸ N ENTER to store the value 200 in memory location N. Whenever you use N in an expression, the calculator will substitute the value 200 until you make a change by storing *another* number in N. Next enter the expression $\dfrac{N(N+1)}{2}$ by typing N × (N + 1) ÷ 2 ENTER. For $N = 200$, you will find that $\dfrac{N(N+1)}{2} = 20100$. Note that there is no distinction made between upper and lower case letters in this case.

The contents of any memory location may be revealed by typing just its letter name and then ENTER. And the TI-92 retains memorized values even when it is turned off, so long as its batteries are good.

5.1.6 Repeated Operations with ANS: As many entry/answer pairs as the history area shows are stored in memory. The last result displayed can be entered on the entry line by pressing 2nd ANS, while the last entry computed is entered on the entry line by pressing 2nd ENTRY. This makes it easy to use the answer from one computation in another computation. For example, press 30 + 15 ENTER so that 45 is the last result displayed. Then press 2nd ANS ÷ 9 ENTER and get 5 because 45 ÷ 9 = 5.

The answer locations are indexed by ans(#), and the entry locations are indexed by entry(#), where # indicates the number of the entry/answer. The pairs are numbered with the most recent computation as 1. Hence the number of a pair changes with each successive computation that is entered. The number of an entry or answer can be found by using the cursor pad (↑) to scroll up to the entry or answer. The number, which is the same for both the entry and the answer, is shown on the bottom of the screen.

To use an earlier answer or entry in a computation, to calculate, say 15 times answer 3 plus 75, press 1 5 × A N S (3) + 7 5 ENTER, using the keyboard to type the letters A, N, and S.

With a function like division, you press the ÷ *after* you enter an argument. For such functions, whenever you would start a new calculation with the previous answer followed by pressing the function key, you may press just the function key. So instead of 2nd ANS ÷ 9 in the previous example, you could have pressed simply ÷ 9 to achieve the same result. This technique also works for these functions: + − × ∧ 2nd x^{-1}.

Here is a situation where this is especially useful. Suppose a person makes $5.85 per hour and you are asked to calculate earnings for a day, a week, and a year. Execute the given keystrokes to find the person's incomes during these periods (results are shown in Figure 5.7).

Pay Period	Keystrokes	Earnings
8-hour day	5.85 × 8 ENTER	$46.80
5-day week	× 5 ENTER	$234
52-week year	× 52 ENTER	$12,168

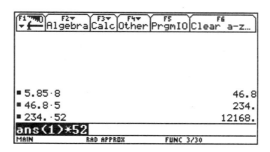

Figure 5.7: ANS variable

5.1.7 *The MATH Menu:* Operators and functions associated with a scientific calculator are available either immediately from the keys of the TI-92 or by the 2nd keys. You have direct access to common arithmetic operations (2nd √ , 2nd x^{-1}, ∧), trigonometric functions (SIN, COS, TAN), and their inverses (2nd SIN^{-1}, 2nd COS^{-1}, 2nd TAN^{-1}), exponential and logarithmic functions (LN, 2nd e^{x}), and a famous constant (2nd π).

A significant difference between the TI-92 graphing calculators and most scientific calculators is that TI-92 requires the argument of a function *after* the function, as you would see in a formula written in your textbook. For example, on the TI-92 you calculate $\sqrt{16}$ by pressing the keys 2nd √ 16) in that order.

Here are keystrokes for basic mathematical operations. Try them for practice on your TI-92.

Expression	Keystrokes	Display
$\sqrt{3^2 + 4^2}$	2nd $\sqrt{}$ 3 ∧ 2 + 4 ∧ 2) ENTER	5.
$2\frac{1}{3}$	2 + 3 2nd x^{-1} ENTER	2.333333333
ln 200	LN 200) ENTER	5.298317367
$2.34 \cdot 10^5$	2.34 × 10 ∧ 5 ENTER	234000.

Technology Tip: Note that if you had set the calculation mode to either AUTO or EXACT (the last line of page 2 of the MODE menu), the TI-92 would display $\frac{7}{3}$ for $2\frac{1}{3}$ and $2\ln(5) + 3\ln(2)$ for ln 200. Thus, you can use either fractions and exact numbers or decimal approximations. The AUTO mode will give exact rational results whenever all of the numbers entered are rational, and decimal approximations for other results.

Additional mathematical operations and functions are available from the MATH menu. Press 2nd MATH to see the various sub-menus. Press 1*[Number]* or just ENTER to see the options available under the Number sub-menu. You will learn in your mathematics textbook how to apply many of them. As an example, calculate the remainder of 437 when divided by 49 by pressing 2nd MATH 1*[Number]* then *either* A*[remain(] or* ↓ ↓ ↓ ↓ ↓ ↓ ↓ ↓ ENTER; finally press 437 , 49) ENTER to see 45. To leave the MATH menu (or any other menu) and take no other action, press 2nd QUIT or just ESC.

Note that you can select a function or a sub-menu from the current menu by pressing either ↓ until the desired item is highlighted and then ENTER, or by pressing the number or letter corresponding to the function or sub-menu. It is easier to press the letter A than to press ↓ nine times to get the remain(function.

Figure 5.8: MATH menu and Number sub-menu

The *factorial* of a non-negative integer is the *product* of *all* the integers from 1 up to the given integer. The symbol for factorial is the exclamation point. So 4! (pronounced *four factorial*) is $1 \cdot 2 \cdot 3 \cdot 4 = 24$. You will learn more about applications of factorials in your textbook, but for now use the TI-92 to calculate 4! Press these keystrokes: 4 2nd MATH 7*[Probability]* 1*[!]* ENTER.

On the TI-92 it is possible to do calculations with complex numbers. To enter the imaginary number i, press 2nd i. For example, to divide $2 + 3i$ by $4 - 2i$, press (2 + 3 2nd i) ÷ (4 − 2 2nd i) ENTER. The result is $0.1 + 0.8i$ (Figure 5.9).

To find the complex conjugate of $4 + 5i$ press 2nd MATH 5*[Complex]* ENTER 4 + 5 2nd i) ENTER (Figure 5.9).

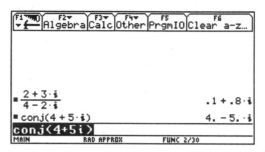

Figure 5.9: Complex number calculations

The TI-92 can also solve for the real and complex roots of an equation. This is done by using the cSolve(function which is not on any of the keys, but can be found in the CATALOG. From the Home screen, pressing 2nd CATALOG gives an alphabetical list of all functions and operations available on the TI-92. You can scroll through the CATALOG page-by-page by pressing 2nd ↓, or if you know what letter the function starts with, pressing the letter moves the cursor to the beginning of the listings for that letter.

The format of cSolve(is cSolve(*expression*, *variable*). For example, to find the zeros of $f(x) = x^3 - 4x^2 + 14x - 20$, from the Home screen press 2nd CATALOG and move the cursor down to cSolve(, then press ENTER. The display will return to the Home screen, with cSolve(on the entry line. To complete the computation, press X ∧ 3 − 4 X ∧ 2 + 14 X − 20 = 0 , X) ENTER. The TI-92 will display the real and complex roots of the equation, as shown in Figure 5.10.

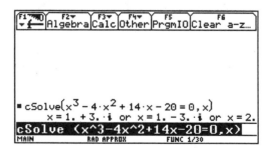

Figure 5.10: cSolve function

All functions and commands found in the CATALOG, can also be used by merely typing the command using the keyboard. Hence, in the Home screen, you could also press C S O L V E (X ∧ 3 − 4 X ∧ 2 + 14 X − 20 = 0 , X) ENTER to find the zeros of $f(x) = x^3 - 4x^2 + 14x - 20$.

5.2 Functions and Graphs

5.2.1 Evaluating Functions: Suppose you receive a monthly salary of $1975 plus a commission of 10% of sales. Let x = your sales in dollars; then your wages W in dollars are given by the equation $W = 1975 + .10x$. If your January sales were $2230 and your February sales were $1865, what was your income during those months?

Here's one method to use your TI-92 to perform this task. Press the ◆ Y= key (above the letter W) or APPS 2*[Y= Editor]* to display the function editing screen (Figure 5.11). You may enter as many as ninety-nine different functions for the TI-92 to use at one time. If there is already a function y1 press ↑ or ↓ as many times as necessary to move the cursor to y1 and then press CLEAR to delete whatever was there. Then enter the expression $1975 + .10x$ by pressing these keys: 1975 + .1 0 X ENTER. Now press ◆ HOME.

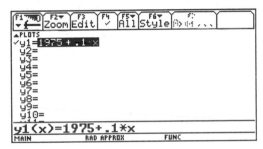

Figure 5.11: Y= screen

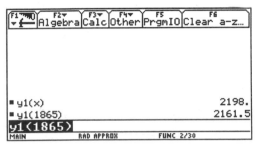

Figure 5.12: Evaluating a function

Assign the value 2230 to the variable x by these keystrokes: 2230 STO▸ X ENTER. Then press the following keystrokes to evaluate y1 and find January's wages: Y 1 (X) ENTER, completes the calculation. It is not necessary to repeat all these steps to find the February wages. Simply press → to begin editing the previous entry, change X to 1865, and press ENTER (see Figure 5.12).

You may also have the TI-92 make a table of values for the function. Press ◆ TblSet to set up the table (Figure 5.13). Move the blinking cursor down to the fourth line beside Independent:, then press → and 2*[ASK]* ENTER. This configuration permits you to input values for x one at a time. Now press ◆ TABLE or APPS 5*[Table]*, enter 2230 in the x column, and press ENTER (see Figure 5.14). Press ↓ to move to the next line and continue to enter additional values for x. The TI-92 automatically completes the table with the corresponding values of y1. Press 2nd QUIT to leave the TABLE screen.

Technology Tip: The TI-92 requires multiplication to be expressed between variables, so xxx does not means x^3, rather it is a new variable named xxx. Thus, you must use either ×'s between the x's or ∧ for powers of x.

Of course, expressed multiplication is not required between a constant and a variable. See your TI-92 manual for more information about the allowed usage of implied multiplication.

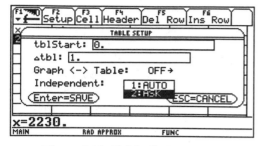

Figure 5.13: Table Setup screen

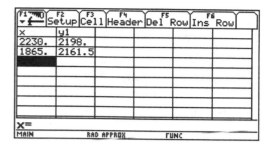

Figure 5.14: Table of values

5.2.2 Functions in a Graph Window: Once you have entered a function in the Y= screen of the TI-92, just press ◆ GRAPH to see its graph. The ability to draw a graph contributes substantially to our ability to solve problems.

For example, here is how to graph $y = -x^3 + 4x$. First press ◆ Y= and delete anything that may be there by moving with the arrow keys to y1 or to any of the other lines and pressing CLEAR wherever necessary. Then, with the cursor on the (now cleared) top line (y1), press (−) X ∧ 3 + 4 X ENTER to enter the function (as in Figure 5.15). Now press ◆ GRAPH and the TI-92 changes to a window with the graph of $y = -x^3 + 4x$.

While the TI-92 is calculating coordinates for a plot, it displays a the word BUSY on the status line.

Technology Tip: If you would like to see a function in the Y= menu and its graph in a graph window, both at the same time, press MODE to open the MODE menu and press F2 to go to the second page. The cursor will be next to Split Screen. Select either TOP-BOTTOM or LEFT-RIGHT by pressing → and 2 or 3, respectively. Now the 2 lines below the Split 1 App line have become readable, since these options apply only when the calculator is in the split screen mode. The Split 1 App will automatically be the screen you were on prior to pressing MODE. You can choose what you want the top or left-hand screen to show by moving down to the Split 1 App line, pressing → and the number of the application you want in that window. The Split 2 App determines what is shown in the bottom or right-hand window. Press ENTER to confirm your choices and your TI-92's screen will now be divided either horizontally or vertically (as you choose). Figure 5.15 shows the graph and the Y= screen with the settings shown in Figure 5.16. The split screen is also useful when you need to do some calculations as you trace along a graph. In split screen mode, one side of the screen will be more heavily outlined. This is the *active screen*, i.e., the screen that you can currently modify. You can change which side is active by using 2nd to access the symbol above the APPS key. For now, restore the TI-92 to Full screen.

Technology Tip: Note that if you set one part of your screen to contain a table and the other to contain a graph, the table will not necessarily correspond to the graph unless you use ◆ TblSet to generate a new table based on the functions(s) being graphed(as in Section 5.2.1).

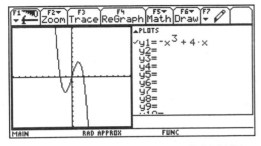

Figure 5.15: Split screen: LEFT-RIGHT

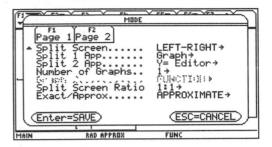

Figure 5.16: MODE settings for Figure 5.15

Your graph window may look like the one in Figure 5.17 or it may be different. Since the graph of $y = -x^3 + 4x$ extends infinitely far left and right and also infinitely far up and down, the TI-92 can display only a piece of the actual graph. This displayed rectangular part is called a *viewing rectangle*. You can easily change the viewing rectangle to enhance your investigation of a graph.

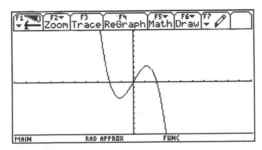

Figure 5.17: Graph of $y = -x^3 + 4x$

The viewing rectangle in Figure 5.17 shows the part of the graph that extends horizontally from −10 to 10 and vertically from −10 to 10. Press ◆ WINDOW to see information about your viewing rectangle. Figure 5.18 shows the WINDOW screen that corresponds to the viewing rectangle in Figure 5.17. This is the *standard viewing rectangle* for the TI-92.

The variables xmin and xmax are the minimum and maximum *x*-values of the viewing rectangle; ymin and ymax are the minimum and maximum *y*-values.

xscl and yscl set the spacing between tick marks on the axes.

xres sets pixel resolution (1 through 10) for function graphs.

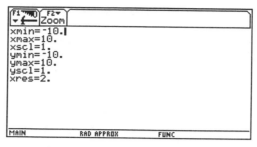

Figure 5.18: Standard WINDOW

Technology Tip: Small xres values improve graph resolution, but may cause the TI-92 to draw graphs more slowly.

Use ↑ and ↓ to move up and down from one line to another in this list; pressing the ENTER key will move down the list. Enter a new value to over-write a previous value and then press ENTER. Remember that a minimum *must* be less than the corresponding maximum or the TI-92 will issue an error message. Also, remember to use the (−) key, not − (which is subtraction), when you want to enter a negative value. Figures 5.17-18, 5.19-20, and 5.21-22 show different WINDOW screens and the corresponding viewing rectangle for each one.

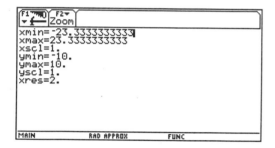

Figure 5.19: Square window

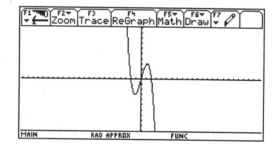

Figure 5.20: Graph of $y = -x^3 + 4x$

To initialize the viewing rectangle quickly to the *standard* viewing rectangle (Figure 5.18), press F2*[Zoom]* 6*[ZoomStd]*. To set the viewing rectangle quickly to a square (Figure 5.19), press F2*[Zoom]* 5*[ZoomSqr]*. More information about square windows is presented later in Section 5.2.4.

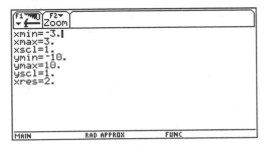

Figure 5.21: Custom window

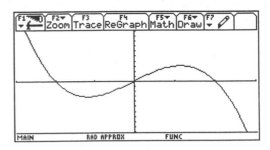

Figure 5.22: Graph of $y = -x^3 + 4x$

Sometimes you may wish to display grid points corresponding to tick marks on the axes. This and other graph format options may be changed while you are viewing the graph by pressing F1 to get the ToolBar menu (Figure 5.23) and then pressing 9*[Format]* to display the Format menu (Figure 5.24) or by pressing ◆ F as indicated on the ToolBar menu in Figure 5.23. Use the cursor pad to move the blinking cursor to Grid; press → 2*[On]* ENTER to redraw the graph. Figure 5.25 shows the same graph as in Figure 5.22 but with the grid turned on.

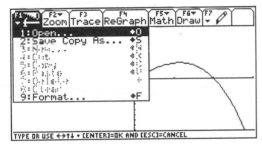

Figure 5.23: ToolBar menu

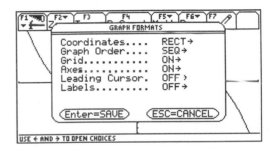

Figure 5.24: Format menu

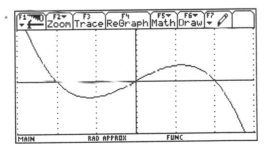

Figure 5.25: Grid turned on for $y = -x^3 + 4x$

In general, you'll want the grid turned *off*, so do that now by pressing ◆ F and turning the Grid option to

OFF, then pressing ENTER.

5.2.3 Graphing Step and Piecewise-Defined Functions: The greatest integer function, written $[[x]]$, gives the greatest *integer* less than or equal to a number x. On the TI-92, the greatest integer function is called floor(and is located under the Number sub-menu of the MATH menu (Figure 5.8). From the Home screen, calculate $[[6.78]] = 6$ by pressing 2nd MATH → 6*[floor(]* 6.78) ENTER.

To graph $y = [[x]]$, go into the Y= menu, move beside y1 and press CLEAR 2nd MATH → 6*[floor(]* X) ENTER ◆ GRAPH. Figure 5.26 shows this graph in a viewing rectangle from −5 to 5 in both directions.

The true graph of the greatest integer function is a step graph, like the one in Figure 5.27. For the graph of $y = [[x]]$, a segment should *not* be drawn between every pair of successive points. You can change this graph from a Line to a Dot graph on the TI-92 by going to the Y= screen, moving up until this function is selected (highlighted) and then pressing F6. This opens the Graph Style menu. Move the cursor down to the second line and press ENTER or press 2; to have the selected graph plotted in Dot style. Now press ◆ GRAPH to see the result.

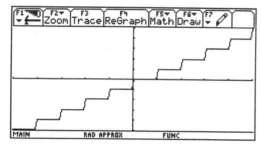   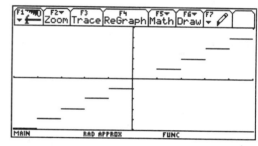

Figure 5.26: **Connected** graph of $y = [[x]]$ Figure 5.27: **Dot** graph of $y = [[x]]$

Technology Tip: When graphing functions in the Dot style, it improves the appearance of the graph to set xres to 1. Figure 5.27 was graphed with xres = 1. Also, the default graph style is Line, so you have to set the style to Dot each time you wish to graph a function in Dot mode.

The TI-92 can graph piecewise-defined functions by using the "when" function. The "when" function is not on any of the keys but can be found in the CATALOG or typed from the keyboard. The format of the when(function is when(*condition, trueResult, falseResult, unknownResult*) where the *falseResult* and *unknownResult* are optional arguments.

For example, to graph the function $f(x) = \begin{cases} x^2 + 2, & x < 0 \\ x - 1, & x \geq 0 \end{cases}$, you want to graph $x^2 + 2$ when the condition $x < 0$ is true and graph $x - 1$ when the condition is false. First, clear any existing functions in the Y= screen. Then move to the y1 line and press W H E N (X 2nd < 0, X ∧ 2 + 1 , X − 1) ENTER (Figure 5.28). Then press ◆ GRAPH to display the graph. Figure 5.29 shows this graph in a viewing rectangle from −5 to 5 in both directions. This was done in Dot style, since the TI-92 will (incorrectly) connect the two sides of the graph at $x = 0$ if the function is graphed in Line style.

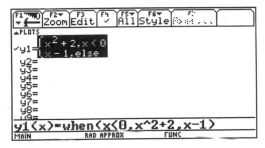

Figure 5.28: Piecewise-defined function

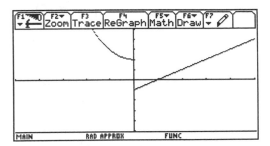

Figure 5.29: Piecewise-defined graph

Other *test* functions, such as $\leq$, $\geq$, and $\neq$ as well as logic operators can be found on the Test sub-menu of the 2nd MATH menu.

5.2.4 Graphing a Circle: Here is a useful technique for graphs that are not functions but can be "split" into a top part and a bottom part, or into multiple parts. Suppose you wish to graph the circle of radius 6 whose equation is $x^2 + y^2 = 36$. First solve for y and get an equation for the top semicircle, $y = \sqrt{36 - x^2}$, and for the bottom semicircle, $y = -\sqrt{36 - x^2}$. Then graph the two semicircles simultaneously.

Use the following keystrokes to draw this circle's graph. First clear any existing functions on the Y= screen. Enter $\sqrt{36 - x^2}$ as y1 and $-\sqrt{36 - x^2}$ as y2 (see Figure 5.30) by pressing 2nd $\sqrt{}$ 36 – X ∧ 2) ENTER (–) 2nd $\sqrt{}$ 36 – X ∧ 2) ENTER. Then press ◆ GRAPH to draw them both (Figure 5.31).

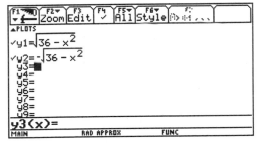

Figure 5.30: Two semicircles

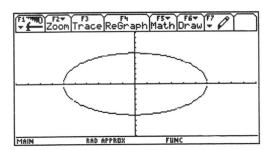

Figure 5.31: Circle's graph - standard WINDOW

Instead of entering $-\sqrt{36 - x^2}$ as y2, you could have entered –y1 as y2 and saved some keystrokes. On the TI-92, try this by going into the Y= screen and pressing ↑ to move the cursor up to y2. Then press CLEAR (–) Y 1 (X) ENTER (Figure 5.32). The graph should be as before.

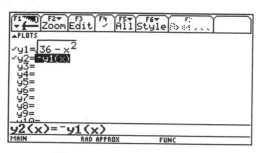

Figure 5.32: Using y1 in y2

If your range were set to a viewing rectangle extending from -10 to 10 in both directions, your graph would look like Figure 5.31. Now this does *not* look a circle, because the units along the axes are not the same. You need what is called a "square" viewing rectangle. Press F2*[Zoom]* 5*[ZoomSqr]* and see a graph that appears more circular.

Technology Tip: Another way to get a square graph is to change the range variables so that the value of ymax $-$ ymin is approximately $\frac{3}{7}$ times xmax $-$ xmin. For example, see the WINDOW in Figure 5.33 to get the corresponding graph in Figure 5.34. This method works because the dimensions of the TI-92's display are such that the ratio of vertical to horizontal is approximately $\frac{3}{7}$.

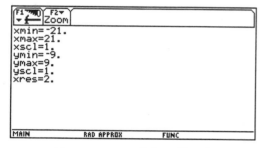

Figure 5.33: $\frac{\text{vertical}}{\text{horizontal}} = \frac{18}{42} = \frac{3}{7}$

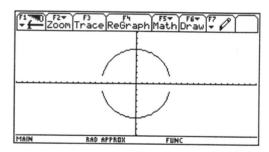

Figure 5.34: A "square" circle

The two semicircles in Figure 5.34 do not meet because of an idiosyncrasy in the way the TI-92 plots a graph.

5.2.5 TRACE: Graph the function $y = -x^3 + 4x$ from Section 5.2.2 using the standard viewing rectangle. (Remember to clear any other functions in the Y= screen.) Press any of the cursor directions ↑ ↓ → ← and see the cursor move from the center of the viewing rectangle. The coordinates of the cursor's location are displayed at the bottom of the screen, as in Figure 5.35, in floating decimal format. This cursor is called a *free-moving cursor* because it can move from dot to dot *anywhere* in the graph window.

Remove the free-moving cursor and its coordinates from the window by pressing ◆ GRAPH, CLEAR, ESC or ENTER. Press the cursor pad again and the free-moving cursor will reappear at the same point you left it.

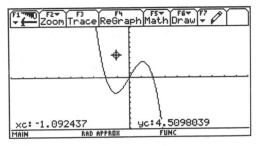

Figure 5.35: Free-moving cursor

Press F3*[TRACE]* to enable the left ← and right → directions to move the cursor along the function. The cursor is no longer free-moving, but is now constrained to the function. The coordinates that are displayed belong to points on the function's graph, so the y-coordinate is the calculated value of the function at the corresponding x-coordinate.

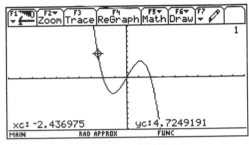

Figure 5.36: TRACE

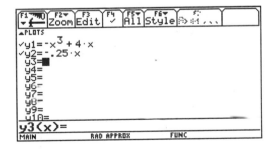

Figure 5.37: Two functions

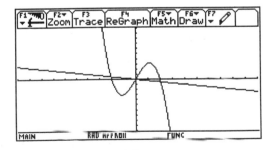

Figure 5.38: $y = -x^3 + 4x$ and $y = -.25x$

Now plot a second function, $y = -.25x$, along with $y = -x^3 + 4x$. Press ◆ Y= and enter $-.25x$ for y2, then press ◆ GRAPH to see both functions (Figure 5.38).

Notice that in Figure 5.37 there are check marks ✓ to the left of *both* y1 and y2. This means that *both* functions will be graphed. In the Y= screen, move the cursor onto y1 and press F4[✓]. The check mark left of y1 should disappear (Figure 5.39). Now press ◆ GRAPH and see that only y2 is plotted (Figure 5.40).

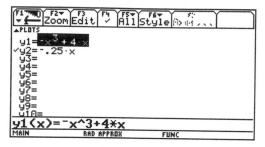

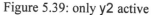

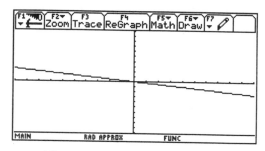

Figure 5.39: only y2 active

Figure 5.40: Graph of $y = -.25x$

Many different functions can be stored in the Y= list and any combination of them may be graphed simultaneously. You can make a function active or inactive for graphing by pressing F4 when the function is highlighted to add a check mark (activate) or remove the check mark (deactivate). Now go back to the Y= screen and do what is needed in order to graph y1 but not y2.

Now activate both functions so that both graphs are plotted. Press F3[TRACE] and the cursor appears first on the graph of $y = -x^3 + 4x$ because it is higher up on the Y= list. You know that the cursor is on this function, y1, because of the numeral 1 that is displayed in the upper right corner of the screen. Press the up ↑ or down ↓ direction to move the cursor vertically to the graph of $y = -.25x$. Now the numeral 2 is shown in the upper right corner of the screen. Next press the left and right arrow keys to trace along the graph of $y = -.25x$. When more than one function is plotted, you can move the trace cursor vertically from one graph to another with the ↑ and ↓ directions.

Technology Tip: By the way, trace the graph of $y = -.25x$ and press and hold either the ← or → direction. The cursor becomes larger and pulses as it moves along the graph. Eventually you will reach the left or right edge of the window. Keep pressing the direction and the TI-92 will allow you to continue the trace by panning the viewing rectangle. Check the WINDOW screen to see that the xmin and xmax are automatically updated.

The TI-92 has a display of 239 horizontal columns of pixels and 103 vertical rows, so when you trace a curve across a graph window, you are actually moving from xmin to xmax in 238 equal jumps, each called Δx. You would calculate the size of each jump to be $\Delta x = \frac{xmax - xmin}{238}$. Sometimes you may want the jumps to be friendly numbers like 0.1 or 0.25 so that, when you trace along the curve, the x-coordinates will be incremented by such a convenient amount. Just set your viewing rectangle for a particular increment Δx by making xmax = xmin + 238 · Δx. For example, if you want xmin = −5 and Δx = 0.3, set xmax = −5 + 238 · 0.3 = 66.4. Likewise, set ymax = ymin + 102 Δy if you want the vertical increment to be some special Δy.

To center your window around a particular point, say (h, k), and also have a certain Δx, set xmin = $h - 119 \cdot \Delta x$ and make xmax = $h + 119 \cdot \Delta x$. Likewise, make ymin = $k - 51 \cdot \Delta y$ and make ymax = $k + 51 \cdot \Delta y$. For example, to center a window around the origin $(0, 0)$, with both horizontal and vertical increments of 0.25, set the range so that xmin = $0 - 119 \cdot 0.25 = -29.75$, xmax = $0 + 119 \cdot 0.25 = 29.75$, ymin = $0 - 51 \cdot 0.25 = -12.75$ and ymax = $0 + 51 \cdot 0.25 = 12.75$.

See the benefit by first plotting $y = x^2 + 2x + 1$ in a standard graphing window. Trace near its y-intercept, which is $(0, 1)$, and move towards its x-intercept, which is $(-1, 0)$. Then press F2[Zoom] 4[ZoomDec] and trace again near the intercepts.

5.2.6 Zoom: Plot again the two graphs, for $y = -x^3 + 4x$ and $y = -.25x$. There appears to be an intersection near $x = 2$. The TI-92 provides several ways to enlarge the view around this point. You can change the viewing rectangle directly by pressing ◆ WINDOW and editing the values of xmin, xmax, ymin, and ymax. Figure 5.42 shows a new viewing rectangle for the range displayed in Figure 5.41. The cursor has been moved near the point of intersection; move your cursor closer to get the best approximation possible for the coordinates of the intersection.

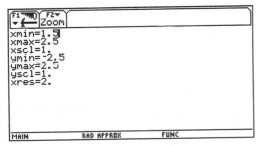

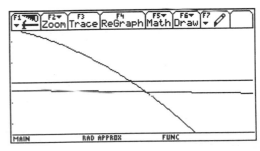

Figure 5.41: New WINDOW

Figure 5.42: Closer view

A more efficient method for enlarging the view is to draw a new viewing rectangle with the cursor. Start again with a graph of the two functions $y = -x^3 + 4x$ and $y = -.25x$ in a standard viewing rectangle. (Press F2[Zoom] 6[ZoomStd] for the standard viewing window.)

Now imagine a small rectangular box around the intersection point, near $x = 2$. Press F2[Zoom] 1[ZoomBox] (Figure 5.43) to draw a box to define this new viewing rectangle. Use the arrow keys to move the cursor, whose coordinates are displayed at the bottom of the window, to one corner of the new viewing rectangle you imagine.

Press ENTER to fix the corner where you moved the cursor; it changes shape and becomes a blinking square (Figure 5.44). Use the arrow keys again to move the cursor to the diagonally opposite corner of the new rectangle (Figure 5.45). (Note that you can use the diagonal directions on the cursor pad for this.) If this box looks all right to you, press ENTER. The rectangular area you have enclosed will now enlarge to fill the graph window (Figure 5.46).

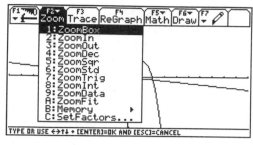

Figure 5.43: F2[*Zoom*] menu

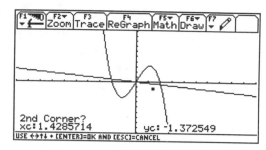

Figure 5.44: One corner selected

You may cancel the zoom any time *before* you press this last ENTER. Press F2[*Zoom*] once more and start over. Press ESC or ◆ GRAPH to cancel the zoom, or press 2nd QUIT to cancel the zoom and return to the Home screen.

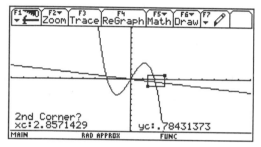

Figure 5.45: Box drawn

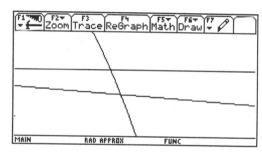

Figure 5.46: New viewing rectangle

You can also quickly magnify a graph around the cursor's location. Return once more to the standard window for the graph of the two functions $y = -x^3 + 4x$ and $y = -.25x$. Press F2[*Zoom*] 2[*ZoomIn*] and then use the cursor pad to move the cursor as close as you can to the point of intersection near $x = 2$ (see Figure 5.47). Then press ENTER and the calculator draws a magnified graph, centered at the cursor's position (Figure 5.48). The range variables are changed to reflect this new viewing rectangle. Look in the WINDOW menu to verify this.

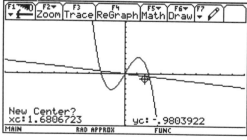

Figure 5.47: Before a zoom in

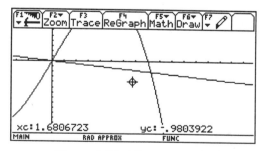

Figure 5.48: After a zoom in

As you see in the F2[Zoom] menu (Figure 5.43), the TI-92 can zoom in (press F2[Zoom] 2) or zoom out (press F2[Zoom] 3). Zoom out to see a larger view of the graph, centered at the cursor position. You can change the horizontal and vertical scale of the magnification by pressing F2[Zoom] C[SetFactors] (see Figure 5.49) and editing xFact and yFact, the horizontal and vertical magnification factors. (The zFact is only used when dealing with 3-dimensional graphs.)

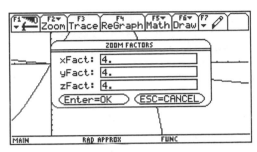

Figure 5.49: ZOOM FACTORS menu

Technology Tip: An advantage of zooming in from square viewing window is that subsequent windows will also be square. Likewise, if you zoom in from a friendly viewing rectangle, the zoomed windows will also be friendly.

The default zoom factor is 4 in both direction. It is not necessary for xFact and yFact to be equal. sometimes, you may prefer to zoom in one direction only, so the other factor should be set to 1. Press ESC to leave the ZOOM FACTORS menu and go back to the graph. (Pressing 2nd QUIT will take you back to the Home screen.)

Technology Tip: The TI-92 remembers the window it displayed before a zoom. So if you should zoom in too much and lose the curve, press F2[Zoom] B[Memory] 1[ZoomPrev] to go back to the window before. If you want to execute a series of zooms but then return to a particular window, press F2[Zoom] B[Memory] 2[ZoomSto] to store the current window's dimensions. Later, press F2[Zoom] B[Memory] 3[ZoomRcl] to recall the stored window.

5.2.7 Relative Minimums and Maximums: Graph $y = -x^3 + 4x$ once again in the standard viewing rectangle. This function appears to have a relative minimum near $x = -1$ and a relative maximum near $x = 1$. You may zoom and trace to approximate these extreme values.

First trace along the curve near the local minimum. Notice by how much the *x*-values and *y*-values change as you move from point to point Trace along the curve until the *y*-coordinate is as *small* as you can get it, so that you are as close as possible to the local minimum, and zoom in (press F2[Zoom] 2[ZoomIn] ENTER or use a zoom box). Now trace again along the curve and, as you move from point to point, see that the coordinates change by smaller amounts than before. Keep zooming and tracing until you find the coordinates of the local minimum point as accurately as you need them, approximately $(-1.15, -3.08)$.

Follow a similar procedure to find the local maximum. Trace along the curve until the y-coordinate is as *great* as you can get it, so that you are as close as possible to the local maximum, and zoom in. The local maximum point on the graph of $y = -x^3 + 4x$ is approximately (1.15, 3.08).

The TI-92 can automatically find the maximum and minimum points. While viewing the graph, press F5*[Math]* to display the Math menu (Figure 5.50). Choose 3*[Minimum]* to calculate the minimum value of the function and 4*[Maximum]* for the maximum. You will be prompted to trace the cursor along the graph first to a point *left* of the minimum/maximum (press ENTER to set this *lower bound*). Note the arrow near the top of the display marking the lower bound (as in Figure 5.51).

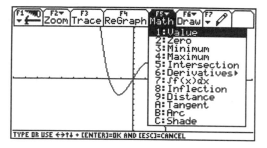

Figure 5.50: Math menu

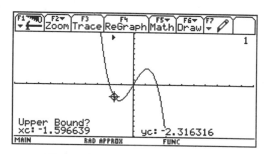

Figure 5.51: Finding a minimum

Now move to a point *right* of the minimum/maximum and set a *upper bound* by pressing ENTER. The coordinates of the relative minimum/maximum point will be displayed (see Figure 5.52). Good choices for the left bound and right bound can help the TI-92 work more efficiently and quickly.

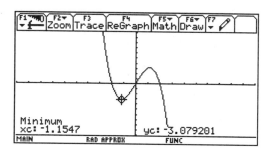

Figure 5.52: Relative minimum on $y = -x^3 + 4x$

Note that if you have more than one graph on the screen, the upper right corner of the TI-83 screen will show the number of the function whose minimum/maximum is being calculated.

5.3 Solving Equations and Inequalities

5.3.1 Intercepts and Intersections: Tracing and zooming are also used to locate an x-intercept of a graph, where a curve crosses the x-axis. For example, the graph of $y = x^3 - 8x$ crosses the x-axis three times (Figure 5.53). After tracing over to the x-intercept point that is farthest to the left, zoom in (Figure 5.54). Continue this process until you have located all three intercepts with as much accuracy as you need. The three x-intercepts of $y = x^3 - 8x$ are approximately -2.828, 0, and 2.828.

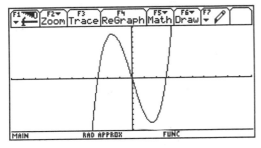

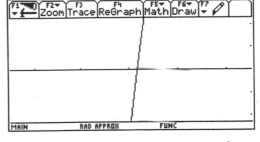

Figure 5.53: Graph of $y = x^3 - 8x$ Figure 5.54: Near an x-intercept of $y = x^3 - 8x$

Technology Tip: As you zoom in, you may also wish to change the spacing between tick marks on the x-axis so that the viewing rectangle shows scale marks near the intercept point. Then the accuracy of your approximation will be such that the error is less than the distance between two tick marks. Change the x-scale on the TI-92 from the WINDOW menu. Move the cursor down to xscl and enter an appropriate value.

The x-intercept of a function's graph is a *zero* of the function, so while viewing the graph, press F5*[Math]* (Figure 5.50) and choose 2*[Zero]* to find a zero of this function. Set a lower bound and upper bound as described in Section 5.2.7. The TI-92 shows the coordinates of the point and indicates that it is a zero (Figure 5.55)

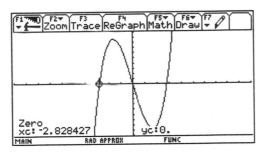

Figure 5.55: A zero of $y = x^3 - 8x$

TRACE and ZOOM are especially important for locating the intersection points of two graphs, say the graphs of $y = -x^3 + 4x$ and $y = -.25x$. Trace along one of the graphs until you arrive close to an intersection point. Then press ↑ or ↓ to jump to the other graph. Notice that the x-coordinate does not change, but the y-coordinate is likely to be different (Figures 5.56 and 5.57).

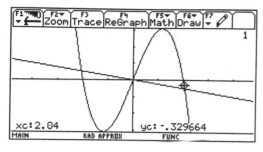

Figure 5.56: Trace on $y = -x^3 + 4x$

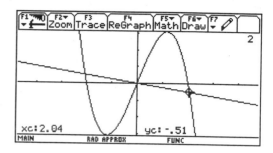

Figure 5.57: Trace on $y = -.25x$

When the two y-coordinates are as close as they can get, you have come as close as you now can to the point of intersection. So zoom in around the intersection point, then trace again until the two y-coordinates are as close as possible. Continue this process until you have located the point of intersection with as much accuracy as necessary.

You can also find the point of intersection of two graphs by pressing F5[Math] 5[Intersection]. Trace with the cursor first along one graph near the intersection and press ENTER; then trace with the cursor along the other graph and press ENTER. Marks + are placed on the graphs at these points. Then set lower and upper bounds for the x-coordinate of the intersection point and press ENTER again. Coordinates of the intersection will be displayed at the bottom of the window (Figure 5.58).

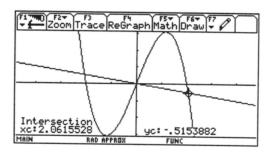

Figure 5.58: An intersection of $y = -x^3 + 4x$ and $y = -.25x$

5.3.2 Solving Equations by Graphing: Suppose you need to solve the equation $24x^3 - 36x + 17 = 0$. First graph $y = 24x^3 - 36x + 17$ in a window large enough to exhibit *all* its x-intercepts, corresponding to all the equation's zeros (roots). Then use trace and zoom, or the TI-92's zero finder, to locate each one. In fact, this equation has just one solution, approximately $x = -1.414$.

Remember that when an equation has more than one x-intercept, it may be necessary to change the viewing rectangle a few times to locate all of them.

The TI-92 has a solve(function. To use this function, you must be in the Home screen. To use the solve(function, press S O L V E (24 X ∧ 3 − 35 X + 17 = 0 , X) ENTER. The TI-92 displays the value of the zero (Figure 5.59). Note that any letter could have been used for the variable. This is the reason that you must indicate to the TI-92 that the variable being used is X.

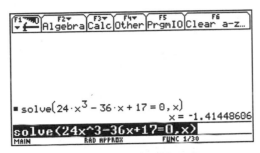

Figure 5.59: solve(function

Technology Tip: To solve an equation like $24x^3 + 17 = 36x$, you may first transform it into standard form, $24x^3 − 36x + 17 = 0$, and proceed as above. However, the solve(function does not require that the function be in standard form. You may also graph the *two* functions $y = 24x^3 + 17$ and $y = 36x$, then zoom and trace to locate their point of intersection.

5.3.3 Solving Systems by Graphing: The solutions to a system of equations correspond to the points of intersection of their graphs (Figure 5.60). For example, to solve the system $y = x^3 + 3x^2 − 2x − 1$ and $y = x^2 − 3x − 4$, first graph them together. Then use zoom and trace or the intersection option in the F5[Math] menu, to locate their point of intersection, approximately (−2.17, 7.25).

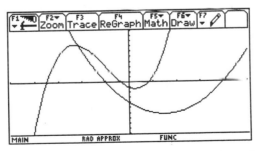

Figure 5.60: Graph of $y = x^3 + 3x^2 − 2x − 1$ and $y = x^2 − 3x − 4$

If you do not use the Intersection option, you must judge whether the two current y-coordinates are sufficiently close for $x = -2.17$ or whether you should continue to zoom and trace to improve the approximation.

The solutions of the system of two equations $y = x^3 + 3x^2 - 2x - 1$ and $y = x^2 - 3x - 4$ correspond to the solutions of the single equation $x^3 + 3x^2 - 2x - 1 = x^2 - 3x - 4$, which simplifies to $x^3 + 2x^2 + x + 3 = 0$. So you may also graph $y = x^3 + 2x^2 + x + 3$ and find its x-intercepts to solve the system or use the solve(function.

5.3.4 Solving Inequalities by Graphing: Consider the inequality $1 - \dfrac{3x}{2} \geq x - 4$. To solve it with your TI-92, graph the two functions $y = 1 - \dfrac{3x}{2}$ and $y = x - 4$ (Figure 5.61). First locate their point of intersection, at $x = 2$. The inequality is true when the graph of $y = 1 - \dfrac{3x}{2}$ lies *above* the graph of $y = x - 4$, and that occurs when $x < 2$. So the solution is the half-line $x \leq 2$, or $(-\infty, 2]$.

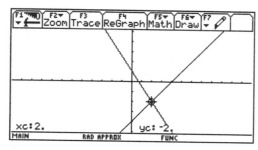

Figure 5.61: Solving $1 - \dfrac{3x}{2} \geq x - 4$

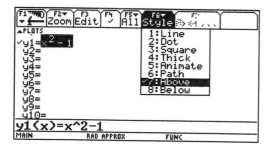

Figure 5.62: Shade **Above** style

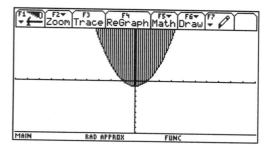

Figure 5.63: Graph of $y \geq x^2 - 1$

The TI-92 is capable of shading the region above or below a graph, or between two graphs. For example, to graph $y \geq x^2 - 1$, first enter the function $y = x^2 - 1$ as y1. Then, highlight y1 and press F6*[Style]* 7*[Above]* (see Figure 5.62). These keystrokes instruct the TI-92 to shade the region above $y = x^2 - 1$. Press ◆ GRAPH to see the graph. The region above the graph will be shaded using the first shading option of vertical lines, as in Figure 5.63.

Now use shading to solve the previous inequality, $1 - \frac{3x}{2} \geq x - 4$. The solution is the region which is *below* the graph of $1 - \frac{3x}{2}$ and *above* $x - 4$. First graph both equations. Then, from the graph screen, press F5*[Math]* C*[Shade]*. The TI-92 will prompt for the function that you want to have the shading *above*. Use ↑ or ↓ to move the cursor to the graph of $x - 4$, then press ENTER. The TI-92 will then prompt for the function that you want to have the shading *below*, so use ↑ or ↓ to move the cursor to the graph of $1 - \frac{3x}{2}$ and press ENTER. The TI-92 will then prompt for the *lower bound* then the *upper bound*, which are the left and right edges, respectively, of the extent of the shading. If you do not enter a lower or upper bound, the values of xmin and xmax will be used. So, in this case, press ENTER twice to set the lower and upper bounds. The shaded area extends left from $x = -2$, hence the solution to $1 - \frac{3x}{2} \geq x - 4$ is the half-line $x \leq 2$, or $(-\infty, 2]$.

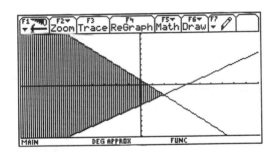

Figure 5.64: Graph of $1 - \frac{3x}{2} \geq x - 4$

5.4   Trigonometry

5.4.1 Degrees and Radians: The trigonometric functions can be applied to angles measured either in radians or degrees, but you should take care that the TI-92 is configured for whichever measure you need. Press MODE to see the current settings. Press ↓ three times and move down to the fourth line of the first page of the mode menu where angle measure is selected. Then press → to display the options. Use ↑ or ↓ to move from one option to the other. Either press the number corresponding to the measure or, when the measure is highlighted, press ENTER to select it. Then press ENTER to confirm your selection and leave the MODE menu.

It's a good idea to check the angle measure setting before executing a calculation that depends on a particular measure. You may change a mode setting at any time and not interfere with pending calculations. From the Home screen, try the following keystrokes to see this in action.

Expression	Keystrokes	Display
$\sin 45°$	MODE ↓ ↓ ↓ → ↓ ENTER ENTER	
	SIN 45) ENTER	.7071067812
$\sin \pi°$	SIN 2nd π) ENTER	.0548036651
$\sin \pi$	MODE ↓ ↓ ↓ → ↑ ENTER ENTER	
	SIN 2nd π) ENTER	0.
$\sin 45$	SIN 45) ENTER	.8509035245
$\sin \dfrac{\pi}{6}$	SIN 2nd π ÷ 6) ENTER	.5

The first line of keystrokes sets the TI-92 in degree mode and calculates the sine of 45 *degrees*. While the calculator is still in degree mode, the second line of keystrokes calculates the sine of π *degrees*, approximately 3.1415°. The third line changes to radian mode just before calculating the sine of π *radians*. The fourth line calculates the sine of 45 *radians* (the calculator remains in radian mode).

The TI-92 makes it possible to mix degrees and radians in a calculation. Execute these keystrokes to calculate $\tan 45° + \sin \frac{\pi}{6}$ as shown in Figure 5.65: TAN 45 2nd MATH 2*[Angle]* 1) + SIN (2nd π ÷ 6) 2nd MATH 2*[Angle]* 2) ENTER. Do you get 1.5 whether your calculator is in set *either* in degree mode *or* in radian mode?

The degree sign can also be entered by pressing 2nd D, which saves keystrokes. There is no corresponding key for the radian symbol.

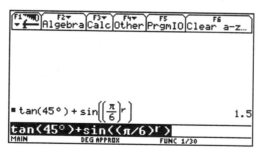

Figure 5.65: Angle measure

Technology Tip: The automatic left parenthesis that the TI-92 places after functions such as sine, cosine, and tangent (as noted in Section 5.1.3) *can* affect the outcome of calculations. In the previous example, the degree sign must be *inside* of the parentheses so that when the TI-92 is in radian mode, it calculates the tangent of 45

TI-92 Graphics Calculator

degrees, rather than converting the tangent of 45 radians into an equivalent number of degrees. Also, the parentheses around the fraction $\frac{\pi}{6}$ are required so that when the TI-92 is in radian mode, it converts $\frac{\pi}{6}$ into radians, rather than converting merely the 6 to radians. Experiment with the placement of parentheses to see how they affect the result of the computation.

5.4.2 Graphs of Trigonometric Functions:
When you graph a trigonometric function, you need to pay careful attention to the choice of graph window. For example, graph $y = \dfrac{\sin 30x}{30}$ in the standard viewing rectangle. Trace along the curve to see where it is. Zoom in to a better window, or use the period and amplitude to establish better WINDOW values.

Technology Tip:.   Since $\pi \approx 3.1$, when in radian mode, set xmin = 0 and xmax = 6.2 to cover the interval from 0 to 2π.

Next graph $y = \tan x$ in the standard window first, then press F2[Zoom] 7[ZoomTrig] to change to a special window for trigonometric functions in which the horizontal increment is $\frac{\pi}{24}$ or 7.5° and the vertical range is from −4 to 4. The TI-92 plots consecutive points and then connects them with a segment, so the graph is not exactly what you should expect. You may wish to change the plot style from Line to Dot (see Section 5.2.3) when you plot the tangent function.

5.5   Scatter Plots

5.5.1 Entering Data:
The table shows the total prize money (in millions of dollars) awarded at the Indianapolis 500 race from 1981 to 1989. (*Source*: Indianapolis Motor Speedway Hall of Fame.)

Year	1981	1982	1983	1984	1985	1986	1987	1988	1989
Prize ($million)	$1.61	$2.07	$2.41	$2.80	$3.27	$4.00	$4.49	$5.03	$5.72

We'll now use the TI-92 to construct a scatter plot that represents these points and to find a linear model that approximates the given data.

The TI-92 holds data in *lists*. You can create as many list names as your TI-92 memory has space to store. Before entering data, clear the data in the lists that you want to use. To delete a list press 2nd VAR-LINK. This will display the list of folders showing the variables defined in each folder. Highlight the name of the list that you wish to delete and press F1[Manage] 1[Delete] ENTER. The TI-92 will ask you to confirm the deletion by pressing ENTER once more.

Now press APPS 6[Data/Matrix Editor] 3[New] ↓ ↓ P R I Z E ENTER to open a new variable called PRIZE (Figure 5.66). Press ENTER to then begin entering the variable values, with the years going in column c1. Instead of entering the full year 198x, enter only x. Here are the keystrokes for the first three years: 1

ENTER 2 ENTER 3 ENTER and so on, then press → to move to the next list. Use the cursor pad to move up to the first row and press 1.61 ENTER 2.07 ENTER 2.41 and so on (see Figure 5.67).

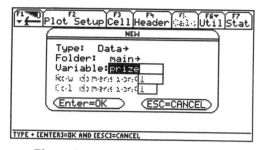

Figure 5.66: Entering a new variable

You may edit statistical data in almost the same way you edit expressions in the Home screen. ← will delete the *entire cell*, not just the character or value to the left of the cursor. Thus, move the cursor to any value you wish to change, then type the correction. To insert or delete a data point, move the cursor over the data point (cell) you wish to add or delete. To insert a cell, move to the cell *below* the place where you want to insert the new cell and press F6*[Util]* 1*[Insert]* 1*[cell]* and a new empty cell is open.

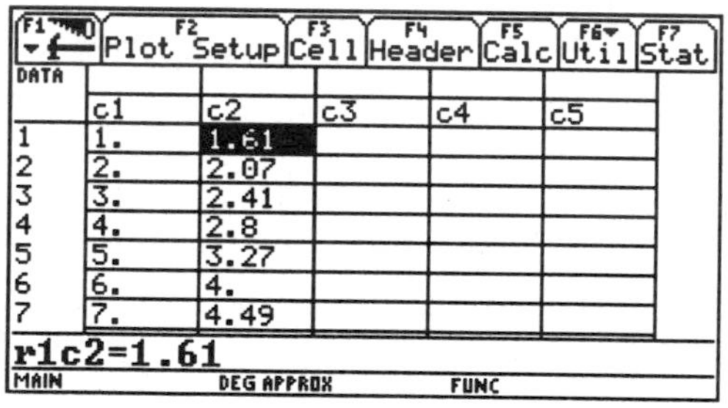

DATA	c1	c2	c3	c4	c5
1	1.	1.61			
2	2.	2.07			
3	3.	2.41			
4	4.	2.8			
5	5.	3.27			
6	6.	4.			
7	7.	4.49			

r1c2=1.61

Figure 5.67: Entering data points

5.5.2 Plotting Data: First check the MODE screen (Figure 5.1) to make sure that you are in FUNCTION graphing mode. With the data points showing, press F2*[Plot Setup]* to display the Plot Setup screen. If no other plots have been entered, Plot 1 is highlighted by default. Press F1*[Define]* to select the options for the plot. Use ↑, ↓, and ENTER to select the Plot Type as Scatter and the Mark as a Box. Use the keyboard to set the independent variable, x, to c1 and the dependent variable, y, to c2 as shown in Figure 5.68, then press ENTER to save the options and press ◆ GRAPH to graph the data points. (Make sure that you have cleared or turned off any functions in the Y= screen, or those functions will be graphed simultaneously.) Figure 5.69 shows this plot in a window from 0 to 10 horizontally and vertically. You may now press F3*[Trace]* to move from data point to data point.

To draw the scatter plot in a window adjusted automatically to include all the data you entered, press F2*[Zoom]* 9 *[ZoomData]*.

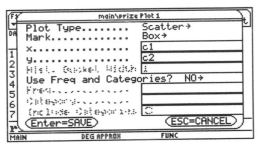

Figure 5.68: Plot1 menu

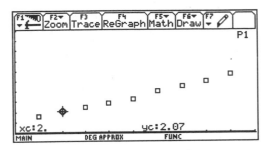

Figure 5.69: Scatter plot

When you no longer want to see the scatter plot, press APPS 6*[Data/Matrix Editor]* 1*[Current]* F2*[Plot Setup]*, highlight Plot 1 and use F4*[✓]* to deselect plot 1. The TI-92 still retains all the data you entered.

5.5.3 Regression Line: The TI-92 calculates slope and *y*-intercept for the line that best fits all the data. After the data points have been entered, while still in the Data/Matrix Editor, press F5*[Calc]*. For the Calculation Type, choose 5*[LinReg]* and set the x variable to c1 and the y variable to c2. In order to have the TI-92 graph the regression equation, set Store RegEQ to as y1(x) as shown in Figure 5.70. Press ENTER and the TI-92 will calculate a linear regression model with the slope named a and the *y*-intercept named b (Figure 5.71). The *correlation coefficient* measures the goodness of fit of the linear regression with the data. The closer the absolute value of the correlation coefficient is to 1, the better the fit; the closer the absolute value of the correlation coefficient is to 0, the worse the fit. The TI-92 displays both the correlation coefficient and the coefficient of determination (R^2).

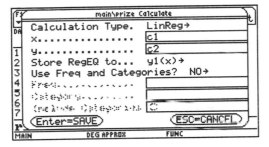

Figure 5.70: Linear regression: Calculate dialog box

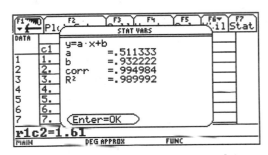

Figure 5.71: Linear regression model

Press ENTER to accept the regression equation and close the STAT VARS screen. To see both the data points and the regression line (Figure 5.72), go to the Plot Setup screen and select Plot1, then press ◆ GRAPH to display the graph.

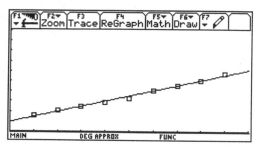

Figure 5.72: Linear regression line

5.5.4 Exponential Growth Model: The table shows the world population (in millions) from 1980 to 1992.

Year	1980	1985	1986	1987	1988	1989	1990	1991	1992
Population (millions)	4453	4850	4936	5024	5112	5202	5294	5384	5478

Clear the previous data by going to the current variable in the Data/Matrix Editor and pressing F1[Tools] 8[Clear Editor] ENTER. Follow the procedure described above to enter the data in order to find an exponential model that approximates the given data. Use 0 for 1980, 5 for 1985, and so on.

The TI-92 will not compute the exponential growth model $y = ae^{cx}$. The exponential regression that The TI-92 will compute is of the form $y = ab^x$. To get this exponential growth model press F5[Calc] and set the Calculation Type to 4[ExpReg], the x variable to c1, and the y variable to c2. Then press ENTER to find the values of a and b (Figure 5.73). In this case, the exponential growth model is $y = 4451(1.017454^x)$. To convert this to the form $y = ae^{cx}$, the required equation is $c = \ln b$, and the exponential growth model in this case is $y = 4451e^{x\ln 1.017454}$ or $y = 4451e^{0.017303t}$.

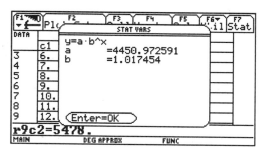

Figure 5.73: Exponential growth model

If you wish to plot and graph the data, follow the method for linear regression. Set an appropriate range for the data and then press ◆ GRAPH. The data will now be plotted in the range. To graph the regression equation also, store the regression equation to a y plot that is free. As in the linear regression model, press ◆ Y= and inactivate or clear any other existing functions, then press ◆ GRAPH to graph the exponential growth model. Note that the exponential regression model does not need to be converted to the form $y = ae^{cx}$ before graphing.

Remember to clear or deselect the plot before viewing graphs of other functions.

5.6   Matrices

5.6.1 Making a Matrix: The TI-92 can work with as many different matrices as the memory will hold..

Here's how to create this 3×4 matrix $\begin{bmatrix} 1 & -4 & 3 & 5 \\ -1 & 3 & -1 & -3 \\ 2 & 0 & -4 & 6 \end{bmatrix}$ in your calculator.

From the Home screen, press APPS 6*[Data/Matrix Editor]* 3*[New]*. Set the Type to Matrix, the Variable to a (this is the 'name' of the matrix), the Row Dimension to 3 and the Col Dimension to 4 (Figure 5.74). Press ENTER to accept these values.

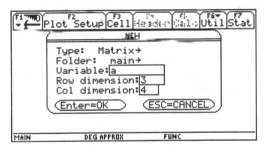

Figure 5.74: Data/Matrix menu   Figure 5.75: Editing a matrix

The display will show the matrix by showing a grid with zeros in the rows and columns specified in the definition of the matrix.

Use the cursor pad or press ENTER repeatedly to move the cursor to a matrix element you want to change. If you press ENTER, you will move right across a row and then back to the first column of the next row. The lower left of the screen shows the cursor's current location within the matrix The element in the second row and first column in Figure 5.75 is highlighted, so the lower left of the window is r2c1 = −1. showing that element's current value. Enter all the elements of matrix a; pressing ENTER after inputing each value.

When you are finished, leave the editing screen by pressing 2nd QUIT or ◆ HOME to return to the Home screen.

5.6.2 Matrix Math: From the Home screen, you can perform many calculations with matrices. To see matrix a, press A ENTER (Figure 5.76).

Perform the scalar multiplication 2 a pressing 2A ENTER. The resulting matrix is displayed on the screen. To create matrix b as 2a press 2 A STO♦ B ENTER (Figure 5.77), or if you do this immediately after calculating 2a, press only STO♦ B ENTER. The calculator will display the matrix.

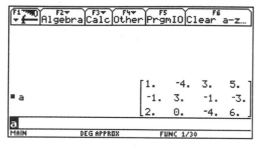

Figure 5.76: Matrix a

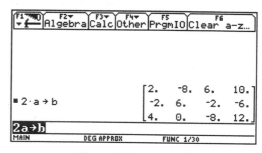

Figure 5.77: Matrix b

To add two matrices, say a and b, create b (with the same dimensions as a) and then press A + B ENTER. Again, if you want to store the answer as a specific matrix, say m, then press STO♦ M. Subtraction is performed in similar manner.

Now create a matrix called c with dimensions of 2×3 and enter the matrix $\begin{bmatrix} 2 & 0 & 3 \\ 1 & -5 & -1 \end{bmatrix}$ as c. For matrix multiplication of c by a, press C × A ENTER. If you tried to multiply a by c, your TI-92 would notify you of an error because the dimensions of the two matrices do not permit multiplication in this way.

The *transpose* of a matrix is another matrix with the rows and columns interchanged. The symbol for the transpose of a is a^T. To calculate a^T, press A 2nd MATH 4*[Matrix]* 1*[^{T}]* ENTER.

5.6.3 Row Operations: Here are the keystrokes necessary to perform elementary row operations on a matrix. Your textbook provides a more careful explanation of the elementary row operations and their uses.

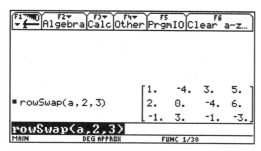

Figure 5.78: Swap rows 2 and 3

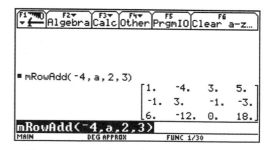

Figure 5.79: Add –4 times row 2 to row 3

To interchange the second and third rows of the matrix a that was defined above, press 2nd MATH 4[Matrix] D[Row ops] 1[rowSwap(] A , 2 , 3) ENTER (see Figure 5.78). The format of this command is rowSwap(*matrix1, rIndex1, rIndex2*).

To add row 2 and row 3 and *store* the results in row 3, press 2nd MATH 4[Matrix] D[Row ops] 2[rowAdd(] A , 2 , 3) ENTER. The format of this command is rowAdd(*matrix1, rIndex1, rIndex2*).

To multiply row 2 by –4 and *store* the results in row 2, thereby replacing row 2 with new values, press 2nd MATH 4[Matrix] D[Row ops] 3[mRow(] (–) 4 , A , 2) ENTER. The format of this command is mRow(*expression, matrix1, index*).

To multiply row 2 by –4 and *add* the results to row 3, thereby replacing row 3 with new values, press 2nd MATH 4[Matrix] D[Row ops] 4[mRowAdd(] (–) 4 , A , 2, 3) ENTER (see Figure 5.79). The format of this command is mRowAdd(*expression, matrix1, Index1, Index2*).

Note that your TI-92 does *not* store a matrix obtained as the result of any row operation. So, when you need to perform several row operations in succession, it is a good idea to store the result of each one in a temporary place.

For example, use row operations to solve this system of linear equations: $\begin{cases} x - 2y + 3z = 9 \\ -x + 3y = -4 \\ 2x - 5y + 5z = 17 \end{cases}$.

First enter this *augmented matrix* as a in your TI-92: $\begin{bmatrix} 1 & -2 & 3 & 9 \\ -1 & 3 & 0 & -4 \\ 2 & -5 & 5 & 17 \end{bmatrix}$. Then return to the Home screen and

store this matrix as e (press A STO♦ E ENTER), so you may keep the original in case you need to recall it.

Here are the row operations and their associated keystrokes. At each step, the result is stored in e and replaces the previous matrix e. The last two steps of the row operations are shown in Figure 5.80.

Row Operations	Keystrokes
add row 1 to row 2	2nd MATH 4 D 2 E , 1 , 2) STO♦ E ENTER
add –2 times row 1 to row 3	2nd MATH 4 D 4 (–) 2 , E , 1 , 3) STO♦ E ENTER
add row 2 to row 3	2nd MATH 4 D 2 E , 2 , 3) STO♦ E ENTER
multiply row 3 by $\frac{1}{2}$	2nd MATH 4 D 3 1 ÷ 2 , E , 3) STO♦ E ENTER

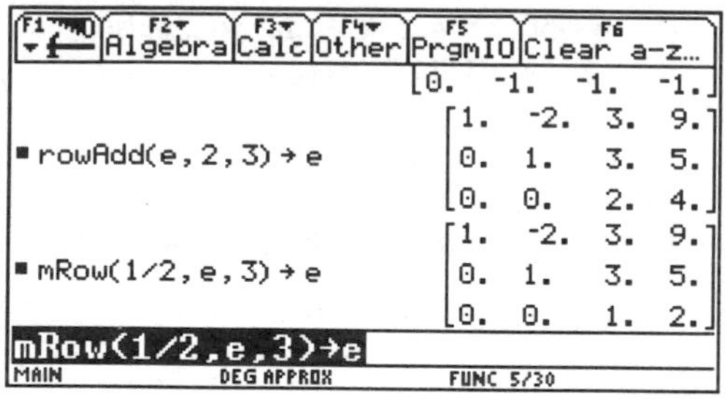

$$\begin{bmatrix} 0. & -1. & -1. & -1. \end{bmatrix}$$

■ rowAdd(e,2,3)→e

$$\begin{bmatrix} 1. & -2. & 3. & 9. \\ 0. & 1. & 3. & 5. \\ 0. & 0. & 2. & 4. \end{bmatrix}$$

■ mRow(1/2,e,3)→e

$$\begin{bmatrix} 1. & -2. & 3. & 9. \\ 0. & 1. & 3. & 5. \\ 0. & 0. & 1. & 2. \end{bmatrix}$$

mRow(1/2,e,3)→e

MAIN   DEG APPROX   FUNC 5/30

Figure 5.80: Final matrix after row operations

Thus $z = 2$, so $y = -1$, and $x = 1$.

Technology Tip: The TI-92 can produce a row-echelon form and the reduced row-echelon form of a matrix. The row-echelon form of matrix **a** is obtained by pressing 2nd MATH 4*[Matrix]* 3*[ref(]* A) ENTER and the reduced row-echelon form is obtained by pressing 2nd MATH 4*[Matrix]* 4*[rref(]* A) ENTER. Note that the row-echelon form of a matrix is not unique, so your calculator may not get exactly the same matrix as you do by using row operations. However, the matrix that the TI-92 produces will result in the same solution to the system.

5.6.4 Determinants and Inverses: Enter this 3×3 square matrix as **a**: $\begin{bmatrix} 1 & -2 & 3 \\ -1 & 3 & 0 \\ 2 & -5 & 5 \end{bmatrix}$. Since this consists of the

first three columns of the matrix **a** that was previously used, you can go to the matrix, move the cursor into the fourth column and press F6*[Util]* 2*[Delete]* 3*[column]*. This will delete the column that the cursor is in.

To calculate its determinant $\begin{vmatrix} 1 & -2 & 3 \\ -1 & 3 & 0 \\ 2 & -5 & 5 \end{vmatrix}$, go to the Home screen and press 2nd MATH 4*[Matrix]* 2*[det(]* A

) ENTER. You should find that the determinant is 2 as shown in Figure 5.81.

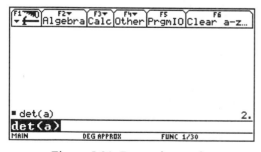

Figure 5.81: Determinant of **a**

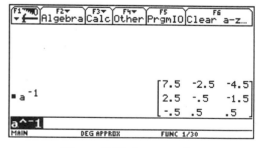

Figure 5.82: Inverse of **a**

TI-92 Graphics Calculator

Since the determinant of the matrix is not zero, it has an inverse matrix. Press A 2nd x^{-1} ENTER to calculate the inverse. The result is shown in Figure 5.82.

Now let's solve a system of linear equations by matrix inversion. Once again, consider $\begin{cases} x-2y+3z = 9 \\ -x+3y = -4 \\ 2x-5y+5z = 17 \end{cases}$.

The coefficient matrix for this system is the matrix $\begin{bmatrix} 1 & -2 & 3 \\ -1 & 3 & 0 \\ 2 & -5 & 5 \end{bmatrix}$ which was entered as matrix a in the previous example. Now enter the matrix $\begin{bmatrix} 9 \\ -4 \\ 17 \end{bmatrix}$ as b. Since b was used before, when we stored 2a as b, press APPS 6[Data/Matrix Editor] 2[Open] → 2[Matrix] ↓ ↓ → and use ↓ to move the cursor to b, then press ENTER twice to go to the matrix previously saved as b, which can be edited. Return to the Home screen (◆ HOME) and press A 2nd x^{-1} × B ENTER to get the answer as shown in Figure 5.83.

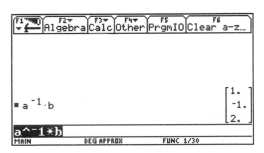

Figure 5.83: Solution matrix

The solution is still $x = 1$, $y = -1$, and $z = 2$.

5.7   Sequences

5.7.1 Iteration with the ANS key: The ANS key enables you to perform *iteration*, the process of evaluating a function repeatedly. As an example, calculate $\dfrac{n-1}{3}$ for $n = 27$. Then calculate $\dfrac{n-1}{3}$ for $n =$ the answer to the previous calculation. Continue to use each answer as n in the *next* calculation. here are keystrokes to accomplish this iteration on the TI-92 calculator. (See the results in Figure 5.84.) Notice that when you use ANS in place of n in a formula, it is sufficient to press ENTER to continue an iteration.

Iteration	Keystrokes	Display
1	27 ENTER	27
2	(2nd ANS − 1) ÷ 3 ENTER	8.666666667
3	ENTER	2.555555556
4	ENTER	.5185185185

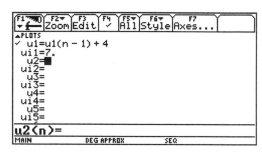

Figure 5.84: Iteration

Press ENTER several more times and see what happens with this iteration. You may wish to try it again with a different starting value.

5.7.2 Arithmetic and Geometric Sequences: Use iteration with the ANS variable to determine the n-th term of a sequence. For example, find the 18th term of an *arithmetic* sequence whose first term is 7 and whose common difference is 4. Enter the first term 7, then start the progression with the recursion formula, 2nd ANS + 4 ENTER. This yields the 2nd term, so press ENTER sixteen more times to find the 18th term. For a *geometric* sequence whose common ratio is 4, start the progression with 2nd ANS × 4 ENTER.

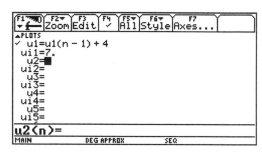

Figure 5.85: Sequential Y= menu

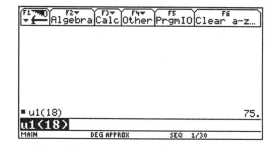

Figure 5.86: Sequence mode

You can also define the sequence recursively with the TI-92 by selecting Sequence in the Graph type on the first page of the MODE menu (see Figure 5.1). Once again, let's find the 18th term of an *arithmetic* sequence whose first term is 7 and whose common difference is 4. Press MODE → 4*[Sequence]* ENTER. Then press

◆ Y= to edit any of the TI-92's sequences, $u1$ through $u99$. Make $u1(n) = u1(n-1) + 4$ and $u1(1) = 7$ by pressing U 1 (N – 1) + 4 ENTER 7 ENTER (Figure 5.85). Press 2nd QUIT to return to the Home screen. To find the 18th term of this sequence, calculate $u1(18)$ by pressing U 1 (18) ENTER (Figure 5.86).

Of course, you could also use the *explicit* formula for the n-th term of an arithmetic sequence $t_n = a + (n-1)d$. First enter values for the variables a, d, and n, then evaluate the formula by pressing A + (N – 1) D ENTER. For a geometric sequence whose n-th term is given by $t_n = a \cdot r^{n-1}$, enter values for the variables a, d, and r, then evaluate the formula by pressing A R ∧ (N – 1) ENTER.

To use the explicit formula in Seq MODE, make $u_1(n) = 7 + (n-1) \cdot 4$ by pressing ◆ Y= then using ↑ to move up to the u1(n) line and pressing CLEAR 7 + (N – 1) × 4 ENTER 2nd QUIT. Once more, calculate $u1(18)$ by pressing U 1 (1 8) ENTER.

5.7.3 Finding Sums of Sequences: You can find the sum of a sequence by combining the features sum(and seq(feature on the LIST sub-menu of the MATH menu. The format of the sum(command is sum(*list*). The format of the seq(command is seq(*expression, variable, low, high, step*) where the *step* argument is optional and the default is for integer values from *low* to *high*. For example, suppose you want to find the sum

$\sum_{n=1}^{12} 4(0.3)^n$. Press 2nd MATH 3*[LIST]* 6*[sum(]* 2nd MATH 3*[LIST]* 1*[seq(]* 4 (. 3) ∧ K , K , 1, 12))

ENTER (Figure 5.87). The seq(command generates a list, which the sum(command then sums. Note that any letter can be used for the variable in the sum, i.e., the K could just have easily been an A or an N.

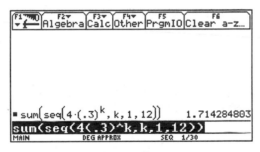

$$\text{Figure 5.87: } \sum_{n=1}^{12} 4(0.3)^n$$

Now calculate the sum starting at $n = 0$ by using →, ←, and ← to edit the range. You should obtain a sum of approximately 5.712848.

5.8   Parametric and Polar Graphs

5.8.1 Graphing Parametric Equations: The TI-92 plots parametric equations as easily as it plots functions. Up to ninety nine pairs of parametric equations can be plotted. In the first page of the MODE menu (Figure 5.1) change the Graph setting to PARAMETRIC. Be sure, if the independent parameter is an angle measure, that the angle measure in the MODE menu has been set to whichever you need, RADIAN or DEGREE.

You can now enter the parametric functions. For example, here are the keystrokes needed to graph the parametric equations $x = \cos^3 t$ and $y = \sin^3 t$. First check that angle measure is in radians. Then press ◆ Y= (COS T)) ∧ 3 ENTER (SIN T)) ∧ 3 ENTER (Figure 5.88).

Press ◆ WINDOW to set the graphing window and to initialize the values of t. In the standard window, the values of t go from 0 to 2π in steps of $\frac{\pi}{24} \approx 0.1309$, with the view from -10 to 10 in both directions. In order to provide a better viewing rectangle press ENTER three times and set the rectangle to go from -2 to 2 horizontally and vertically (Figure 5.89). Now press ◆ GRAPH to draw the graph (Figure 5.90).

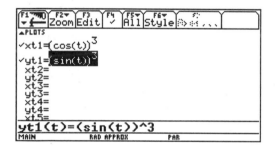
Figure 5.88: $x = \cos^3 t$ and $y = \sin^3 t$

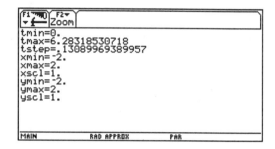
Figure 5.89: Parametric WINDOW menu

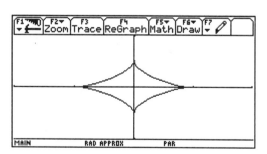
Figure 5.90: Parametric graph of $x = \cos^3 t$ and $y = \sin^3 t$

You may Zoom and Trace along parametric graphs just as you did with function graphs. However, unlike with function graphs, the cursor will not move to values outside of the t range, so ← will not work when t = 0, and → will not work when t = 2π. As you trace along this graph, notice that the cursor moves in the *counter-clockwise* direction as t increases.

5.8.2 Rectangular-Polar Coordinate Conversion: The Angle sub-menu of the MATH menu provides a function for converting between rectangular and polar coordinate systems. These functions use the current angle measure setting, so it is a good idea to check the default angle measure before any conversion. Of course, you may override the current angle measure setting, as explained in Section 5.4.1. For the following examples, the TI-92 is set to radian measure.

Given the rectangular coordinates $(x, y) = (4, -3)$, convert to polar coordinates (r, θ) in the Home screen by pressing 2nd MATH 2*[Angle]* 5*[R ▶Pr(]* 4 , (−) 3) ENTER. The value of r is displayed; now press 2nd MATH 2*[Angle]* 6*[R ▶Pθ(]* 4, (−) 3) ENTER to display the value of θ (Figure 5.91). The polar coordinates are approximately $(5, -0.6435)$.

Suppose $(r, \theta) = (3, \pi)$. Convert to rectangular coordinates (x, y) by pressing 2nd MATH 2*[Angle]* 3*[P ▶Rx(]* 3 , 2nd π) ENTER. The x-coordinate is displayed. Press 2nd MATH 2*[Angle]* 4*[P ▶Ry(]* 3 , 2nd π) ENTER to display the y-coordinate (Figure 5.92). The rectangular coordinates are $(-3, 0)$.

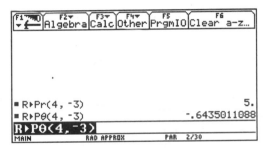
Figure 5.91: Rectangular to polar coordinates

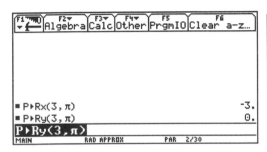
Figure 5.92: Polar to rectangular coordinates

5.8.3 Graphing Polar Equations: The TI-92 graphs polar functions in the form $r = f(\theta)$. In the Graph line of the MODE menu, select POLAR for polar graphs. You may now graph up to ninety nine polar functions at a time. Be sure that the angle measure has been set to whichever you need, RADIAN or DEGREE. Here we will use radian measure.

For example, to graph $r = 4\sin\theta$, press ◆ Y= for the polar graph editing screen. Then enter the expression $4\sin\theta$ by pressing 4 SIN θ) ENTER. The θ key is on the lower right of the keyboard, near the ENTER key.

Choose a good viewing rectangle and an appropriate interval and increment for θ. In Figure 5.93, the viewing rectangle is roughly "square" and extends from −14 to 14 horizontally and from −6 to 6 vertically. (Refer back to the Technology Tip in Section 5.2.4.)

Figure 5.93 shows *rectangular* coordinates of the cursor's location on the graph. You may sometimes wish to trace along the curve and see *polar* coordinates of the cursor's location. The first line of the Graph Format menu (Figure 5.24) has options for displaying the cursor's position in rectangular (RECT) or polar (POLAR) form.

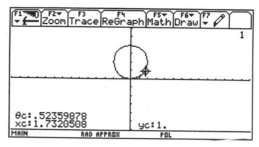

Figure 5.93: Polar graph of $r = 4\sin\theta$

5.9   Probability

5.9.1 Random Numbers: The command rand(generates numbers. You will find this command in the **Probability** sub-menu of the MATH menu in the Home screen. Press 2nd MATH 7*[Probability]* 4*[rand()*) ENTER to generate a random number between 0 and 1. Press ENTER to generate another number; keep pressing ENTER to generate more of them.

If you need a random number between, say, 0 and 10, then press 10 2nd MATH 7*[Probability]* 4*[rand()*) ENTER. To get a random number between 5 and 15, press 5 + 10 2nd MATH 7*[Probability]* 4*[rand()*) ENTER.

If you need the random number to be an *integer* between 1 and 10 (inclusive), press 2nd MATH 7*[Probability]* 4*[rand()* 10) ENTER. For a random negative integer between −1 and −10 (inclusive), press 2nd MATH 7*[Probability]* 4*[rand()* (−) 10) ENTER

5.9.2 Permutations and Combinations: To calculate the number of permutations of 12 objects taken 7 at a time, $_{12}P_7$, press 2nd MATH 7*[Probability]* 2*[nPr()* 12 , 7) ENTER (Figure 5.94). Thus $_{12}P_7 = 3,991,680$.

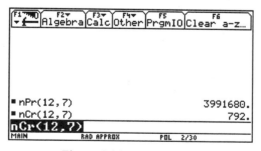

Figure 5.94: $_{12}P_7$ and $_{12}C_7$

TI-92 Graphics Calculator

For the number of combinations of 12 objects taken 7 at a time, $_{12}C_7$, press 2nd MATH 7*[Probability]* 3*[nCr]* 12 , 7) ENTER (Figure 5.94). Thus $_{12}C_7 = 792$.

5.9.3 Probability of Winning: A state lottery is configured so that each player chooses six different numbers from 1 to 40. If these six numbers match the six numbers drawn by the State Lottery Commission, the player wins the top prize. There are $_{40}C_6$ ways for the six numbers to be drawn. If you purchase a single lottery ticket, your probability of winning is 1 in $_{40}C_6$. Press 1 ÷ 2nd MATH 7*[Probability]* 3*[nCr]* 40 , 6) ENTER to calculate your chances, but don't be disappointed.

5.10 Programming

5.10.1 Entering a Program: The TI-92 is a programmable calculator that can store sequences of commands for later replay. Here's an example to show you how to enter a useful program that solves quadratic equations by the quadratic formula.

Press APPS 7*[Program Editor]* to access the programming menu. The TI-92 has space for many programs, each named by a name you give it. To create a new program now, start by pressing APPS 7*[Program Editor]* 3*[New]*.

Set the Type to Program and the Folder to main (unless you have another folder in which you want to have the program). Enter a descriptive title for the program in the Variable line. Name this program Quadrat and press ENTER twice to go to the program editor. The program name and the beginning and ending commands of the program are automatically displayed with the cursor on the first line after Prgm, the begin program command.

In the program, each line begins with a colon : supplied automatically by the calculator. Any command you could enter directly in the TI-92's Home screen can be entered as a line in a program. There are also special programming commands.

Input the program Quadrat by pressing the keystrokes given in the listing below. You may interrupt program input at any stage by pressing 2nd QUIT. To return later for more editing, press APPS 7*[Program Editor]* 2*[Open]*, move the cursor down to the Variable list, highlight this program's name, and press ENTER twice.

Each time you press ENTER while writing a program, the TI-92 *automatically* inserts the : character at the beginning of the next line.

The instruction manual for your TI-92 gives detailed information about programming. Refer to it to learn more about programming and how to use other features of your calculator.

Note that this program makes use of the TI-92's ability to compute complex numbers. Make sure that Complex Format on the MODE screen (Figure 5.1) is set to RECTANGULAR.

Enter the program Quadrat by pressing the given keystrokes. A space entered by using the spacebar on the keyboard is indicated by ␣.

Program Line	Keystrokes

: Input "Enter a", a   ⬆ I N P U T ␣ 2nd " ⬆ E N T E R ␣ A 2nd " , A ENTER

     displays the words ENTER A on the TI-92 screen and waits for you to input a value that will be assigned to the variable A

: Input "Enter b", b   ⬆ I N P U T ␣ 2nd " ⬆ E N T E R ␣ B 2nd " , B ENTER

: Input "Enter c", c   ⬆ I N P U T ␣ 2nd " ⬆ E N T E R ␣ C 2nd " , C ENTER

: $b^2 - 4*a*c \rightarrow d$ B ∧ 2 − 4 × A × C STO✦ D ENTER

     calculates the discriminant and stores its value as d

: $(-b+\sqrt{(d)})/(2a) \rightarrow m$ (((−) B + 2nd √ D)) ÷ (2 A) STO✦ M ENTER

     calculates one root and stores it as m

: $(-b-\sqrt{(d)})/(2a) \rightarrow n$ (((−) B − 2nd √ D)) ÷ (2 A) STO✦ N ENTER

     calculates the other root and stores it as n

: If d<0 Then   ⬆ I F ␣ D 2nd < 0 ␣ ⬆ T H E N ENTER

     tests to see if the discriminant is negative;

: Goto Label1   ⬆ G O T O ␣ ⬆ L A B E L 1 ENTER

     if the discriminant is negative, jumps to the line Label1 below

: EndIf   ⬆ E N D ⬆ I F ENTER

     if the discriminant is not negative, continues on to the next line

: If d=0 Then   ⬆ I F ␣ D = 0 ␣ ⬆ T H E N ENTER

     tests to see if the discriminant is zero;

: Goto Label2   ⬆ G O T O ␣ ⬆ L A B E L 2 ENTER

     if the discriminant is zero, jumps to the line Label 2 below

: EndIf   ⬆ E N D ⬆ I F ENTER

     if the discriminant is not zero, continues on to the next line

: Disp "Two real roots",m,n   ⬆ D I S P ␣ 2nd " ⬆ T W O ␣ R E A L ␣ R O O T S 2nd " , M , N ENTER

     displays the message "Two real roots" and both roots

: Stop	⬆ S T O P ENTER

stops program execution

: Lbl Label1	⬆L B L ⌴ ⬆L A B E L 1 ENTER

jumping point for the Goto command above

: Disp "Complex roots",m,n	⬆ D I S P ⌴ 2nd " ⬆ C O M P L E X ⌴ R O O T S
	2nd " , M , N ENTER

displays the message "Complex roots" and both roots

: Stop	⬆ S T O P ENTER

: Lbl Label2	⬆L B L ⌴ ⬆L A B E L 2 ENTER

:Disp "Double root", m	⬆ D I S P ⌴ 2nd " ⬆ D O U B L E ⌴ R O O T
	2nd " , M ENTER

displays the message "Double root" and the solution (root)

When you have finished, press 2nd QUIT to leave the program editor and move on.

If you want to remove a program from memory, press 2nd VAR-LINK, use the cursor pad to highlight the name of the program you want to delete, then press F1*[Manage]* 1*[Delete]* ENTER and then ENTER again to confirm the deletion from the calculator's memory.

Technology Tip: The program uses the variables a, b, c, d, m, and n. Note that any previous values for these variables, including matrices, will be replaced by the values used by the program. The TI-92 does not distinguish between A and a in these uses. Note that you will have to clear the variables (using 2nd VAR-LINK) in order to use these names again in the current folder. From the Home screen, F6 will clear all 1-character variables. Another way to deal with this is to create a new folder. From the Home screen, press F4*[Other]* B*[NewFold]* and type the name of the new folder. The work you do from that point on will be in the new folder, as indicated by the folder name in the lower left corner of the Status line. You can change folders from

the MODE menu or, from the Home screen, by typing setFold(*foldername*), where *foldername* is the existing folder that you wish to be in.

5.10.2 Executing a Program: To execute the program you have entered, go to the Home screen and type the name of the program, including the parentheses and then press ENTER to execute it. If you have forgotten its name, press 2nd VAR-LINK to list all the variables that exist. The programs will have PRGM after the name. You can execute the program from this screen by highlighting the name and then pressing ENTER. The screen will return to the Home screen and you will have to enter the closing parenthesis) and press ENTER to execute the program.

The program has been written to prompt you for values of the coefficients *a*, *b*, and *c* in a quadratic equation

$ax^2 + bx + c = 0$. Input a value, then press **ENTER** to continue the program.

If you need to interrupt a program during execution, press **ON**.

After the program has run, the TI-92 will display the appropriate message and the root(s). The TI-92 will be on the Program I/O screen *not* the Home screen. The **F5** key toggles between the Home screen and the Program I/O screen or you can use **2nd QUIT**, ◆ **HOME** to go to the Home screen, or the **APPS** menu to go any screen.

The instruction manual for your TI-92 gives detailed information about programming. Refer to it to learn more about programming and how to use other features of your calculator.

5.11 Differentiation

5.11.1 Limits: Suppose you need to find this limit: $\lim\limits_{x \to 0} \dfrac{\sin 4x}{x}$. Plot the graph of $f(x) = \dfrac{\sin 4x}{x}$ in a convenient viewing rectangle that contains the point where the function appears to intersect the line $x = 0$ (because you want the limit as $x \to 0$). Your graph should support the conclusion that $\lim\limits_{x \to 0} \dfrac{\sin 4x}{x} = 4$ (Figure 5.95).

To test whether the conclusion that $\lim\limits_{x \to \infty} \dfrac{2x - 1}{x + 1} = 2$ is reasonable, evaluate $f(x) = \dfrac{2x - 1}{x + 1}$ for several large positive values of x (since you want the limit as $x \to \infty$). For example, evaluate $f(100)$, $f(1000)$, and $f(10,000)$. Another way to test the conclusion is to examine the graph of $f(x) = \dfrac{2x - 1}{x + 1}$ in a viewing rectangle that extends over large values of x. See, as in Figure 5.96 (where the viewing rectangle extends horizontally from 0 to 90), whether the graph is asymptotic to the horizontal line $y = 2$. Enter $\dfrac{2x - 1}{x + 1}$ for y1 and 2 for y2.

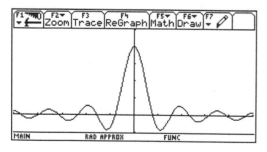

Figure 5.95: Checking $\lim\limits_{x \to 0} \dfrac{\sin 4x}{x} = 4$

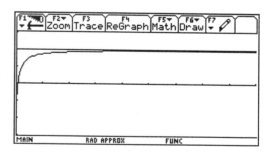

Figure 5.96: Checking $\lim\limits_{x \to \infty} \dfrac{2x - 1}{x + 1} = 2$

5.11.2 Numerical Derivatives: The derivative of a function f at x can be defined as the limit of the slopes of secant lines, so $f'(x) = \lim_{\Delta x \to 0} \dfrac{f(x + \Delta x) - f(x - \Delta x)}{2\Delta x}$. And for small values of Δx, the expression $\dfrac{f(x + \Delta x) - f(x - \Delta x)}{2\Delta x}$ gives a good approximation to the limit.

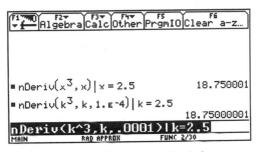

Figure 5.97: Using nDeriv(

The TI-92 has a function , nDeriv(, which is available in the Calculus sub-menu of the MATH menu, that will calculate the *symmetric difference*, $\dfrac{f(x + \Delta x) - f(x - \Delta x)}{2\Delta x}$. So, to find a numerical approximation to $f'(2.5)$ when $f(x) = x^3$ and with $\Delta x = 0.001$, go to the Home screen and press 2nd MATH A[Calculus] A[nDeriv(] X ∧ 3 , X) 2nd | X = 2.5 ENTER as shown in Figure 5.97. The format of this command is nDeriv(*expression, variable, Δx*), where the optional argument Δx controls the accuracy of the approximation. The added expression, 2nd | X = 2.5 give the value of x at which the derivative is evaluated. The | is found on the keyboard above the K, so press 2nd K to enter it. If no value for Δx is provided, the TI-92 automatically uses $\Delta x = 0.001$. If no value for x is given, the TI-92 will give the symmetric difference as a function of x. The same derivative is also approximated in Figure 5.97 using $\Delta x = 0.0001$. For most purposes, $\Delta x = 0.001$ gives a very good approximation to the derivative. Note that in Figure 5.97 any letter can be used for the variable.

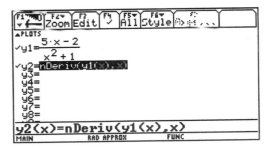

Figure 5.98: Entering $f(x)$ and $f'(x)$

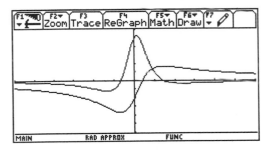

Figure 5.99: Graphs of $f(x)$ and $f'(x)$

Technology Tip: It is sometimes helpful to plot both a function and its derivative together. In Figure 5.99, the function $f(x) = \dfrac{5x-2}{x^2+1}$ and its numerical derivative (actually, an approximation to the derivative given by the symmetric difference) are graphed on viewing window that extends from -6 to 6 vertically and horizontally. You can duplicate this graph by first entering $\dfrac{5x-2}{x^2+1}$ for y1 and then entering its numerical derivative for y2 by pressing 2nd MATH A[Calculus] A[nDeriv(] Y 1 (X) , X) ENTER (Figure 5.98).

Graphing the derivative will be quite slow. Making the xres value larger on the WINDOW screen will speed up the plotting of the graph.

Technology Tip: To approximate the *second* derivative $f''(x)$ of a function $y = f(x)$ or to plot the second derivative, first enter the expression for y1 and its derivative for y2 as above. Then enter the second derivative for y3 by pressing 2nd MATH A[Calculus] A[nDeriv(] Y 2 (X) , X) ENTER.

You may also approximate a derivative while you are examining the graph of a function. When you are in a graph window, press F5[Math] 6[Derivatives] 1[dy/dx], then use the cursor pad to trace along the curve to a point where you want the derivative or enter a value and press ENTER. For example, with the TI-92 in Function graphing mode, graph the function $f(x) = \dfrac{5x-2}{x^2+1}$ in the standard viewing rectangle. Then press F5[Math] 6[Derivatives] 1[dy/dx]. The coordinates of the point in the center of the of the range will appear. To find the numerical derivative at $x = -2.3$, press -2.3 ENTER. Figure 5.100 shows the derivative at that point to be about -0.7746922.

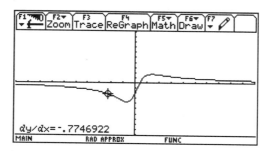

Figure 5.100: Derivative of $f(x) = \dfrac{5x-2}{x^2+1}$ at $x = -2.3$

If more than one function is graphed you can use $\uparrow$ and $\downarrow$ to scroll between the functions.

Note that different options are available from pressing F5[Math] 6[Derivatives] depending on whether the function(s) being graphed are in FUNCTION, PARAMETER, or POLAR mode.

5.11.3 Newton's Method: With the TI-92, you may iterate using Newton's method to find the zeros of a

function. Recall that Newton's Method determines each successive approximation by the formula
$x_{n+1} = x_n - \dfrac{f(x_n)}{f'(x_n)}$.

As an example of the technique, consider $f(x) = 2x^3 + x^2 - x + 1$. Enter this function as y1 and graph it in the standard viewing window. A look at its graph suggests that it has a zero near $x = -1$, so start the iteration by going to the Home screen and storing -1 as x. Then press these keystrokes: X − Y 1 (X) ÷ 2nd MATH A[Calculus] A[nDeriv(] Y 1 (X) , X) STO• X ENTER ENTER (Figure 5.101) to calculate the first two iterations of Newton's method. Press ENTER repeatedly until two successive approximations differ by less than some predetermined value, say 0.0001. Note that each time you press ENTER, the TI-92 will use the *current* value of x, and that value is changing as you continue the iteration.

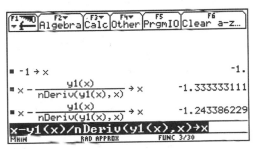

Figure 5.101: Newton's method

Technology Tip: Newton's Method is sensitive to your initial value for x, so look carefully at the function's graph to make a good first estimate. Also, remember that the method sometimes fails to converge!

You may want to write a short program for Newton's Method. See your calculator's manual for further information.

5.12   Integration

5.12.1 Approximating Definite Integrals: The TI-92 has a function, nInt(,which is available in the Calculus sub-menu of the MATH menu, that will approximate a definite integral. For example, to find a numerical approximation to $\int_0^1 \cos x^2 dx$ go to the Home screen and press 2nd MATH A[Calculus] B[nInt(] COS X ∧ 2) , X , 0 , 1) ENTER (Figure 5.102). The format of this command is nInt(*expression, variable, lower limit, upper limit*). The algorithm that the TI-92 uses to calculate the numerical integral is adaptive, and has an accuracy goal of six significant digits. If it seems that this goal has not been achieved, the calculator will display the warning "Questionable accuracy."

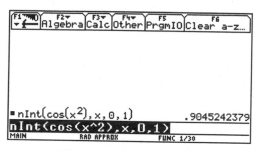

Figure 5.102: Using nInt(

5.12.2 Areas: You may approximate the area under the graph of a function $y = f(x)$ between $x = A$ and $x = B$ with your TI-92. To do this you use the F5*[Math]* menu when you have a graph displayed. For example, here are the keystrokes for finding the area under the graph of the function $y = \cos x^2$ between $x = 0$ and

$x = 1$. The area is represented by the definite integral $\int_0^1 \cos x^2 dx$. First clear any existing graphs and then press COS X $\wedge$ 2) ENTER followed by ◆ GRAPH to draw the graph . The range in Figure 5.103 extends from −5 to 5 horizontally and from −2 to 2 vertically. Now press F5*[Math]* 7*[∫f(x)dx]*. The TI-92 will prompt you for the lower and upper limits which are entered by pressing 0 ENTER 1 ENTER. The region between the graph and the x-axis from the lower limit to the upper limit is shaded and the approximate value of the integral is displayed (Figure 5.104).

Technology Tip: If the function takes on negative values between the lower and upper limits, the value that the TI-92 displays it the value of the integral, *not* the area of the shaded region.

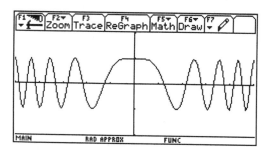

Figure 5.103: Graph of $y = \cos x^2$

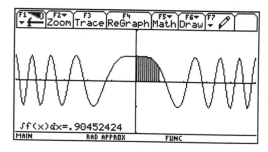

Figure 5.104: Graph and area

Technology Tip: Suppose that you want to find the area between two functions, $y = f(x)$ and $y = g(x)$ from $x = A$ and $x = B$. If $f(x) \geq g(x)$ for $A \leq x \leq B$, then graph the expression $f(x) - g(x)$ and use the method above to find the required area.

Chapter 6

Casio fx-7700GE/9700GE

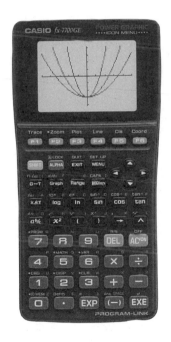
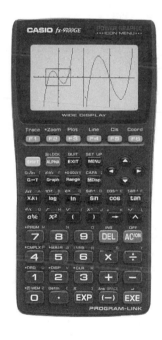

6.1   Getting started with the Casio *fx*-7700GE or *fx*-9700GE

Note: There are some differences between the Casio *fx*-7700GE and the Casio *fx*-9700GE, but for the most part they operate identically. We will point out the differences as needed. The Casio *fx*-9700GE has a wider screen than the Casio *fx*-7700GE and displays more significant digits. Illustrations in this chapter are primarily for the Casio *fx*-7700GE whenever the differences between the two calculators are not significant. Likewise, when menu names differ only slightly (for example, REC on the Casio *fx*-7700GE is RECT on the Casio *fx*-9700GE), we will use the Casio *fx*-7700GE's names. We will refer to both calculators by "Casio *fx*-7700/9700GE" whenever the *same* sequence of keystrokes works on *both* calculators.

6.1.1 Basics: Press the AC$^{/ON}$ key to begin using your Casio *fx*-7700/9700GE. The main menu screen will appear on your calculator. (See Figure 6.1 for the Casio *fx*-7700GE and Figure 6.2 for the Casio *fx*-9700GE.)

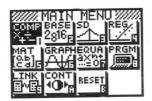

Figure 6.1: Casio *fx*-7700GE main menu

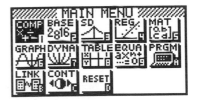

Figure 6.2: Casio *fx*-9700GE main menu

If you need to adjust the display contrast, select CONT icon from the main menu by using the arrow keys and press EXE, or by selecting the appropriate letter, A for the Casio *fx*-7700GE and C for the Casio *fx*-9700GE. Press ◄ (the *left* arrow key) to lighten and ► (the *right* arrow key) to darken. When you have finished with the calculator, turn it off to conserve battery power by pressing SHIFT and then OFF.

Technology Tip: To return to the main menu you can press the MENU key. In general, whenever you need to return to the main menu, you can press the MENU key. Note that to enter any mode from the main menu, you just need to select the appropriate number or letter. Thus, you can jump quickly to the contrast screen by pressing MENU A on the Casio *fx*-7700GE or MENU C on the Casio *fx*-9700GE.

Check your Casio *fx*-7700/9700GE's calculation mode settings by selecting COMP from the main menu (Menu 1). The following screen showing the COMP settings will appear. (See Figure 6.3 for the Casio *fx*-7700GE and Figure 6.4 for the *fx*-9700GE).

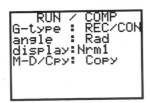

Figure 6.3: Casio *fx*-7700GE COMP settings

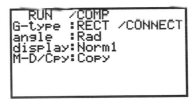

Figure 6.4: Casio *fx*-9700GE COMP settings

Press **SHIFT SET UP** to change the GRAPH TYPE, DRAW TYPE, and M-DISP/COPY. Use the up and down arrow keys to scroll through the three categories. When the cursor is at GRAPH TYPE press F1 to select the rectangular coordinates mode. When the cursor is at DRAW TYPE press F1 to select the connected mode. When the cursor is at M-DISP/COPY press F1 to select M-DISP. (Note while both Figure 6.3 and Figure 6.4 display Copy, this was in order to copy the image of the screen into this document.) Press **EXIT** to return to the COMP screen and then press **AC**/ON to clear the screen. To set the unit of angle measurement to radians, press **SHIFT DRG** (located at the 1 key). Press **F2 EXE** to select the radian mode. You can return to the COMP screen by pressing the **MENU** key and selecting the COMP icon. Press **AC**/ON to clear the screen to start calculating.

6.1.2 Editing: One advantage of the Casio *fx*-7700/9700GE is that up to seven lines are visible at one time, so you can *see* a long calculation. For example, enter the calculation mode (**MENU** 1) and type this sum (Figure 6.5):

$$1 + 2 + 3 + 4 + 5 + 6 + 7 + 8 + 9 + 10 + 11 + 12 + 13 + 14 + 15 + 16 + 17 + 18 + 19 + 20$$

Then press **EXE** to see the answer too.

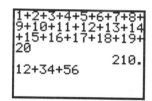

Figure 6.5: Calculation screen

Often we do not notice a mistake until we see how unreasonable an answer is. The Casio *fx*-7700/9700GE permit you to redisplay an entire calculation, edit it easily, then execute the *corrected* calculation.

Suppose you had typed 12 + 34 + 56 as in Figure 6.5 but had *not yet* pressed EXE, when you realize that 34 should have been 74. Simply press ◄ (the *left* arrow key) as many times as necessary to move the blinking cursor left to 3, then type 7 to write over it. On the other hand, if 34 should have been 384, move the cursor back to 4, press **SHIFT INS** (the cursor changes to a blinking frame) and then type 8 (inserts at the cursor position and the other characters are pushed to the right). If the 34 should have been 3 only, move the cursor to 4, and press **DEL** to delete it.

Technology Tip: To move quickly to the *beginning* of an expression you are currently editing, press ▲ (the *up* arrow key); to jump to the *end* of that expression, press ▼ (the *down* arrow key).

Even if you had pressed **EXE**, you may still edit the previous expression. Press the *left* or *right* arrow key to *redisplay* the last expression that we entered. Now you can change it. If you press ◄, the cursor will be at the *end* of the previous expression; if you press ► the cursor will appear at the *beginning*. Even if you have already pressed some keys since the last EXE, but *not* EXE again, you can still recall the previous expression by first pressing AC/ON to clear the screen and then pressing ◄ or ►.

Technology Tip: When you need to evaluate a formula for different values of a variable, use the editing feature to simplify the process. For example, suppose you want to find the balance in an investment account if there is now $5000 in the account and interest is compounded annually at the rate of 8.5%. The formula for the balance is $P = \left(1 + \frac{r}{n}\right)^{nt}$, where P = principal, r = rate of interest (expressed as a decimal), n = number of times interest is compounded each year, and t = number of years. In our example, this becomes $5000(1+.085)^t$. Here are the keystrokes for finding the balance after $t = 3$, 5, and 10 years. Figure 6.6 shows the first set of keystrokes and the result.

Years	*Keystrokes*	*Balance*
3	5000 (1 + .085) ∧ 3 EXE	$6386.45
5	◄ ◄ 5 EXE	$7518.28
10	◄ ◄ 10 EXE	$11,304.92

Figure 6.6: Editing expressions

Then to find the balance from the same initial investment but after 5 years when the annual interest rate is 7.5%, press the keys to change the last calculation above: ◄ ◄ DEL ◄ 5 ◄ ◄ ◄ ◄ ◄ 7 EXE.

6.1.3 Key Functions: Most keys on the Casio *fx*-7700/9700GE offer access to more than one function, just as the keys on a computer keyboard can produce more than one letter ("g" and "G") or even quite different characters ("5" and "%"). The primary function of a key is indicated on the key itself, and you access that function by a simple press on the key.

To access the *second* function indicated to the *left* above a key, first press SHIFT (the cursor changes to a blinking **S** and a menu appears at the bottom of the screen) and *then* press the key. For example to calculate $\sqrt{25}$, press SHIFT $\sqrt{}$ 25 EXE.

When you want to use a letter or other character printed to the *right* above a key, first press ALPHA (the cursor changes to a blinking **A** and a menu appears at the bottom of the screen) and then the key. For example, to use the letter K in a formula, press ALPHA K. If you need several letters in a row, press SHIFT A-LOCK, which is like the CAPS LOCK key on a computer keyboard, and then press all the letters you want. Remember to press ALPHA when you are finished and want to restore the keys to their primary functions.

6.1.4 Order of Operations:
The Casio *fx*-7700/9700GE performs calculations according to the standard algebraic rules. Working outwards from inner parentheses, calculations are performed from left to right. Powers and roots are evaluated first, followed by multiplications and divisions, and then additions and subtractions.

Enter these expressions to practice using your Casio *fx*-7700/9700GE.

Expression	Keystrokes	Display
$7 - 5 \cdot 3$	7 − 5 × 3 EXE	−8
$(7 - 5) \cdot 3$	(7 − 5) × 3 EXE	6
$120 - 10^2$	120 − 10 SHIFT x² EXE	20
$(120 - 10)^2$	(120 − 10) SHIFT x² EXE	12100
$\dfrac{24}{2^3}$	24 ÷ 2 ∧ 3 EXE	3
$\left(\dfrac{24}{2}\right)^3$	(24 ÷ 2) ∧ 3 EXE	1728
$(7 - -5) \cdot -3$	(7 − −5) × − 3 EXE	−36

6.1.5 Algebraic Expressions and Memory:
Your calculator can evaluate expressions such as $\dfrac{N(N+1)}{2}$ after you have entered a value for N. Suppose you want $N = 200$. Press 200 → ALPHA N EXE to store the value 200 in memory location N. Whenever you use N in an expression, the calculator will substitute the value 200 until you make a change by storing *another* number in N. Next enter the expression $\dfrac{N(N+1)}{2}$ by typing

ALPHA N (ALPHA N + 1) ÷ 2 EXE. For $N = 200$, you will find that $\dfrac{N(N+1)}{2} = 20100$.

The contents of any memory location may be revealed by typing just its letter name and then EXE. And the Casio *fx*-7700/9700GE retains memorized values even when it is turned off, so long as its batteries are good.

6.1.6 Repeated Operations with Ans:
The result of your *last* calculation is always stored in memory location Ans and replaces any previous result. This makes it easy to use the answer from one computation in another computation. For example, press 30 + 15 EXE so that 45 is the last result displayed. Then press SHIFT Ans ÷ 9 EXE and get 5 because $45 \div 9 = 5$.

With a function like division, you press the ÷ *after* you enter an argument. For such functions, whenever you start a new calculation with the previous answer followed by pressing the function key, you may press just the function key. So instead of SHIFT Ans ÷ 9 in the previous example, you could have pressed simply ÷ 9 to achieve the same result. This technique also works for these functions: $+ \quad - \quad \times \quad \wedge \quad x^2 \quad x^{-1}$.

Here is a situation where this is especially useful. Suppose a person makes \$5.85 per hour and you are asked to calculate earnings for a day, a week, and a year. Execute the given keystrokes to find the person's incomes during these periods (results are shown in Figure 6.7 for the Casio *fx*-7700GE and Figure 6.8 for the Casio *fx*-9700GE).

Pay Period	Keystrokes	Earnings
8-hour day	5.85 × 8 EXE	$46.80
5-day week	SHIFT Ans × 5 EXE	$234
52-week year	× 52 EXE	$12,168

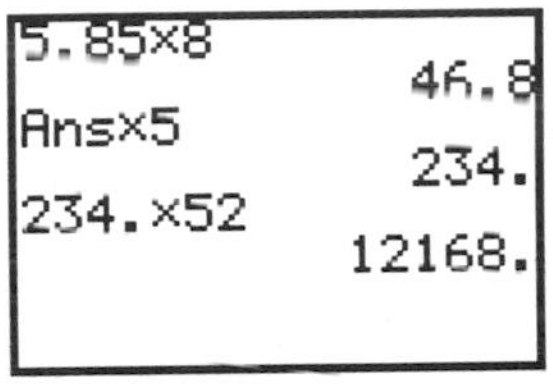

Figure 6.7: Casio *fx*-7700GE
SHIFT Ans key

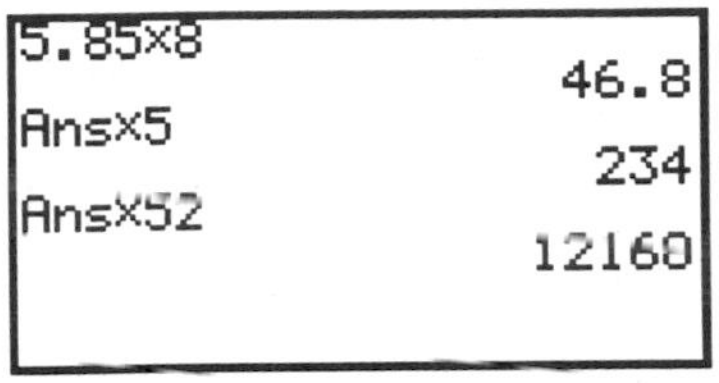

Figure 6.8: Casio *fx*-9700GE
SHIFT Ans key

In general, the Casio *fx*-7700/9700GE does not distinguish between the negative sign and the subtraction operator. But when you enter –4 as the *first* number in a calculation, you must use the negative key (–) rather than the – key. Press these keys for an illustration: 8 EXE – 5 EXE (–) 5 EXE.

6.1.7 The MATH Menu: Operators and functions associated with a scientific calculator are available either immediately from the keys of the Casio *fx*-7700/9700GE or by the SHIFT keys. You have direct access to common arithmetic operations (x^2, SHIFT $\sqrt{}$, SHIFT x^{-1}, ∧), trigonometric functions (sin, cos, tan), and their inverses (SHIFT $\sin^{-1}$, SHIFT $\cos^{-1}$, SHIFT $\tan^{-1}$), exponential and logarithmic functions (log, SHIFT 10^X, ln, SHIFT e^X), and a famous constant (SHIFT π).

A significant difference between the Casio *fx*-7700/9700GE graphing calculators and most scientific calculators is that the Casio *fx*-7700/9700GE requires the argument of a function *after* the function, as you would see in a formula written in your textbook. For example, on the Casio *fx*-7700/9700GE you calculate $\sqrt{16}$ by pressing the keys $\sqrt{}$ 16 in that order.

The Casio *fx*-7700/9700GE has a special fraction key $a\frac{b}{c}$ for entering fractions and mixed numbers. To enter a fraction such as $\frac{2}{5}$, press 2 $a\frac{b}{c}$ 5 EXE. To enter a mixed number like $2\frac{3}{4}$, press 2 $a\frac{b}{c}$ 3 $a\frac{b}{c}$ 4 EXE. Press $a\frac{b}{c}$ to toggle between the mixed number and its decimal equivalent; press SHIFT $\frac{d}{c}$ and see $2\frac{3}{4}$ as an improper fraction, $\frac{11}{4}$.

Here are keystrokes for basic mathematical operations. Try them for practice on your Casio *fx*-7700/9700GE.

Note that the Casio *fx*-9700GE will display more digits for the second and third expressions.

Expression	Keystrokes	Display
$\sqrt{3^2 + 4^2}$	SHIFT $\sqrt{\ }$ (3 x^2 + 4 x^2) EXE	5
$2\frac{1}{3}$	2 $a\frac{b}{c}$ 1 $a\frac{b}{c}$ 3 EXE $a\frac{b}{c}$	2.333333333
log 200	LOG 200 EXE	2.301029996
$2.34 \cdot 10^5$	2.34 × SHIFT 10^x 5 EXE	234000

Additional mathematical operations and functions are available from the MATH menu. Press SHIFT MATH to see the six categories of mathematical functions. They are listed across the bottom of the Casio *fx*-7700/9700GE's screen and correspond to the function keys, F1 to F6 (Figure 6.9).

You will learn in your mathematics textbook how to apply many of them. As an example, with the MATH menu on display, calculate $|-5|$ by pressing F3 (for access to numerical functions) and then F1 (−) 5 EXE (Figure 6.10). To clear a menu from the screen, press EXIT. In this case, you press EXIT twice, the first time to clear the NUM menu to get back to the MATH menu and then to clear the MATH menu.

Figure 6.9: MATH menu

Figure 6.10: MATH NUM menu

The *factorial* of a non-negative integer is the *product* of *all* the integers from 1 up to the given integer. The symbol for factorial is the exclamation point. So 4! (pronounced *four factorial*) is $1 \cdot 2 \cdot 3 \cdot 4 = 24$. You will learn more about applications of factorials in your textbook, but for now use the Casio *fx*-7700/9700GE to calculate 4! Press these keystrokes: 4 SHIFT MATH F2 F1 EXE.

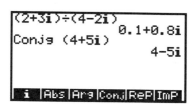

Figure 6.11: Complex number calculations

Casio fx-9700GE: On the Casio *fx*-9700GE it is possible to do calculations with complex numbers. Press SHIFT CMPLX to activate the complex number calculation menu at the bottom of the screen. For example, to divide $2 + 3i$ by $4 - 2i$, press (2 + 3 F1 *[i]*) ÷ (4 – 2 F1 *[i]*) EXE. The result is $0.1 + 0.8i$ (Figure 6.11).

To find the complex conjugate of $4 + 5i$ press F4 *[Conj]* (4 + 5 F1 *[i]*) EXE (Figure 6.11).

Casio fx-9700GE: The Casio *fx*-9700GE can also solve for the real and complex roots of a quadratic or cubic function in the equation mode (MENU 9). The functions must be in the form $f(x) = ax^3 + bx^2 + cx + d$ or $f(x) = ax^2 + bx + c$ where $a \neq 0$. Enter the equation mode and press F2 *[POLY]*. You are then prompted for the degree (either 2 or 3) of your polynomial (Figure 6.12).

Figure 6.12: Prompt for degree

For example, to find all the zeros of $f(x) = x^3 - 4x^2 + 14x - 20$ select F2 *[3]* at the prompt and then enter the coefficients into the table by pressing 1 EXE (–) 4 EXE 14 EXE (–) 20 EXE (Figure 6.13). If you had not pressed EXE yet, you can change a coefficient by pressing AC$^{/ON}$ and entering a new value. Otherwise, move your cursor over the coefficient with your arrow keys and enter a new value. Now press F1 *[SOLV]* and the calculator will display the solutions (Figure 6.14).

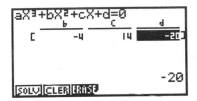

Figure 6.13: Entering the coefficients

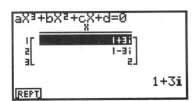

Figure 6.14: Zeros of $f(x) = x^3 - 4x^2 + 14x - 20$

To perform a new calculation press F1 *[REPT]*. If you are computing the roots of another cubic function then you can either edit the existing coefficients or press F2 *[CLER]* to reset all the coefficients to zero. If you are computing the roots of a quadratic function, press F3 *[ERASE]* F1 *[YES]* F1 *[2]* and then enter in the coefficients of the quadratic and proceed.

Note that it may take considerable time for the calculation result of a cubic equation to appear on the display. Failure of a result to appear immediately does not mean that the unit is not functioning properly.

6.2   Functions and Graphs

6.2.1 Evaluating Functions: Suppose you receive a monthly salary of $1975 plus a commission of 10% of sales. Let x = your sales in dollars; then your wages W in dollars are given by the equation $W = 1975 + .10x$. If your January sales were $2230 and your February sales were $1865, what was your income during those months?

Here's one method to use your Casio *fx*-7700/9700GE to perform this task. First be in the COMP mode and press AC/ON to clear your screen. Then set x = 2230 by pressing 2230 → X,θ,T. (The X,θ,T key lets you enter a variable x without having to use the ALPHA key.) Then press SHIFT ↵ to allow another expression to be input on a single command line. Finally, enter the expression $1975 + .10x$ by pressing these keys: 1975 + .10 X,θ,T. Now press EXE to calculate the answer (Figure 6.15).

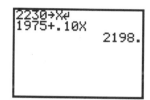

Figure 6.15: Evaluating a function

It is not necessary to repeat all these steps to find the February wages. Simply press ▶ to recall the entire previous line, change 2230 to 1865, and press EXE.

Another method is to use the function memory. The Casio *fx*-7700/9700GE can store up to six functions. First be in the COMP mode and press AC$^{/ON}$ to clear your screen. Press SHIFT F-MEM to display the function memory menu at the bottom of the screen. Enter the expression $1975 + .10x$ by pressing these keys: 1975 + .10 X,θ,T. Then store this as function memory number 1 by pressing F1 *[STO]* 1 (Figure 6.16). Press AC/ON to clear the screen, leaving the function memory menu. Then set x = 2230 by pressing 2230 → X,θ,T EXE. Recall the entire expression by pressing F2 *[RCL]* 1, and then press EXE to calculate the answer (Figure 6.17). To find the February wages, set x = 1865 and then evaluate the function for the new value by pressing F2 *[RCL]* 1 EXE. By pressing F2 1 you recall the entire expression. You can also press F3 1 to get the variable f1.

Figure 6.16: Storing a function   Figure 6.17: Evaluating a function

In general, to *store* a function enter it first in the calculation screen, but do *not* press EXE. If the function memory menu is active, press F1 and an integer from 1 to 6, otherwise you must press SHIFT F-MEM before pressing F1. To *recall* a function from the calculation screen press F2 and the integer corresponding to the function you want. Press F4 for a list of functions currently in the function memory.

Casio fx-9700GE: With the Casio *fx-9700GE*, you can also use your calculator's TABLE mode to create a table of values for a function. (Unfortunately, this feature is *not* available on the Casio *fx-7700GE*.) From the main menu select the TABLE icon (MENU 8) to get the TABLE&GRAPH screen (Figure 6.18).

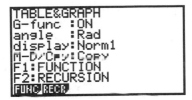

Figure 6.18: TABLE&GRAPH screen

Then press F1 *[FUNC]*. If there is no function stored in memory, you should get a prompt to input the function formula. If there is already a function, its numeric table appears on the display. In this case, you should press F1 *[NEW]* and then F1 *[YES]* to proceed. Enter the function $1975 + .10x$ by pressing these keys: 1975 + .10 X,θ,T (Figure 6.19). Press F2 *[RANGE]* to set the conditions for the x-variable when generating a function table. Start is the starting value of the x-variable, End is the ending value of the x-variable, and pitch is the change of the x-variable. In this case set Start to 2230, End to 2230 and pitch to 0 (Figure 6.20).

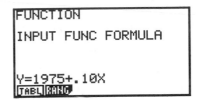

Figure 6.19: Inputting the function

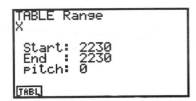

Figure 6.20: TABLE Range

Press F1 *[TABLE]* to see the table (Figure 6.21). We will now add $x = 1865$ to the table by pressing F3 *[ROW]* F3 *[ADD]* 1865 EXE (Figure 6.22).

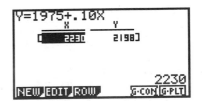

Figure 6.21: Initial table of values

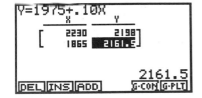

Figure 6.22: Evaluating a function in a table

Press EXIT to get the previous menu at the bottom of the screen. To add more entries use the ROW and ADD commands as described above. If you wish to change the function press F2 *[EDIT]* to edit the function.

Technology Tip: The Casio *fx*-7700/9700GE does not require multiplication to be expressed between variables, so *xxx* means x^3. It is often easier to press two or three *x*'s together than to search for the square key or the powers key. Of course, expressed multiplication is also not required between a constant and a variable. Hence to enter $2x^3 + 3x^2 - 4x + 5$ in the Casio *fx*-7700/9700GE, you might save keystrokes and press just these keys: 2 X,θ,T X,θ,T X,θ,T + 3 X,θ,T X,θ,T − 4 X,θ,T + 5

6.2.2 Functions in a Graph Window: On the Casio *fx*-7700/9700GE, you can easily generate the graph of a function. The ability to draw a graph contributes substantially to our ability to solve problems.

For example, here is how to graph $y = -x^3 + 4x$ from the COMP mode. First press Graph and then (−) X,θ,T ∧ 3 + 4 X,θ,T to enter the function (as in Figure 6.23). Now press EXE and the Casio *fx*-7700/9700GE changes to a window with the graph of $y = -x^3 + 4x$.

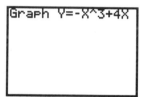

Figure 6.23: Graph command in COMP mode

While the Casio *fx*-7700/9700GE is busy calculating coordinates for a plot, it displays a solid square at the top right of the graph window. When you see this indicator, even though the screen does not change, you know that the calculator is working.

Switch back and forth between the graph window and the home screen by pressing G↔T.

The graph window on the Casio *fx*-7700GE may look like the one in Figure 6.24, while the graph window on the Casio *fx*-9700GE may look like the one in Figure 6.25. Since the graph of $y = -x^3 + 4x$ extends infinitely far left and right and also infinitely far up and down, the Casio *fx*-7700/9700GE can only display only a piece of the actual graph. This displayed rectangular part is called a *viewing rectangle*.

You can easily change the viewing rectangle to enhance your investigation of a graph. For example, press any of the arrow keys to pan the graph window in the corresponding direction. If you press the down arrow, for example, the window will pan down so that you may look at points below the current window.

The viewing rectangle for the Casio *fx*-7700GE in Figure 6.24 shows the part of the graph that extends horizontally form −4.7 to 4.7 and vertically from −3.1 to 3.1, while the viewing rectangle for the Casio *fx*-9700GE in Figure 6.25 shows the part of the graph that extends horizontally form −6.3 to 6.3 and vertically from −3.706 to 3.706. Press RANGE to see information about your viewing rectangle. Figures 6.26 and 6.27 show the RANGE screen that corresponds to the viewing rectangles in Figures 6.24 and 6.25, respectively. These are the *standard* viewing rectangles for the Casio *fx*-7700GE and Casio *fx*-9700GE, respectively.

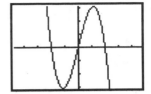

Figure 6.24: Casio *fx*-7700GE

graph of $y = -x^3 + 4x$

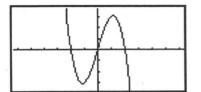

Figure 6.25: Casio *fx*-9700GE

graph of $y = -x^3 + 4x$

The variables Xmin and Xmax are the minimum and maximum *x*-values of the viewing rectangles; Ymin and Ymax are the minimum and maximum *y*-values.

Xscl and Yscl set the spacing between tick marks on the axes. (These appear as Xscale and Yscale on the Casio *fx*-9700GE.)

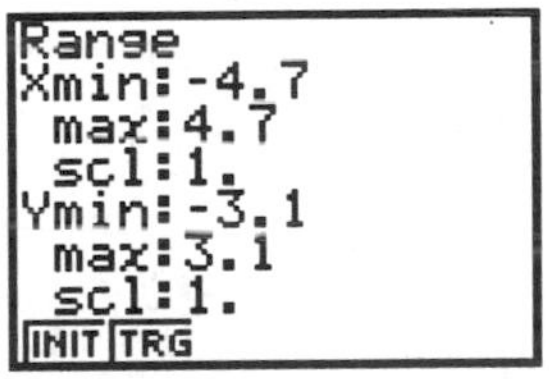

```
Range
Xmin:-4.7
 max:4.7
 scl:1.
Ymin:-3.1
 max:3.1
 scl:1.
INIT TRG
```

Figure 6.26: Casio *fx*-7700GE
standard RANGE

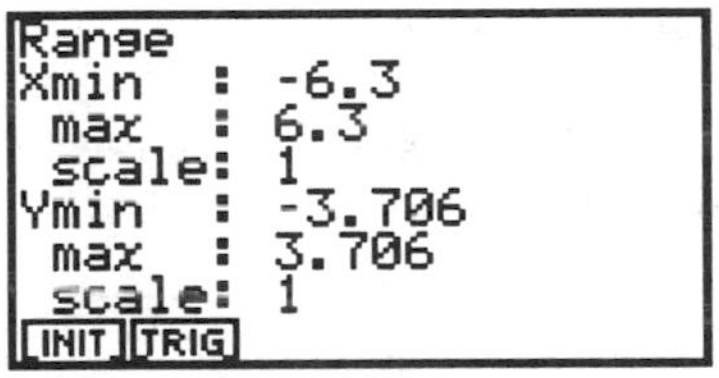

```
Range
Xmin  : -6.3
 max  : 6.3
 scale: 1
Ymin  : -3.706
 max  : 3.706
 scale: 1
INIT TRIG
```

Figure 6.27: Casio *fx*-9700GE
standard RANGE

Use the arrow keys ▲ and ▼ to move up and down from one line to another in this list; pressing the EXE key will move down the list. Enter a new value to over-write a previous value and then press EXE. When you enter a new value in the RANGE menu be sure that your cursor is at the far left. Note that the DEL key does not work when you are in the RANGE menu. To leave the RANGE menu, press the EXIT key. Finally, press EXE to redraw the graph. The following figures show different RANGE screens and the corresponding viewing rectangle for each one. Note that these figures are from the Casio *fx*-7700GE; figures for the Casio *fx*-9700GE are similar.

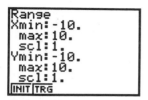

Figure 6.28: −10 to 10 in both directions

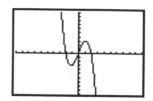

Figure 6.29: Graph of $y = -x^3 + 4x$

Figure 6.30: Custom window

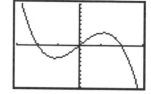

Figure 6.31: Graph of $y = -x^3 + 4x$

To initialize the viewing rectangle quickly to the *standard* viewing rectangle, press RANGE F1 EXIT. Then press EXE to redraw the graph.

As you pan over the graph by pressing the arrow keys, the RANGE to dimensions are updated automatically. More information about viewing rectangles is presented later in Section 6.2.4.

Technology Tip: Clear any graphs drawn in the COMP mode by pressing F5 EXE when the Casio *fx*-7700/9700GE is showing the graph screen or SHIFT F5 EXE when it is displaying the calculation screen. (When you press SHIFT the graph commands are displayed at bottom, and you select the clear screen command Cls.)

If you are going to use a function later, or if you need to perform some calculations before returning to its graph, save it in the Casio *fx*-7700/9700GE's function memory as described in Section 6.2.1.

Technology Tip: It's a good idea to reserve at least one function memory location, say f1, for temporary storage of functions, and use the remaining locations for longer-term storage.

Another procedure for graphing on the Casio *fx*-7700/9700GE is to use the GRAPH mode. To enter the GRAPH mode select the GRAPH icon (MENU 6) from the main menu. The top of the screen should say GRAPH FUNC: RECT. If not, press SHIFT SET UP F1 to specify the rectangular mode for drawing. Press EXIT to return to the main GRAPH screen. Erase any existing functions by pressing F1 *[STO]*, scrolling to the desired function, and then pressing F6 *[SET]*. You can store up to 20 functions in memory. To enter $y = -x^3 + 4x$, press (−) X,θ,T ∧ 3 + 4 X,θ,T F1 *[STO]*, move your cursor to Y1, and then F6 *[SET]*. If you have a Casio *fx*-7700GE, the screen in Figure 6.32 will appear. If you have a Casio *fx*-9700GE, the screen in Figure 6.33 will appear. Now press F6 *[DRW]* to draw the graph.

Figure 6.32: Casio *fx*-7700GE
graph function screen

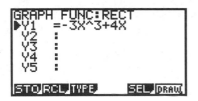

Figure 6.33: Casio *fx*-9700GE
graph function screen

As before, you can switch back and forth between the graph window and the graph function screen by pressing G↔T.

For editing functions or graphing multiple functions it is preferable to use the GRAPH mode to draw functions. Unless otherwise noted, we will be using the GRAPH mode for our graphs.

To change a function in the GRAPH mode, recall it to the edit line by pressing F2 *[RCL]*, then move the cursor to the function you want remove and press F6 *[SET]*.

6.2.3 *Graphing Step and Piecewise–Defined Functions:*   The greatest integer function, written $[[x]]$, gives the greatest *integer* less than or equal to a number x. On the Casio *fx*-7700/9700GE, the greatest integer function is called Intg and is located as F5 under the NUM sub-menu of the MATH menu (see Figure 6.10). Calculate $[[6.78]] = 6$ in the COMP mode by pressing SHIFT MATH F3 F5 6.78 EXE.

To graph $y = [[x]]$, enter the GRAPH mode (MATH 6) and press SHIFT MATH F3 F5 X,θ,T SHIFT QUIT F1 F6 F6. Figure 6.34 shows this graph in a viewing rectangle from −5 to 5 in both directions.

The true graph of the greatest integer function is a step graph, like the one in Figure 6.35. For the graph of $y = [[x]]$, a segment should *not* be drawn between every pair of successive points. You can change from CONNECT line to PLOT graph on the Casio *fx*-7700/9700GE by pressing SHIFT SET UP to enter the GRAPH menu. Use the arrow keys to scroll to DRAW TYPE and change to plot by pressing F2. Press EXIT and then F6 to draw the new graph (Figure 6.35).

In general, you'll want your graph to be connected, so do that by returning to the set up menu, scrolling down to DRAW TYPE, and pressing F1.

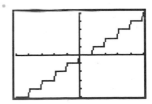

Figure 6.34: Connected graph of $y = [[x]]$

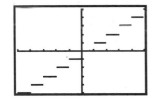

Figure 6.35: Plotted graph of $y = [[x]]$

It is not possible to graph piecewise–defined functions on the Casio *fx*-7700/9700GE.

6.2.4 Graphing a Circle: Here is a useful technique for graphs that are not functions but can be "split" into a top part and a bottom part, or into multiple parts. Suppose you wish to graph the circle of radius 6 whose equation is $x^2 + y^2 = 36$. First solve for y and get an equation for the top semicircle, $y = \sqrt{36 - x^2}$, and for the bottom semicircle, $y = -\sqrt{36 - x^2}$. Then graph the two semicircles simultaneously.

Use the following keystrokes to draw this circle's graph in the GRAPH screen. Store $\sqrt{36 - x^2}$ as Y1 by pressing SHIFT $\sqrt{\ }$ (36 – X,θ,T x^2) F1 F6. Then store $-\sqrt{36 - x^2}$ as Y2 by pressing (–) SHIFT $\sqrt{\ }$ (36 – X,θ,T x^2) F1 ▼ F6 (Figure 6.36). Next press F6 to graph both halves (Figure 6.37). Make sure that the RANGE is set large enough to display a circle of radius 6.

Figure 6.36: Two semicircles

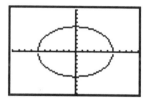

Figure 6.37: One view of circle's graph

Instead of entering $-\sqrt{36 - x^2}$ as Y2, you could have entered –Y1 as Y2 and saved some keystrokes. On the Casio *fx*-7700GE, try this by going into the GRAPH screen and pressing (–) SHIFT VAR F1 *[GRP]* F1 *[Y]* 1 EXIT EXIT F1 *[STO]* ▼ F6 *[SET]*. On the Casio *fx*-9700GE, press (–) SHIFT VAR F3 *[GRPH]* F1 *[Y]* 1 EXIT EXIT F1 *[STO]* ▼ F6 *[SET]*. The graph should be as before. The VAR menu (displayed along the bottom of the screen in Figure 6.38 for the Casio *fx*-7700GE and in Figure 6.39 for the Casio *fx*-9700GE) enables you to recall graphic functions and other information from memory.

Figure 6.38: Casio *fx*-7700GE
GRAPH mode

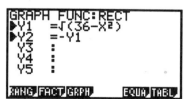

Figure 6.39: Casio *fx*-7700GE
GRAPH mode

If your range were set to a viewing rectangle extending from –10 to 10 in both directions, your graph would look like Figure 6.37. Now this does *not* look a circle, because the units along the axes are not the same. You need what is called a "square" viewing rectangle. The Casio *fx*-7700/9700GE's standard viewing rectangle is square but too small to display a circle of radius 6, so you should double the dimensions of the standard viewing window. On the Casio *fx*-7700GE, change the range to extend horizontally from −9.4 to 9.4 and vertically from −6.2 to 6.2 (Figure 6.40). On the Casio *fx*-9700GE, change the range to extend horizontally from −12.6 to 12.6 and vertically from −7.412 to 7.412 (Figure 6.42). The graphs for the better circles are shown in Figure 6.41 for the Casio *fx*-7700GE and in Figure 6.43 for the Casio *fx*-9700GE.

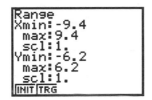

Figure 6.40: Casio *fx*-7700GE
twice standard range

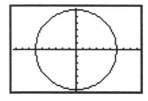

Figure 6.41: Casio *fx*-7700GE
better circle

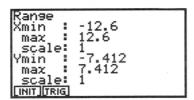

Figure 6.42: Casio *fx*-9700GE
twice standard range

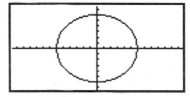

Figure 6.43: Casio *fx*-9700GE
better circle

Technology Tip for the Casio fx-7700GE: Another way to get an approximately square graph on the Casio *fx*-7700GE is to change the range variables so that the value of Ymax − Ymin is $\frac{2}{3}$ times Xmax − Xmin. For example, use the RANGE values in Figure 6.44 to get the corresponding graph in Figure 6.45. This method works because the dimensions of the Casio *fx*-7700GE's display are such that the ratio of vertical to horizontal is approximately $\frac{2}{3}$.

Figure 6.44: Casio *fx*-7700GE
$\frac{\text{vertical}}{\text{horizontal}} = \frac{16}{24} = \frac{2}{3}$

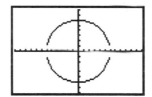

Figure 6.45: Casio *fx*-7700GE
"square" circle

Technology Tip for the Casio fx-9700GE: Another way to get an approximately square graph on the Casio

fx-9700GE is to change the range variables so that the value of Ymax − Ymin is $\frac{3}{5}$ times Xmax − Xmin. For example, use the RANGE values in Figure 6.46 to get the corresponding graph in Figure 6.47. This method works because the dimensions of the Casio fx-9700GE's display are such that the ratio of vertical to horizontal is approximately $\frac{3}{5}$.

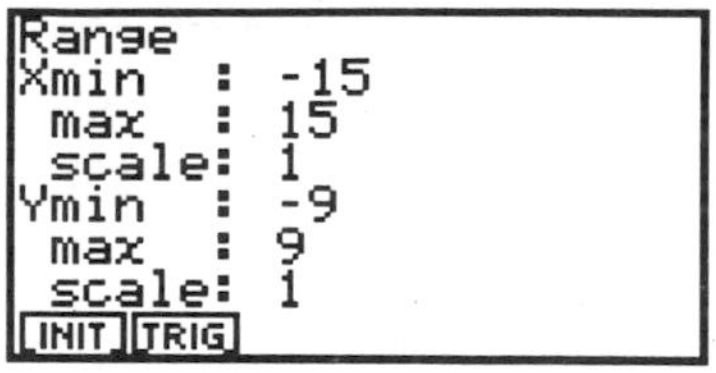

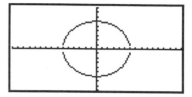

Figure 6.46: Casio fx-9700GE
$$\frac{\text{vertical}}{\text{horizontal}} = \frac{18}{30} = \frac{3}{5}$$

Figure 6.47: Casio fx-9700GE
"square" circle

The two semicircles in Figures 6.45 and 6.47 do not meet because of an idiosyncrasy in the way the Casio fx-7700/9700GE plots a graph.

6.2.5 TRACE: In the graph mode (MENU 6) graph the function $y = -x^3 + 4x$ from Section 6.2.2 using the standard viewing rectangle. (Remember to clear or cancel any other functions in the graph function screen.) When the graph window is displayed, press F1 to enable ◄ (the *left* arrow key) and ► (the *right* arrow key) to trace along the function. The coordinates that are displayed belong to points on the function's graph, so the y-coordinate is the calculated value of the function at the corresponding x-coordinate (Figure 6.48). (Note that on the Casio fx-9700GE the x-coordinate of the trace begins at the left-most value of the rectangle so the cursor will not appear until it is traced onto the screen.)

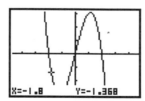

Figure 6.48: Trace

Press F6 to cycle between the x-coordinate alone, the y-coordinate alone, and both coordinates.

Now plot a second function, $y = -.25x$, along with $y = -x^3 + 4x$. From the above graph window, return to the GRAPH screen by pressing G↔T, and enter the second function as Y2 by pressing (−) .25 X,θ,T F1 ▼ F6. (See Figure 6.49 for the Casio fx-7700GE and Figure 6.51 for the Casio fx-9700GE.) Finally, press F6 to draw both functions. (See Figure 6.50 for the Casio fx-7700GE and Figure 6.52 for the Casio fx-9700GE.)

Figure 6.49: Casio *fx*-7700GE
two functions in GRAPH mode

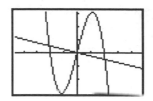

Figure 6.50: Casio *fx*-7700GE
graph of $y = -.25x$ and $y = -x^3 + 4x$

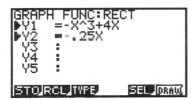

Figure 6.51: Casio *fx*-9700GE
two functions in GRAPH mode

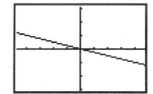

Figure 6.52: Casio *fx*-9700GE
graph of $y = -.25x$ and $y = -x^3 + 4x$

Notice that in Figure 6.49 the equal signs next to Y1 and Y2 are *both* highlighted, while in Figure 6.51 there is a ▶ to the left of *both* Y1 and Y2. This means that *both* functions will be graphed. In the GRAPH screen, press F5 *[SEL]* F2 *[CAN]* EXIT. On the Casio *fx*-7700GE the equal sign next to Y1 should no longer be highlighted (Figure 6.53), while on the Casio *fx*-9700GE the ▶ should no longer be next to Y1 (Figure 6.55). Now press F6 and see that only Y2 is plotted (See Figure 6.54 for the Casio *fx*-7700GE and Figure 6.56 for the Casio *fx*-9700GE.)

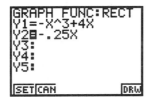

Figure 6.53· Casio *fx*-7700GE
only Y2 active

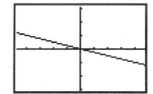

Figure 6.54: Casio *fx*-7700GE
graph of $y = -.25x$

Up to twenty different functions can be stored in list of functions in the GRAPH screen and any combination of them may be graphed simultaneously. You can make a function active or inactive for graphing pressing the F5 to activate the select feature, scrolling to the desired function, and pressing F1 to set or F2 to cancel. Now go back to the GRAPH screen and do what is need in order to graph Y1 but not Y2.

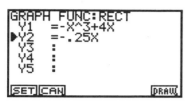

Figure 6.55: Casio *fx*-9700GE
only Y2 active

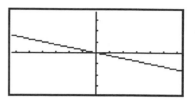

Figure 6.56: Casio *fx*-9700GE
graph of $y = -.25x$

Now activate both functions so that both graphs are plotted. Press F1 to trace and the cursor will be on the graph of $y = -x^3 + 4x$ because it is higher up on the list of active functions in the GRAPH screen. Press the up ▲ or down ▼ arrow key to move the cursor vertically to the graph of $y = -.25x$. Next press the left and right arrow keys to trace along the graph of $y = -.25x$. When more than one function is plotted, you can move the trace cursor vertically from one graph to another in this way.

Technology Tip for the Casio fx-9700GE: On the Casio *fx*-9700GE, when there is more than one function plotted, the function of the graph being graphed or traced can be displayed. In SET UP, move your cursor down to GRAPH FUNC and press F1 *[ON]* to turn the display on or F2 *[OFF]* to turn it off.

Technology Tip: By the way, trace the graph of $y = -.25x$ and press and hold either ◄ or ►. Eventually you will reach the left or right edge of the window. Keep pressing the arrow key and the Casio *fx*-7700/9700GE will allow you to continue the trace by panning the viewing rectangle. Check the RANGE screen to see that the Xmin and Xmax are automatically updated.

Technology Tip for the Casio fx-7700GE: The Casio *fx*-7700GE has a display of 95 horizontal columns of pixels and 63 vertical rows, so when you trace a curve across a graph window, you are actually moving from Xmin to Xmax in 94 equal jumps, each called Δx. You would calculate the size of each jump to be $\Delta x = \dfrac{\text{Xmax} - \text{Xmin}}{94}$. Sometimes you may want the jumps to be friendly numbers like 0.1 or 0.25 so that, when you trace along the curve, the x-coordinates will be incremented by such a convenient amount. Just set your viewing rectangle for a particular increment Δx by making Xmax = Xmin + 94 · Δx. For example, if you want Xmin = –5 and $\Delta x = 0.3$, set Xmax = –5 + 94 · 0.3 = 23.2.

On the Casio *fx*-7700GE, to center your window around a particular point, say (h, k), and also have a certain Δx, set Xmin = $h - 47 \cdot \Delta x$ and make Xmax = $h + 47 \cdot \Delta x$. Likewise, make Ymin = $k - 31 \cdot \Delta y$ and make Ymax = $h + 31 \cdot \Delta x$. For example, to center a window around the origin $(0, 0)$, with both horizontal and vertical increments of 0.25, set the range so that Xmin = 0 – 47 · 0.25 = –11.75, Xmax = 0 + 47 · 0.25 = 11.75, Ymin = 0 – 31 · 0.25 = –7.75 and Ymax = h + 31 · 0.25 = 7.75.

The Casio *fx*-7700GE's standard viewing window is already a friendly viewing rectangle, centered at the origin $(0, 0)$ with $\Delta x = \Delta y = 0.1$

Technology Tip for the Casio fx-9700GE: On the other hand, the Casio *fx*-9700GE has a display of 127 horizontal columns of pixels and 63 vertical rows, so when you trace a curve across a graph window, you are

Casio *fx*-7700/9700GE Power Graphic Calculators

actually moving from Xmin to Xmax in 126 equal jumps. Hence, you would calculate the size of each jump to be $\Delta x = \dfrac{\text{Xmax} - \text{Xmin}}{126}$. Sometimes you may want the jumps to be friendly numbers like 0.1 or 0.25 so that, when you trace along the curve, the x-coordinates will be incremented by such a convenient amount. Just set your viewing rectangle for a particular increment Δx by making Xmax = Xmin + 126 · Δx. For example, if you want Xmin = −5 and $\Delta x = 0.3$, set Xmax = −5 + 126 · 0.3 = 38.3.

On the Casio fx-9700GE, to center your window around a particular point, say (h, k), and also have a certain Δx, set Xmin = $h − 63 \cdot \Delta x$ and make Xmax = $h + 63 \cdot \Delta x$. Likewise, make Ymin = $k − 31 \cdot \Delta y$ and make Ymax = $h + 31 \cdot \Delta x$. For example, to center a window around the origin $(0, 0)$, with both horizontal and vertical increments of 0.25, set the range so that Xmin = 0 − 63 · 0.25 = −15.75, Xmax = 0 + 63 · 0.25 = 15.75, Ymin = 0 − 31 · 0.25 = −7.75 and Ymax = 0 + 31 · 0.25 = 7.75.

The Casio fx-9700GE's standard viewing window is a square window, centered at the origin $(0, 0)$ with $\Delta x = 0.1$.

See the benefit by first plotting $y = x^2 + 2x + 1$ in a window that extends from −10 to 10 in both directions. Trace near its y-intercept, which is $(0, 1)$, and move towards its x-intercept, which is $(−1, 0)$. Then initialize the range to the standard window and trace again near the intercepts.

6.2.6 ZOOM: Plot again the two graphs, for $y = -x^3 + 4x$ and $y = -.25x$. There appears to be an intersection near $x = 2$. The Casio fx-7700/9700GE provides several ways to enlarge the view around this point. You can change the viewing rectangle directly by pressing RANGE and editing the values of Xmin, Xmax, Ymin, and Ymax. Figure 6.57 shows a new viewing rectangle for the range extending from 1.5 to 2.5 horizontally and from −2.5 to 2.5 vertically.

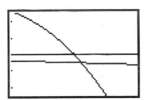

Figure 6.57: Closer view

Trace along the graphs until coordinates of a point that is close to the intersection are displayed.

A more efficient method for enlarging the view is to draw a new viewing rectangle with the cursor. Start again with a graph of the two functions $y = -x^3 + 4x$ and $y = -.25x$ in a standard viewing rectangle.

Now imagine a small rectangular box around the intersection point, near $x = 2$. Press F2 *[ZOOM]* to activate the ZOOM menu at the bottom of the screen. (See Figure 6.58 for the Casio fx-7700GE and Figure 6.59 for The Casio fx-9700GE.)

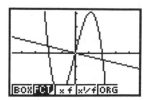

Figure 6.58: Casio *fx*-7700GE
Zoom menu

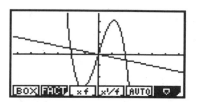

Figure 6.59: Casio *fx*-9700GE
Zoom menu

Now press F1 *[BOX]* to draw a box to define this new viewing rectangle. On the Casio *fx*-7700GE you should press SHIFT QUIT to remove the menu from the bottom of the screen. Use the arrow keys to move the cursor, which is now free-moving and whose coordinates are displayed at the bottom of the window, to one corner of the new viewing rectangle you imagine (Figure 6.60).

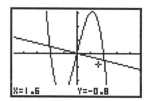

Figure 6.60: One corner selected

Press EXE to fix the corner where you moved the cursor. Use the arrow keys again to move the cursor to the diagonally opposite corner of the new rectangle (Figure 6.61). If this box looks all right to you, press EXE. The rectangular area you have enclosed will now enlarge to fill the graph window (Figure 6.62).

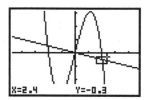

Figure 6.61: Box drawn

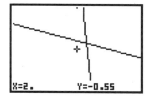

Figure 6.62: New viewing rectangle

You may cancel the zoom any time *before* you press this last EXE. Press another function key such as F1 to cancel the zoom and initiate a trace instead, or press F2 to zoom again and start over. Even if you did execute the zoom, you may still return to the original viewing rectangle and start over on the Casio *fx*-7700GE by pressing F2 *[Zoom]* F5 *[ORG]* and on the Casio *fx*-9700GE by pressing F2 *[Zoom]* F6 *[•]* F1 *[ORG]*.

Casio fx-9700GE: The Casio *fx*-9700GE has a split screen feature that enables you to see two views of a graph simultaneously. (This feature is *not* available on the Casio *fx*-7700GE.) In SET UP, move the cursor down to DUAL GRAPH and toggle it on. Now when you zoom, the left window displays the original graph and the right window displays the result of the zoom (see Figure 6.63).

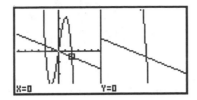

Figure 6.63: DUAL GRAPH

In the Casio *fx*-9700GE's DUAL GRAPH mode, only the left side can be acted on. So to achieve another zoom, first press F6 F2 *[CHNG]* to exchange the left and right windows. When you press RANGE, you will find *two* ranges that can be changed independently. The F6 key toggles between the left side range and the right side range.

Technology Tip for the Casio fx-9700GE: Use the G↔T key to toggle the Casio *fx*-9700GE from the dual graph to full-screen left side to full-screen right side to the GRAPH FUNC screen.

The Casio *fx*-7700/9700GE can quickly magnify a graph around the cursor's location. Return once more to the standard range for the graph of the two functions $y = -x^3 + 4x$ and $y = -.25x$. Trace along the graphs to move the cursor as you can to the point of intersection near $x = 2$ (Figure 6.64). Then press F2 F3 *[× f]* and the calculator draws a magnified graph, centered at the cursor's position (Figure 6.65). The range values are changed to reflect this new viewing rectangle. Look in the RANGE menu to check.

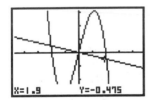

Figure 6.64: Before a zoom in

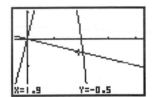

Figure 6.65: After a zoom in

As you see in the Zoom menu (press F2), the Casio *fx*-7700/9700GE can zoom in (press F2 F3 *[× f]*) or zoom out (press F2 F4 *[× 1/f]*). Zoom out to see a larger view of the graph, centered at the cursor position. You can change the horizontal and vertical scale of the magnification by pressing F2 F2 and editing Xfact and Yfact, the horizontal and vertical magnification factors.

Technology Tip: An advantage of zooming in from the default viewing rectangle is that subsequent windows will also be square. Likewise, if you zoom in from a friendly viewing rectangle, the zoomed windows will also be friendly.

The default zoom factor is 2 in both directions (press F1 *[INIT]* in the Zoom Factor menu). It is not necessary for Xfact and Yfact to be equal. Sometimes, you may prefer to zoom in one direction only, so the other factor should be set to 1, Press EXIT to leave the Zoom Factor menu.

Technology Tip: If you should zoom in too much and lose the curve, zoom back to the original viewing rectangle and start over. Or use the arrow keys to pan over if you think the curve is not too far away. You can also just initialize the range to the Casio *fx*-7700/9700GE's standard window.

Technology Tip for the Casio fx-9700GE: The Casio *fx*-9700GE can automatically select the necessary *vertical* range for a function. For auto scaling, press F2 *[Zoom]* F5 *[AUTO]*. Take care, because sometimes when you are graphing two functions together, the calculator will auto scale for one function in such a way that the other function will no longer be visible. For example, plot the two functions $y = -x^3 + 4x$ and $y = -.25x$ in the Casio *fx*-9700GE's standard viewing rectangle, then auto scale and trace along both functions.

6.2.7 Relative Minimums and Maximums: Graph $y = -x^3 + 4x$ once again in the standard viewing rectangle. This function appears to have a relative minimum near $x = -1$ and a relative maximum near $x = 1$. You may zoom and trace to approximate these extreme values.

First trace along the curve near the local minimum. Notice by how much the *x*-values and *y*-values change as you move from point to point Trace along the curve until the *y*-coordinate is as *small* as you can get it, so that you are as close as possible to the local minimum, and zoom in (press F2 F3 or use a zoom box). Now trace again along the curve and, as you move from point to point, see that the coordinates change by smaller amounts than before. Keep zooming and tracing until you find the coordinates of the local minimum point as accurately as you need them, approximately $(-1.15, -3.08)$.

Follow a similar procedure to find the local maximum. Trace along the curve until the *y*-coordinate is as *great* as you can get it, so that you are as close as possible to the local maximum, and zoom in. The local maximum point on the graph of $y = -x^3 + 4x$ is approximately $(1.15, 3.08)$.

Casio fx-9700GE: The Casio *fx*-9700GE can automatically find the maximums and minimums for functions drawn in the GRAPH mode. After graphing $y = -x^3 + 4x$, press SHIFT G-SOLV to activate the graph solve menu at the bottom of the screen (Figure 6.66). (Note that the features in the G-SOLV menu is *not* available on the Casio *fx*-7700GE.)

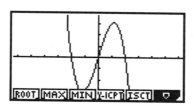

Figure 6.66: G-SOLVE menu

After activating the graph solve menu, press F3 *[MIN]* to calculate the minimum (Figure 6.67). Then find the maximum by pressing SHIFT G-SOLV F2 *[MAX]* (Figure 6.68).

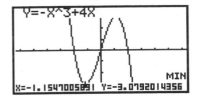

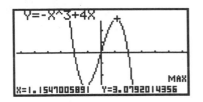

Figure 6.67: Minimum of $y = -x^3 + 4x$ Figure 6.68: Maximum of $y = -x^3 + 4x$

Note that if you have more than one graph on the screen, the calculator will pause until you specify the graph whose maximum or minimum you want to calculate. As you use the up and down arrow keys to move between graphs, press EXE when the equation you want to evaluate appears on the screen.

If your graph has more than one maximum or minimum, you can use the left and right arrows to move between them.

6.3   Solving Equations and Inequalities

6.3.1 Intercepts and Intersections: Tracing and zooming are also used to locate an x-intercept of a graph, where a curve crosses the x-axis. For example, the graph of $y = x^3 - 8x$ crosses the x-axis three times (Figure 6.69). After tracing over to the x-intercept point that is farthest to the left, zoom in (Figure 6.70). Continue this process until you have located all three intercepts with as much accuracy as you need. The three x-intercepts of $y = x^3 - 8x$ are approximately −2.828, 0, and 2.828.

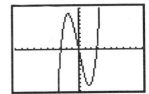

Figure 6.69: Graph of $y = x^3 - 8x$ Figure 6.70: Near an x-intercept of $y = x^3 - 8x$

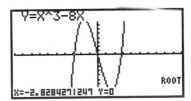

Figure 6.71: A root of $y = x^3 - 8x$

Technology Tip: As you zoom in, you may also wish to change the spacing between tick marks on the *x*-axis so that the viewing rectangle shows scale marks near the intercept point. Then the accuracy of your approximation will be such that the error is less than the distance between two tick marks. Change the *x*-scale on the Casio 7700/9700 from the RANGE menu. Move the cursor down to Xscale and enter an appropriate value.

Casio fx-9700GE: The *x*-intercept of a function's graph is a *root* of the equation $f(x) = 0$, and the Casio *fx*-9700GE can automatically search for the roots. (This feature is *not* available on the Casio *fx*-7700GE.) First plot the function in GRAPH mode and then activate the graph solve menu by pressing SHIFT G-SOLV. (Refer back to figure 6.66.) Then press F1 *[ROOT]* to locate an *x*-intercept on the graph in the current window (Figure 6.71). The calculator searches from left to right to find an *x*-intercept in the current window; press the right arrow key to search for the next *x*-intercept to the right.

TRACE and ZOOM are especially important for locating the intersection points of two graphs, say the graphs of $y = -x^3 + 4x$ and $y = -.25x$. Trace along one of the graphs until you arrive close to an intersection point. Then press ▲ or ▼ to jump to the other graph. Notice that the *x*-coordinate does not change, but the *y*-coordinate is likely to be different (Figures 6.72 and 6.73).

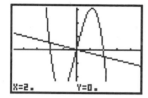

Figure 6.72: Trace on $y = -x^3 + 4x$

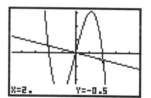

Figure 6.73: Trace on $y = -.25x$

When two *y*-coordinates are as close as they can get, you have come as close as you now can to the point of intersection. So zoom in around the intersection point, then trace again until the two *y*-coordinates are as close as possible. Continue this process until you have located the point of intersection with as much accuracy as necessary.

Technology Tip: Press F6 a couple of times to display only the *y*-coordinate. Then while tracing towards an intersection, it's easier to see where the *y*-coordinates are closest.

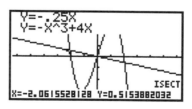

Figure 6.74: An intersection of $y = -x^3 + 4x$ and $y = -.25x$

Technology Tip for the Casio fx-9700GE: While the graphs are displayed, automate the Casio *fx*-9700GE to

Casio *fx*-7700/9700GE Power Graphic Calculators

search for points of intersection by pressing SHIFT G-SOLV F5 *[ISCT]*. (This feature is *not* available on the Casio *fx*-7700GE.) If more than two functions are being plotted, the calculator will ask you to specify the two whose intersection you seek. The calculator searches from left to right to find an intersection point in the current window; press the right arrow to continue the search for the *next* intersection point. Figure 6.74 shows one of the intersection points.

6.3.2 Solving Equations by Graphing: Suppose you need to solve the equation $24x^3 - 36x + 17 = 0$. First graph $y = 24x^3 - 36x + 17$ in a window large enough to exhibit *all* its x-intercepts, corresponding to all the equation's roots. Then use trace and zoom or the Casio *fx*-9700GE's ROOT command, to locate each one. In fact, this equation has just one solution, approximately $x = -1.414$.

Remember that when an equation has more than one root, it may be necessary to change the viewing rectangle a few times to locate all of them.

Technology Tip: To solve an equation like $24x^3 + 17 = 36x$, you may first transform it into standard form, $24x^3 - 36x + 17 = 0$, and proceed as above. However, you may also graph the *two* functions $y = 24x^3 + 17$ and $y = 36x$, then zoom and trace to locate their point of intersection.

6.3.3 Solving Systems by Graphing: The solutions to a system of equations correspond to the points of intersection of their graphs (Figure 6.75). For example, to solve the system $y = x^3 + 3x^2 - 2x - 1$ and $y = x^2 - 3x - 4$, first graph them together. Then use zoom and trace or the Casio *fx*-9700GE's ISCT command, to locate their point of intersection, approximately $(-2.17, 7.25)$.

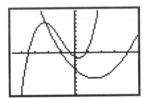

Figure 6.75: Graph of $y = x^3 + 3x^2 - 2x - 1$ and $y = x^2 - 3x - 4$

You must judge whether the two current y-coordinates are sufficiently close for $x = -2.17$ or whether you should continue to zoom and trace to improve the approximation.

The solutions of the system of two equations $y = x^3 + 3x^2 - 2x - 1$ and $y = x^2 - 3x - 4$ correspond to the solutions of the single equation $x^3 + 3x^2 - 2x - 1 = x^2 - 3x - 4$, which simplifies to $x^3 + 2x^2 + x + 3 = 0$. So you may also graph $y = x^3 + 2x^2 + x + 3$ and find its x-intercepts to solve the system.

6.3.4 Solving Inequalities by Graphing: Consider the inequality $1 - \dfrac{3x}{2} \geq x - 4$. To solve it with your Casio fx-7700/9700GE, graph the two functions $y = 1 - \dfrac{3x}{2}$ and $y = x - 4$ (Figure 6.76). First locate their point of intersection, at $x = 2$. The inequality is true when the graph of $y = 1 - \dfrac{3x}{2}$ lies *above* the graph of $y = x - 4$, and that occurs when $x < 2$. So the solution is the half-line $x \leq 2$, or $(-\infty, 2]$.

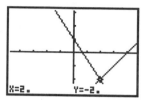

Figure 6.76: Solving $1 - \dfrac{3x}{2} \geq x - 4$

The Casio fx-7700/9700GE is capable of graphing inequalities of the form $y \leq f(x)$, $y < f(x)$, $y \geq f(x)$, or $y > f(x)$. For example, to graph $y \geq x^2 - 1$ in GRAPH mode, press F3 *[TYPE]* F4 *[INEQ]*. Input $x^2 - 1$. Now when you press F1 *[STO]* to store this expression, several inequality options appear (Figure 6.77). In this case, we want to select F5 *[Y≥]*. Press F6 to draw the graph (Figure 6.78).

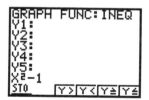

Figure 6.77: Inequality options

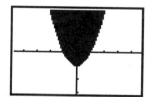

Figure 6.78: Graph of $y \geq x^2 - 1$

Next press F1 to trace along the boundaries of the inequality. Notice that the Casio fx-7700/9700GE displays coordinates appropriately as inequalities. Zooming is also available for inequality graphs.

Solve a system of inequalities, such as $1 - \dfrac{3x}{2} \geq y$ and $y > x - 4$, by plotting the two inequality graphs simultaneously. First, clear the graph window and reset the range to a convenient window. Input $1 - \dfrac{3x}{2}$ as an inequality type and store it as Y1 by pressing F1 *[STO]* F6 *[Y≤]*. Likewise, input $x - 4$ as an inequality type and store it by pressing F1 *[STO]* ▼ F3 *[Y>]*. Now press F6 to draw the two inequalities as in Figure 6.79.

Technology Tip: Since you can change the mode of the Casio fx-7700/9700GE at any time, you can graph inequalities and equations together at the same time. Simply change to inequality type before entering an inequality, and change to rectangular type before entering an equation.

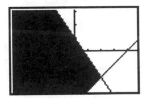

Figure 6.79: Graphs of $1 - \dfrac{3x}{2} \geq y$ and $y > x - 4$

6.4   Trigonometry

6.4.1 Degrees and Radians: The trigonometric functions can be applied to angles measured either in radians or degrees, but you should take care that the Casio *fx*-7700/9700GE is configured for whichever measure you need. From the main menu, enter the COMP mode to see the current settings. To change your angle setting, press SHIFT DRG to activate the angular unit menu at the bottom of the screen. Press F1 for degree measure or F2 for radian measure, then press EXE to make it so.

It's a good idea to check the angle measure setting before executing a calculation that depends on a particular measure. You may change a mode setting at any time and not interfere with pending calculations. Try the following keystrokes to see this in action.

Expression	Keystrokes	Display
$\sin 45°$	SHIFT DRG F1 EXE	
	sin 45 EXE	0.7071067812
$\sin \pi°$	sin SHIFT π EXE	0.05480366515
$\sin \pi$	SHIFT DRG F2 EXE	
	sin SHIFT π EXE	0
$\sin 45$	sin 45 EXE	0.8509035245
$\sin \dfrac{\pi}{6}$	sin (SHIFT π ÷ 6) EXE	0.5

The first line of keystrokes sets the Casio 7700/9700 in degree mode and calculates the sine of 45 *degrees*. While the calculator is still in degree mode, the second line keystrokes calculates the sine of π *degrees*, approximately 3.1415°. The third line changes to radian mode just before calculating the sine of π *radians*. The fourth line calculates the sine of 45 *radians* (the calculator remains in radian mode).

The Casio *fx*-7700/9700GE makes it possible to mix degrees and radians in a calculation. Execute these keystrokes to calculate $\tan 45° + \sin \frac{\pi}{6}$ as shown in Figure 6.80: tan 45 SHIFT DRG F4 + sin (SHIFT π ÷ 6)

F5 EXE. Do you get 1.5 whether your calculator is in set *either* in degree mode *or* in radian mode?

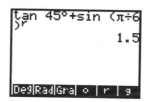

Figure 6.80: Angle measure

6.4.2 Graphs of Trigonometric Functions: When you graph a trigonometric function, you need to pay careful attention to the choice of graph window. For example, graph $y = \dfrac{\sin 30x}{30}$ in the standard viewing rectangle. Trace along the curve to see where it is. Zoom in to a better window, or use the period and amplitude to establish a better WINDOW.

Technology Tip: In the RANGE menu of the Casio *fx*-7700GE the viewing rectangle can be set by pressing F2 to a special window for trigonometric functions so that horizontal range is from -2π to 2π in radian mode or from $-360°$ to $360°$ in degree mode and the vertical range is from -1.6 to 1.6. On the Casio *fx*-9700GE horizontal range is from -3π to 3π in radian mode or from $-540°$ to $540°$ in degree mode and the vertical range is from -1.6 to 1.6.

6.5   Scatter Plots

6.5.1 Entering Data: The table shows the total prize money (in millions of dollars) awarded at the Indianapolis 500 race from 1981 to 1989. (*Source*: Indianapolis Motor Speedway Hall of Fame.)

Year	1981	1982	1983	1984	1985	1986	1987	1988	1989
Prize ($million)	$1.61	$2.07	$2.41	$2.80	$3.27	$4.00	$4.49	$5.03	$5.72

We'll now use the Casio *fx*-7700/9700GE to construct a scatter plot that represents these points and to find a linear model that approximates the given data.

Figure 6.81: REG settings

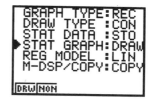

Figure 6.82: Setting up REG

In the main menu select the REG icon (MENU 4) to enter the regression mode. You want to set the mode to the settings shown in Figure 6.81. In particular you want S-data set to STO and S-graph set to DRAW for data storage and graphing. If you need to change the set up, press SHIFT SET UP. For example, to set STAT GRAPH to DRAW press ▼ ▼ ▼ F1 *[DRAW]* (Figure 6.82). Press EXIT AC/ON to enter the regression screen.

Instead of entering the full year 198x we will enter only x. Since the data will be plotted as it is entered, choose an appropriate viewing rectangle, say from 0 to 10 horizontally and vertically. First clear any existing statistical data by pressing F2 *[EDIT]* and then F1 *[DEL]* as many times as necessary to delete any existing entries. Press EXIT when you are done. Here are the keystrokes to enter the data for the first three years: 1 F3 1.61 F1 2 F3 2.07 F1 3 F3 2.41 F1. Continue to enter all the given data .

You can go back and forth from your graph to the REG screen by pressing the G↔T. You may edit statistical data in the same way you edit expressions in the home screen. Press F2 to display the data table (Figure 6.83). Move the cursor to the *x* and *y* value for any data point you wish to change, then type the correction and press EXE. To insert or delete data, move the cursor to the *x* or *y* value for any data point you wish to add or delete. Press F2 and a new data point is created; press F1 and the data point is deleted. Press EXIT to return to the regression screen.

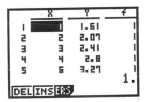

Figure 6.83: Editing data points

6.5.2 Plotting Data: Once all the data points have been entered, press SHIFT CLR F2 *[SCL]* EXE to clear statistical memory. Then press F6 *[CALC]* to calculate the statistics associated with the data. Recalculation is necessary whenever you edit data or change the graph window. Figure 6.84 shows the scatter plot in a viewing rectangle extending form 0 to 10 along the horizontal and vertical axes.

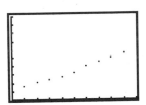

Figure 6.84: Scatter plot

6.5.3 Regression Line: The Casio *fx*-7700/9700GE calculates and the slope and *y*-intercept for the line that best fits all the data. From the plot of the data, return to the regression screen by pressing G↔T. Press F6 *[CALC]* to calculate a linear regression model. As you can see in Figure 6.85, the Casio *fx*-7700/9700GE names the *y*-intercept A and calls the slope B. The number r (between −1 and 1) is called the *correlation coefficient* and measures the goodness of fit of the linear regression with the data. The closer the absolute value of r is to 1, the better the fit; the closer the absolute value of r is to 0, the worse the fit. Press the function key F1 EXE for A, F2 EXE for B, and F3 EXE for r.

Graph the line $y = A + Bx$ by pressing GRAPH SHIFT F4 1 EXE. See how well this line fits with your data points (Figure 6.86).

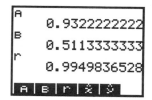

Figure 6.85: Linear regression model

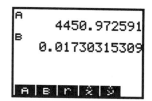

Figure 6.86: Linear regression line

6.5.4 Exponential Growth Model: The table shows the world population (in millions) from 1980 to 1992.

Year	1980	1985	1986	1987	1988	1989	1990	1991	1992
Population (millions)	4453	4850	4936	5024	5112	5202	5294	5384	5478

In the regression mode, press SHIFT SET UP. Scroll to STAT GRAPH and press F2 *[NON]*. Then move to REG MODEL and press F3 *[EXP]*, so the set up screen looks like Figure 6.87. EXIT the screen and follow the procedure described above and enter the data to find an exponential model that approximates the given data. Use 0 for 1980, 5 for 1985, and so on. You may find it easier to EDIT the existing data rather than entering the new data. Press EXIT F6 *[CALC]*.

Now press F6 *[REG]* to compute the exponential growth model $y = ae^{bx}$. Press F1 *[A]* EXE F2 *[B]* EXE to find the values of A and B (Figure 6.88). In this case, the exponential growth model is $y = 4451e^{0.017303t}$.

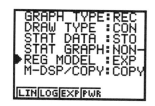

Figure 6.87

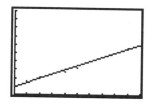

Figure 6.88

6.6 Matrices

6.6.1 Making a Matrix: The Casio *fx*-7700GE can display and use 5 different matrices (Mat A through Mat E), while the Casio *fx*-9700GE can work with 26 different matrices (Mat A through Mat Z). Here's how to create this 3×4 matrix $\begin{bmatrix} 1 & -4 & 3 & 5 \\ -1 & 3 & -1 & -3 \\ 2 & 0 & -4 & 6 \end{bmatrix}$ in your calculator as Mat A.

Enter the matrix mode by selecting the MAT icon from the main menu. Then press F4 *[LIST]* for the matrix list. The display will show the dimension of each matrix if the matrix exists; otherwise, it will display None (Figure 6.89). Move the cursor to Mat A and press F2 *[DIM]* 3 EXE 4 EXE to enter its dimensions of 3 rows by 4 columns. Return to the matrix list by pressing EXIT once, then press F1 *[EDIT]* to edit Mat A.

Use the arrow keys or press EXE repeatedly to move the cursor to a matrix element you want to change. If you press EXE, you will move right across a row and then back to the first column of the next row. The element in the second row and first column in Figure 6.90 is highlighted, so that the element's current value is displayed at the bottom right corner of the screen. Continue to enter all the elements of Mat A; press EXE after inputting each value.

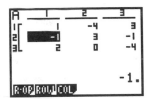

Figure 6.89: Matrix list   Figure 6.90: Editing a matrix

When you are finished, leave the editing screen by pressing EXIT once to return to the matrix list.

6.6.2 Matrix Math: You can perform many calculations with matrices in the matrix mode. To calculate the scalar multiplication 2 Mat A, enter the matrix calculation screen and press 2 F1 *[Mat]* ALPHA A EXE. The resulting matrix is displayed on the screen and is stored in matrix memory as Mat Ans (Figure 6.91). If you would rather have the matrix stored as specific matrix, say Mat C, you should press 2 F1 *[Mat]* ALPHA A → F1 *[Mat]* ALPHA C EXE (Figure 6.92).

To add two matrices, say Mat A and Mat B, create Mat B (with the same dimensions as Mat A) and then press F1 *[Mat]* ALPHA A + F1 *[Mat]* ALPHA B EXE. Again, if you want to store the answer as a specific matrix, say Mat C, then press → F1 *[Mat]* ALPHA C before executing the above command. Subtraction is performed in similar manner.

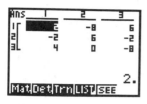

Figure 6.91: 2 Mat A

Figure 6.92: Matrix calculation

Now set the dimensions of **Mat C** as 2×3 and enter the matrix $\begin{bmatrix} 2 & 0 & 3 \\ 1 & -5 & -1 \end{bmatrix}$ as **Mat C**. For matrix multiplication of **Mat C** by **Mat A**, press F1 *[Mat]* ALPHA C × F1 *[Mat]* ALPHA A EXE. If you tried to multiply **Mat A** by **Mat C**, your calculator would signal an error because the dimensions of the two matrices do not permit multiplication in this way.

The *transpose* of a matrix is another matrix with the rows and columns interchanged. To calculate the transpose of **Mat A** , press F3 *[Trn]* F1 *[Mat]* ALPHA A EXE.

6.6.3 Row Operations: Here are the keystrokes necessary to perform elementary row operations on a matrix. Your textbook provides a more careful explanation of the elementary row operations and their uses.

Enter the editing screen for **Mat A**. Press F1 to activate the row operations menu at the bottom of the calculator screen (Figure 6.93).

After you select a row operation, your calculator will prompt you through it. For example, to interchange the second and third rows of **Mat A** defined above, press F1 2 EXE 3 EXE, while the calculator prompts for the row numbers (Figure 6.94). The format of this command is Swap Row m ↔ Row n.

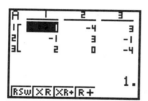

Figure 6.93: Row operations menu

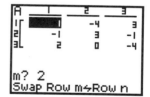

Figure 6.94: Swap rows 2 and 3

To add row 2 and row 3 and *store* the results in row 3, press F4 2 EXE 3 EXE. The format of this command is Row m + Row n → Row n.

To multiply row 2 by –4 and *store* the results in row 2, thereby replacing row 2 with new values, press F2 (-) 4 EXE 2 EXE. The format of this command is k × Row m → Row m.

To multiply row 2 by –4 and *add* the results to row 3, thereby replacing row 3 with new values, press F3 (-) 4 EXE 2 EXE 3 EXE. The format of this command is k × Row m + Row n → Row n.

Casio *fx*-7700/9700GE Power Graphic Calculators

Note that as you perform row operations on the Casio *fx*-7700/9700GE, your old matrix is replaced by the new matrix, so if you want to keep the original matrix in case you need it, you should save it under another name.

For example, use row operations to solve this system of linear equations: $\begin{cases} x - 2y + 3z = 9 \\ -x + 3y = -4 \\ 2x - 5y + 5z = 17 \end{cases}$.

First enter this *augmented matrix* as Mat A in your Casio *fx*-7700/9700GE: $\begin{bmatrix} 1 & -2 & 3 & 9 \\ -1 & 3 & 0 & -4 \\ 2 & -5 & 5 & 17 \end{bmatrix}$. Next store this

matrix as Mat E (press EXIT a couple of times to go back to the matrix home screen, then press F1 *[Mat]* ALPHA A → F1 *[Mat]* ALPHA E EXE, as in Figure 6.95), so you may keep the original in case you need to recall it.

Figure 6.95

We now edit Mat E. Here are the row operations and their associated keystrokes. At each step, the result is stored as Mat E and replaces the previous Mat E. The completion of the row operations is shown in Figure 6.96. First press F1 to begin performing row operations.

Row Operations	Keystrokes
add row 1 to row 2	F4 1 EXE 2 EXE
add −2 times row 1 to row 3	F3 (−) 2 EXE 1 EXE 3 EXE
add row 2 to row 3	F4 2 EXE 3 EXE
multiply row 3 by $\frac{1}{2}$	F2 1 ÷ 2 EXE 3 EXE

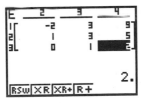

Figure 6.96

Thus $z = 2$, so $x = 1$ and $y = -1$.

6.6.4 Determinants and Inverses: Enter this 3×3 square matrix as Mat A: $\begin{bmatrix} 1 & -2 & 3 \\ -1 & 3 & 0 \\ 2 & -5 & 5 \end{bmatrix}$. To calculate its

determinant $\begin{vmatrix} 1 & -2 & 3 \\ -1 & 3 & 0 \\ 2 & -5 & 5 \end{vmatrix}$, go to the matrix home screen and press F2 *[Det]* F1 *[Mat]* ALPHA A EXE. You

should find that the determinant is 2 as shown in Figure 6.97.

Since the determinant of the matrix is not zero, it has an inverse matrix. Press F1 *[Mat]* ALPHA A SHIFT x^{-1} EXE to calculate the inverse. The result is shown in Figure 6.98.

Figure 6.97: Determinant of Mat A

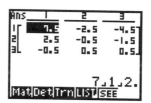

Figure 6.98: Inverse of Mat A

Now let's solve a system of linear equations by matrix inversion. Once again, consider $\begin{cases} x - 2y + 3z = 9 \\ -x + 3y = -4 \\ 2x - 5y + 5z = 17 \end{cases}$.

The coefficient matrix for this system is the matrix $\begin{bmatrix} 1 & -2 & 3 \\ -1 & 3 & 0 \\ 2 & -5 & 5 \end{bmatrix}$ which was entered in the previous example.

Now enter the matrix $\begin{bmatrix} 9 \\ -4 \\ 17 \end{bmatrix}$ as Mat B. Then in matrix mode, press F1 *[Mat]* ALPHA A SHIFT x^{-1} × F1 *[Mat]*

ALPHA B EXE to get the answer as shown in Figure 6.99.

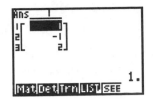

Figure 6.99: Solution matrix

The solution is still $x = 1$, $y = -1$, and $z = 2$.

6.7   Sequences

6.7.1 Iteration with the Ans key: Compute the following in the COMP mode (MENU 1). The Ans feature enables you to perform iterations to evaluate a function repeatedly. As an example, calculate $\frac{n-1}{3}$ for $n = 27$.

Then calculate $\frac{n-1}{3}$ for n = the answer to the previous calculation. Continue to use each answer as n in the *next* calculation. here are keystrokes to accomplish this iteration on the Casio *fx*-7700/9700GE calculator. (See the results in Figure 6.100.) Notice that when you use Ans in place of n in a formula, it is sufficient to press EXE to continue an iteration.

Iteration	Keystrokes	Display
1	27 EXE	27
2	(SHIFT Ans – 1) ÷ 3 EXE	8.666666667
3	EXE	2.555555556
4	EXE	0.5185185185
5	EXE	–0.1604938272

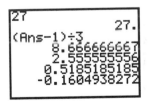

Figure 6.100: Iteration

Press EXE several more times and see what happens with this iteration. You may wish to try it again with a different starting value.

6.7.2 Arithmetic and Geometric Sequences: Use iteration with the Ans variable to determine the n-th term of a sequence. For example, find the 18th term of an *arithmetic* sequence whose first term is 7 and whose common difference is 4. Enter the first term 7, then start the progression with the recursion formula, SHIFT Ans + 4 EXE. This yields the 2nd term, so press EXE sixteen more times to find the 18th term. For a *geometric* sequence whose common ratio is 4, start the progression with SHIFT Ans × 4 EXE.

Of course, you could also use the *explicit* formula for the n-th term of an arithmetic sequence $t_n = a + (n-1)d$. First enter values for the variables a, d, and n, then evaluate the formula by pressing ALPHA A + (ALPHA N − 1) ALPHA D EXE. For a geometric sequence whose n-th term is given by $t_n = a \cdot r^n$, enter values for the variables a, d, and n, then evaluate the formula by pressing ALPHA A ALPHA R ∧ (ALPHA N − 1) EXE.

Casio fx-9700GE: You can also define the sequence recursively with the Casio *fx*-9700GE in the table mode. (This feature is *not* available on the Casio *fx*-7700GE.) From the main menu enter the table mode by selecting the TABLE icon (MENU 8)and then press F2 *[RECR]*. Next press F6 *[TYPE]* F2 *[a$_{n+1}$]* to select the recursion type. Once again, let's find the 18th term of an *arithmetic* sequence whose first term is 7 and whose common difference is 4. Input the recursion formula $a_{n+1} = a_n + 4$ by pressing F4 *[a$_n$]* + 4 (Figure 6.101). Now make $a_1 = 7$ (because the first term is a_1 where $n = 1$) and display a table that contains the 16th term a_{16} to the 20th term a_{20} by pressing F2 *[RANG]* F6 *[a$_1$]* 16 EXE 20 EXE 7 F1 *[TABL]* (Figures 6.102 and 6.103).

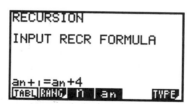

Figure 6.101: Recursion formula

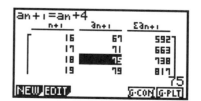

Figure 6.102: TABLE Range   Figure 6.103: $a_{18} = 75$

To use the explicit formula in a Casio *fx*-9700GE recursion table, make $a_n = 7 + (n-1) \cdot 4$ by starting a new table by pressing F1 *[NEW]* F1 *[YES]*. Now press F6 *[TYPE]* F1 *[aₙ]* 7 + (F3 *[n]* – 1) × 4. Once more, calculate term a_{18} by pressing F2 *[RANG]* 18 EXE 18 F1 *[TABL]* (Figures 6.104 – 6.106).

Figure 6.104: Explicit formula

Technology Tip for the Casio fx-9700GE: A table whose starting and ending range values are the same has just one entry. So to display a single *n*-th term series, set both the starting and ending range values to *n*.

There are more detailed instructions for using the table recursion mode in the Casio *fx*-9700 manual.

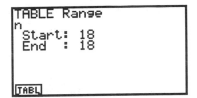

Figure 6.105: TABLE Range

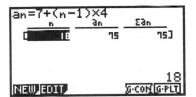

Figure 6.106: $a_{18} = 75$

6.7.3 Finding Sums of Sequences (Casio fx-9700GE): You can use recursion option in the table mode of the Casio *fx*-9700GE to find the sum of a sequence. For example, suppose you want to find the sum $\sum_{n=1}^{12} 4(0.3)^n$. Erase any existing formula by pressing F1 *[NEW]* F1 *[YES]*. Now press F6 *[TYPE]* F1 *[aₙ]* 4 × (.3) ∧ F3 *[n]*. Now you must set the range from 1 to 12 by pressing F2 *[RANG]* 1 EXE 12. Press F1 *[TABL]*. The last entry in the table is the sum (Figure 6.107).

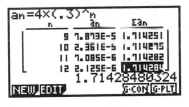

Figure 6.107: $\sum_{n=1}^{12} 4(0.3)^n$

Now calculate the sum starting at $n = 0$ by editing the range. You should obtain a sum of 5.712848.

6.8 Parametric and Polar Graphs

6.8.1 Graphing Parametric Equations: The Casio *fx*-7700/9700GE plots parametric equations as easily as it plots functions. Enter the graph mode by selecting the **GRAPH** icon from the main menu. Change to parametric mode by pressing **F3** *[TYP]* **F3** *[PRM]*. Be sure, if the independent parameter is an angle measure, that the angle measure has been set to whichever you need, **Rad** or **Deg**.

You can now enter the parametric functions. For example, here are the keystrokes need to graph the parametric equations $x = \cos^3 t$ and $y = \sin^3 t$. First check that angle measure is in radians. Then in the graph screen press (**cos** **X,θ,T**) $\wedge$ 3 **F4** *[,]* (**sin** **X,θ,T**) $\wedge$ 3 (Figure 6.108). Now, when you press the variable key **X,θ,T**, you get a **T** because the calculator is in parametric mode. Store the function by pressing **F1** *[STO]* **F6** *[SET]* (Figure 6.109).

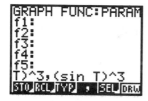

Figure 6.108: Graph screen

Figure 6.109: $x = \cos^3 t$ and $y = \sin^3 t$

Press **RANGE F1** *[INIT]* to set the standard graphing window and to initialize the values of **T**. Press **RANGE** again to see that the values of **T** go from 0 to 2π in steps of $\frac{2\pi}{100} \approx 0.062832$. In order to provide a better viewing rectangle press **RANGE** twice and set the rectangle to go from -2 to 2 horizontally and vertically. **EXIT** back to the graph screen and press **F6** *[DRW]* to draw the graph (Figure 6.110).

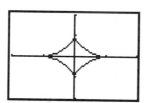

Figure 6.110: Parametric graph of $x = \cos^3 t$ and $y = \sin^3 t$

You may **ZOOM** and **TRACE** along parametric graphs just as you did with function graphs. As you trace along this graph, notice that the cursor moves in the counterclockwise direction as **T** increases.

Note that you can also graph parametric equations in the COMP mode (**MENU 1**) by using the **GRAPH** key.

First clear any existing graph(s) from the graph window by pressing SHIFT F5 *[CLS]* EXE. After setting the graph type as parametric and choosing an appropriate RANGE, press GRAPH (cos X,θ,T) ∧ 3 SHIFT , (sin X,θ,T) ∧ 3 EXE.

6.8.2 Rectangular-Polar Coordinate Conversion:
Conversion between rectangular and polar coordinate systems is accomplished directly through keystrokes on the Casio *fx*-7700/9700GE. These functions use the current angle measure setting, so it is a good idea to check the default angle measure before any conversion. Of course, you may override the current angle measure setting, as explained in Section 6.4.1. For the following examples, the Casio *fx*-7700/9700GE is set to radian measure.

The Casio *fx*-7700/9700GE uses the variables I and J to store the results of a conversion. So going from rectangular to polar coordinates, you get $(r, \theta) = (I, J)$. Going from polar to rectangular, you get $(x, y) = (I, J)$.

We perform these calculations in the COMP mode. To convert between rectangular and polar coordinates, activate the coordinate menu at the bottom of the screen by pressing SHIFT MATH F5 *[COR]*.

Given the rectangular coordinates $(x, y) = (4, -3)$, convert to polar coordinates (r, θ) by pressing F1 *[Pol]* 4 SHIFT , − 3) EXE. The value of *r* is displayed; now press ALPHA J EXE to display the value of θ (Figure 6.111). The polar coordinates are approximately $(5, -0.6435)$.

Suppose $(r, \theta) = (3, \pi)$. Convert to rectangular coordinates (x, y) by pressing F2 *[Rec]* 3 SHIFT , SHIFT π) EXE. The *x*-coordinate is displayed; press ALPHA J EXE to display the *y*-coordinate (Figure 6.112). The rectangular coordinates are $(-3, 0)$.

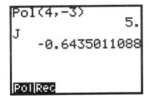

Figure 6.111: Rectangular to polar coordinates   Figure 6.112: Polar to rectangular coordinates

6.8.3 Graphing Polar Equations:
The Casio *fx*-7700/9700GE graphs polar functions in the form $r = f(\theta)$. Enter the graph mode by selecting the GRAPH icon from the main menu (MENU 6). Change to polar mode by pressing F3 *[TYP]* F2 *[POL]*. Be sure that the angle measure has been set to whichever you need, Rad or Deg. Here we will use radian measure. Press RANGE F1 *[INIT]* to initialize the graph window so θ goes from 0 to 2π.

For example, to graph $r = 4\sin\theta$, enter 4 sin X,θ,T in the graph screen. Now, when you press the variable key X,θ,T, you get a θ because the calculator is in polar mode. Store the function by pressing F1 *[STO]* F6 *[SET]*.

Choose a good viewing rectangle and an appropriate interval and increment for θ. In Figure 6.113 for the Casio *fx*-7700GE, the viewing rectangle is roughly "square" and extends from −6 to 6 horizontally and from −4 to 4 vertically. In Figure 6.114 for the Casio *fx*-9700GE, the viewing rectangle is roughly "square" and extends from −10 to 10 horizontally and from −6 to 6 vertically. (Refer back to the Technology Tips in Section 6.2.4.)

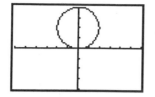

Figure 6.113: Casio *fx*-7700GE
polar graph of $r = 4 \sin \theta$

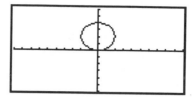

Figure 6.114: Casio *fx*-9700GE
polar graph of $r = 4 \sin \theta$

Trace along this graph to see the polar coordinates of the cursor's location displayed at the bottom of the window. Zooming works just the same as before.

Note that you can also graph polar equations in the COMP mode (MENU 1) by using the GRAPH key. First clear any existing graph(s) from the graph window by pressing SHIFT F5 *[CLS]* EXE. After setting the graph type as polar and choosing an appropriate RANGE, press GRAPH 4 sin X,θ,T EXE.

6.9   Probability

6.9.1 Random Numbers: The command Rn# generates a number between 0 and 1. You will find this command in the PRB (probability) sub-menu of the MATH menu. In the COMP mode (MENU 1) press SHIFT MATH F2 *[PRB]* to activate the probability menu at the bottom of the screen. We will assume that this menu is active for the rest of this section. Then press F4 *[Rn#]* EXE to generate a random number. Press EXE to generate another number; keep pressing EXE to generate more of them.

If you need a random number between, say, 0 and 10, then press 10 F4 *[Rn#]* EXE. To get a random number between 5 and 15, press 5 + 10 F4 *[Rn#]* EXE.

6.9.2 Permutations and Combinations: To calculate the number of permutations of 12 objects taken 7 at a time, $_{12}P_7$, press 12 F2 *[nPr]* 7 EXE (Figure 6.115). Thus $_{12}P_7 = 3,991,680$.

For the number of combinations of 12 objects taken 7 at a time, $_{12}C_7$, press 12 F3 *[nCr]* 7 EXE (Figure 6.115). Thus $_{12}C_7 = 792$.

Figure 6.115: $_{12}P_7$ and $_{12}C_7$

6.9.3 Probability of Winning: A state lottery is configured so that each player chooses six different numbers from 1 to 40. If these six numbers match the six numbers drawn by the State Lottery Commission, the player wins the top prize. There are $_{40}C_6$ ways for the six numbers to be drawn. If you purchase a single lottery ticket, your probability of winning is 1 in $_{40}C_6$. Press 1 ÷ 40 F3 [nCr] 6 EXE to calculate your chances, but don't be disappointed.

6.10   Programming

6.10.1 Entering a Program: The Casio *fx*-7700/9700GE is a programmable calculator that can store sequences of commands for later replay. Here's an example to show you how to enter a useful program that solves quadratic equations by the quadratic formula.

Select the PRGM icon from the main menu (MENU 8 on the Casio *fx*-7700GE and MENU A on your Casio *fx*-9700). On the Casio *fx*-9700 press F1 *[PRGM]*. You should now have a program list on your calculator. (See Figure 6.116 for the Casio *fx*-7700 and Figure 6.117 for the Casio *fx*-9700). The Casio *fx*-7700/9700GE has space for up to 38 programs, each named by a number or letter. If a program location is not used, the word empty appears to the right of its name in the list.

Figure 6.116: Casio *fx*-7700GE
program list

Figure 6.117: Casio *fx*-9700GE
program list

Press the up or down arrow keys to move the cursor to an empty program area; you may also press the key corresponding to a program's name and jump directly there. For example, to go to program 5, press 5; to edit program B, press ALPHA B.

When the cursor is blinking next to the program area you've chosen, press EXE to write a new program in that area or to edit a program that is already there.

Now enter a descriptive title, so press SHIFT A-LOCK and name this program QUADRATIC. Press ALPHA to cancel the alpha lock. Then press EXE to begin writing the actual program. If you do not enter a title, the first line of the program appears in the program list.

Any command you could enter directly in the Casio *fx*-7700/9700GE's computation screen can be entered as a line in a program. There are also special programming commands.

Each time you press EXE while writing a program, the Casio *fx*-7700/9700GE *automatically* inserts the ↵ character at the end of the previous line. For simplicity, since this happens every time you press EXE, the ↵

character is not shown in the program listing below.

Note that while entering a program the program menu can be activated by pressing SHIFT PRM (Figure 6.118). This menu is used to access a variety of commands that are need for writing a program.

Figure 1.118: Program menu

The instruction manual for you Casio *fx*-7700/9700GE gives detailed information about programming. Refer to it to learn more about programming and how to use other features of your calculator.

Enter the program QUADRATIC by pressing the given keystrokes.

Program Line	Keystrokes
"ENTER A"? → A	SHIFT A-LOCK F2 E N T E R SPACE A F2 SHIFT PRGM F4 → ALPHA A EXE

displays the words ENTER A on the screen and waits for you to input a value that will be assigned to the variable A

"ENTER B"? → B	SHIFT A-LOCK F2 E N T E R SPACE B F2 SHIFT PRGM F4 → ALPHA B EXE
"ENTER C"? → C	SHIFT A-LOCK F2 E N T E R SPACE C F2 SHIFT PRGM F4 → ALPHA C EXE
$B^2 - 4AC$ → D	ALPHA B x^2 – 4 ALPHA A ALPHA C → ALPHA D EXE

calculates the discriminate and stores its value as D

$D < 0 \Rightarrow$ Goto 1	ALPHA D F2 F4 0 EXIT F1 F1 F2 1 EXE

tests to see if the discriminant is negative;

if the discriminant is negative, jumps to the line Lbl 1 below; if the discriminant is not negative, continues on to the next line

$D = 0 \Rightarrow$ Goto 2   ALPHA D EXIT F2 F1 0 EXIT F1 F1 F2 2 EXE

> tests to see if the discriminant is zero;

> if the discriminant is zero, jumps to the line Lbl 2 below; if the discriminant is not zero, continues on to the next line

"TWO REAL ROOTS"   SHIFT A-LOCK F2 T W O SPACE R E A L SPACE R O O T S F2 ALPHA EXE

$(-B + \sqrt{D}) \div (2A) \rightarrow M : M\llcorner$ ((−) ALPHA B + SHIFT √ ALPHA D) ÷ (2 ALPHA A) → ALPHA M EXIT F6 ALPHA M F5

> calculates one root and stores it as M, then displays it and pauses

$(-B - \sqrt{D}) \div (2A) \rightarrow N : N$ ((−) ALPHA B − SHIFT √ ALPHA D) ÷ (2 ALPHA A) → ALPHA N EXIT F6 ALPHA N EXE

Goto 3   F1 F2 3 EXE

> jumps to the end of the program

Lbl 1   F3 1 EXE

> jumping point for the Goto command above

"COMPLEX ROOTS"   SHIFT A-LOCK F2 C O M P L E X SPACE R O O T S F2 ALPHA EXE

> displays a message in case the roots are complex numbers

"REAL PART"   SHIFT A-LOCK F2 R E A L SPACE P A R T F2 ALPHA EXE

$-B \div (2A) \rightarrow R : R\llcorner$ (−) ALPHA B ÷ (2 ALPHA A) → ALPHA R F6 ALPHA R F5

> calculates and displays the real part $-\dfrac{b}{2a}$ of the complex roots

"IMAGINARY PART"   SHIFT A-LOCK F2 I M A G I N A R Y SPACE P A R T F2 ALPHA EXE

$\sqrt{-D} \div (2A) \rightarrow I : I$ SHIFT √ (−) ALPHA D ÷ (2 ALPHA A) → ALPHA I F6 ALPHA I EXE

> calculates and displays the imaginary part $\dfrac{\sqrt{-D}}{2a}$ of the complex roots; since D is negative, we must use $-D$ as the radicand

Goto 3   F1 F2 3 EXE

Lbl 2	F3 2 EXE
"DOUBLE ROOT"	SHIFT A-LOCK F2 D O U B L E SPACE R O O T F2 ALPHA EXE

displays a message in case there is a double root

−B ÷ (2A) → M : M	(−) ALPHA B ÷ (2 ALPHA A) → ALPHA M EXIT F6 ALPHA M EXE

the quadratic formula reduces to $-\dfrac{b}{2a}$ when D is zero

Lbl 3	F1 F3 3

When you have finished, press **MENU** to leave the program editor and move on.

If you want to clear a program, enter the program editor again. Move to the program you want to delete, and when the cursor is blinking next to its name, press **F2** to remove it from the calculator's memory.

Casio fx-9700GE: Here, for the Casio *fx*-9700GE, is an alternative version of the program QUADRATIC that also finds roots of quadratic equations while taking advantage of that calculator's ability to work with complex numbers.

Program Line	*Keystrokes*
"ENTER A"? → A	SHIFT A-LOCK F2 E N T E R SPACE A F2 SHIFT PRGM F4 → ALPHA A EXE

displays the words ENTER A on the Casio *fx*-9700GE screen and waits for you to input a value that will be assigned to the variable A

"ENTER B"? → B	SHIFT A-LOCK F2 E N T E R SPACE B F2 SHIFT PRGM F4 → ALPHA B EXE
"ENTER C"? → C	SHIFT A-LOCK F2 E N T E R SPACE C F2 SHIFT PRGM F4 → ALPHA C EXE
B^2 − 4AC → D	ALPHA B x^2 − 4 ALPHA A ALPHA C → ALPHA D EXE

calculates the discriminant and stores its value as D

(−B + √D) ÷ (2A) → M	((−) ALPHA B + SHIFT √ ALPHA D) ÷ (2 ALPHA A) → ALPHA M EXIT F6 ALPHA M EXE

calculates one root and stores it as M

(−B − √D) ÷ (2A) → N	((−) ALPHA B − SHIFT √ ALPHA D) ÷ (2 ALPHA A) → ALPHA N EXIT F6 ALPHA N EXE

D < 0 $\Rightarrow$ Goto 1	ALPHA D F2 F4 0 EXIT F1 F1 F2 1 EXE

tests to see if the discriminant is negative;

if the discriminant is negative, jumps to the line Lbl 1 below; if the discriminant is not negative, continues on to the next line

D = 0 $\Rightarrow$ Goto 2	ALPHA D EXIT F2 F1 0 EXIT F1 F1 F2 2 EXE

tests to see if the discriminant is zero;

if the discriminant is zero, jumps to the line Lbl 2 below; if the discriminant is not zero, continues on to the next line

"TWO REAL ROOTS"	SHIFT A-LOCK F2 T W O SPACE R E A L SPACE R O O T S F2 ALPHA EXE
M⏴	ALPHA M EXIT F5

displays one root and pauses

N	ALPHA N EXE
Goto 3	F1 F2 3 EXE

jumps to the end of the program

Lbl 1	F3 1 EXE

jumping point for the Goto command above

"COMPLEX ROOTS"	SHIFT A-LOCK F2 C O M P L E X SPACE R O O T S F2 ALPHA EXE

displays a message in case the roots are complex numbers

M⏴	ALPHA M EXIT F5
N	ALPHA N EXE
Goto 3	F1 F2 3 EXE
Lbl 2	F3 2 EXE
"DOUBLE ROOT"	SHIFT A-LOCK F2 D O U B L E SPACE R O O T F2 ALPHA EXE

displays a message in case there is a double root

M	ALPHA M EXE
Lbl 3	F1 F3 3

6.10.2 Executing a Program: To run the program you have entered, enter the COMP mode (MENU 1) and then press SHIFT PRGM F3 and the number or letter that it was named; finally, press EXE to run it. If you have forgotten its name, you must go back to the program editor to find the program, then press F1 to run it.

The program has been written to prompt you for values of the coefficients a, b, and c in a quadratic equation $ax^2 + bx + c = 0$. Input a value, then press EXE to continue the program.

If you need to interrupt a program during execution, press AC$^{/ON}$.

The program has been written to prompt you for values of the coefficients a, b, and c in a quadratic equation $ax^2 + bx + c = 0$. Input a value, then press EXE to continue the program.

If you need to interrupt a program during execution, press AC$^{/ON}$.

6.11   Differentiation

6.11.1 Limits: Suppose you need to find this limit: $\lim\limits_{x \to 0} \dfrac{\sin 4x}{x}$. Plot the graph of $f(x) = \dfrac{\sin 4x}{x}$ in a convenient viewing rectangle that contains the point where the function appears to intersect the line $x = 0$ (because you want the limit as $x \to 0$). Your graph should support the conclusion that $\lim\limits_{x \to 0} \dfrac{\sin 4x}{x} = 4$ (Figure 6.119).

To test whether the conclusion that $\lim\limits_{x \to \infty} \dfrac{2x-1}{x+1} = 2$ is reasonable, evaluate $f(x) = \dfrac{2x-1}{x+1}$ for several large positive values of x (since you want the limit as $x \to \infty$). For example, evaluate $f(100)$, $f(1000)$, and $f(10,000)$. Another way to test the conclusion is to examine the graph of $f(x) = \dfrac{2x-1}{x+1}$ in a viewing rectangle that extends over large values of x. See, as in Figure 6.120 (where the viewing rectangle extends horizontally from 0 to 100), whether the graph is asymptotic to the horizontal line $y = 2$. Enter $\dfrac{2x-1}{x+1}$ for Y1 and 2 for Y2.

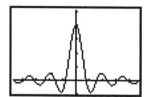

Figure 6.119: Checking $\lim\limits_{x \to 0} \dfrac{\sin 4x}{x} = 4$

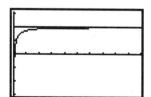

Figure 6.120: Checking $\lim\limits_{x \to \infty} \dfrac{2x-1}{x+1} = 2$

Casio *fx*-7700/9700GE **Power Graphic Calculators**

6.11.2 Numerical Derivatives: The derivative of a function *f* at *x* can be defined as the limit of the slopes of secant lines, so $f'(x) = \lim_{\Delta x \to 0} \dfrac{f(x + \Delta x) - f(x - \Delta x)}{2\Delta x}$. And for small values of Δx, the expression $\dfrac{f(x + \Delta x) - f(x - \Delta x)}{2\Delta x}$ gives a good approximation to the limit.

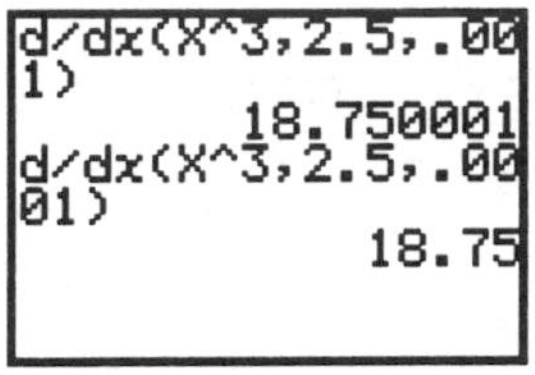

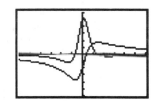

Figure 6.121: Using *d/dx*

The Casio *fx*-7700/9700GE has a function which is accessed by pressing SHIFT *d/dx* that will calculate the symmetric difference, $\dfrac{f(x + \Delta x) - f(x - \Delta x)}{2\Delta x}$. So to find a numerical approximation to $f'(2.5)$ when $f(x) = x^3$ and with $\Delta x = 0.001$, enter the calculation mode (MENU 1) and press SHIFT *d/dx* x,θ,T ∧ 3 SHIFT , 2.5 SHIFT , .001) EXE as shown in Figure 6.121. The format of this command is d/dx(*expression, variable, value,* Δx). The same derivative is also approximated in Figure 6.121 using $\Delta x = 0.0001$. For most purposes, $\Delta x = 0.001$ gives a very good approximation to the derivative.

Technology Tip: It is sometimes helpful to plot both a function and its derivative together. In Figure 6.123, the function $f(x) = \dfrac{5x - 2}{x^2 + 1}$ and its numerical derivative (actually, an approximation to the derivative given by the symmetric difference) are graphed on viewing window that extends from −6 to 6 vertically and horizontally. In the graph mode (MENU 6), enter $\dfrac{5x - 2}{x^2 + 1}$ for Y1. In the edit line of the graph mode, press SHIFT *d/dx* SHIFT VAR F1 *[GRP]* (F3 *[GRPH]* on the Casio *fx*-9700GE) F1 *[Y]* 1 SHIFT , x,θ,T SHIFT , .001) EXIT EXIT (Figure 6.122). Then store this as Y2.

Figure 6.122: Entering $f(x)$ and $f'(x)$

Figure 6.123: Graphs of $f(x)$ and $f'(x)$

Casio fx-9700GE: The Casio *fx*-9700GE can compute the derivative of a point on a graph drawn in the graph mode. For example, enter the graph mode (MENU 6) and graph the function $f(x) = \dfrac{5x-2}{x^2+1}$ in the standard viewing rectangle. Then press SHIFT G-SOLV F6 [⚫] F3 [d/dx]. The coordinates for the left-most x-value of the range will appear, along with the derivative at that point above the y-coordinate. You can use the left and right arrow keys to move to a specific point, say $x = -2.3$. Figure 6.124 shows the derivative at that point to be about -0.774692.

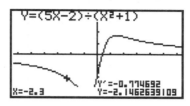

Figure 6.124: Derivative of $f(x) = \dfrac{5x-2}{x^2+1}$ at $x = -2.3$

If more than one function is graphed you can use ▲ and ▼ to scroll between the functions.

6.11.3 Newton's Method: With the Casio *fx*-7700/9700GE, you may iterate using Newton's method to find the zeros of a function. Recall that Newton's Method determines each successive approximation by the formula $x_{n+1} = x_n - \dfrac{f(x_n)}{f'(x_n)}$.

As an example of the technique, consider $f(x) = 2x^3 + x^2 - x + 1$. In the calculation mode (MENU 1) enter this function in the function memory as f1 (refer back to Section 6.2.1). Set the range to the standard viewing window and clear any graphs from the window (SHIFT F5 [CLS] EXE). Graph the function in the calculation mode by pressing GRAPH SHIFT F-MEM F3 [fn] 1 EXE. A look at the graph suggests that it has a zero near $x = -1$, so start the iteration by storing -1 as x. Then press these keystrokes: x,θ,T – SHIFT F-MEM F3 [fn] 1 ÷ SHIFT d/dx F3 [fn] 1 SHIFT , x,θ,T SHIFT , .001) → x,θ,T EXE EXE (Figure 6.125) to calculate the first two iterations of Newton's method. Press EXE repeatedly until two successive approximations differ by less than some predetermined value, say 0.0001. Note that each time you press EXE, the Casio will use the *current* value of x, and that value is changing as you continue the iteration.

Figure 6.125: Newton's method

Technology Tip: Newton's Method is sensitive to your initial value for x, so look carefully at the function's graph to make a good first estimate. Also, remember that the method sometimes fails to converge!

You may want to write a short program for Newton's Method. See your calculator's manual for further information.

6.12   Integration

6.12.1 Approximating Definite Integrals: The Casio *fx*-7700/9700GE has a function which is accessed by pressing SHIFT ∫dx that will approximate a definite integral. For example, to find a numerical approximation to $\int_0^1 \cos x^2\, dx$ first enter the calculation mode (MENU 1) and then press SHIFT ∫dx cos X,θ,T x^2 SHIFT , 0 SHIFT , 1 EXE (first two lines in Figure 6.126). The Casio *fx*-7700/9700GE uses a method known as Simpson's Rule to perform the calculation. The format of this command is ∫(*expression, lower limit, upper limit, n*), where the last optional entry n is an integer from 1 to 9 which gives the number of intervals 2^n used in computing the integral. If n is not specified the calculator will automatically assign a value. The last two lines in Figure 6.126 shows the calculation when n is 7.

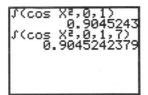

Figure 6.126: Using ∫dx

6.12.2 Areas: You may approximate the area under the graph of a function $y = f(x)$ between $x = A$ and $x = B$ with your Casio *fx*-7700/9700GE. To do this you must be in the calculation mode (MENU 1) and access the command by pressing SHIFT G-∫dx. For example, here are the keystrokes for finding the area under the graph of the function $y = \cos x^2$ between $x = 0$ and $x = 1$. The area is represented by the definite integral $\int_0^1 \cos x^2\, dx$. Press SHIFT G-∫dx cos X,0,T x^2 SHIFT , 0 SHIFT , 1 (Figure 6.127) followed by EXE to draw the graph. Notice that these keystrokes are nearly identical to the keystrokes used above in computing the integral; the difference is from using the command G-∫dx instead of ∫dx. As you could add a third value n to your command before executing to specify the number of intervals used in computing the integral. The range in Figure 6.128 extends from −5 to 5 horizontally and from −2 to 2 vertically. If you need to change your range press the RANGE key, enter your range, press EXIT to enter the calculation screen, and then press EXE to redraw.

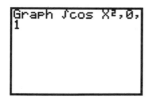

Figure 6.127: Using G-∫dx

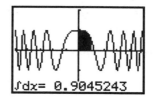

∫dx= 0.9045243

Figure 6.128: Graph and area

Technology Tip: Suppose that you want to find the area between two functions, $y = f(x)$ and $y = g(x)$ from $x = A$ and $x = B$. If $f(x) \geq g(x)$ for $A \leq x \leq B$, then graph the expression $f(x) - g(x)$ in the manner described above to find the required area.

Casio *fx*-7700/9700GE Power Graphic Calculators

Chapter 7

Casio CFX-9800G

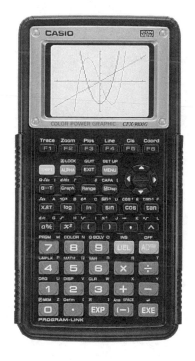

7.1   Getting started with the Casio *CFX*-9800G

7.1.1 Basics: Press the AC/^{ON} key to begin using your Casio *CFX*-9800G. The main menu screen will appear on your calculator (Figure 7.1).

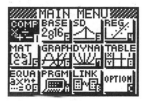

Figure 7.1: Casio *CFX*-9800G main menu

If you need to adjust the display contrast, select the OPTION icon from the main menu by using the arrow keys and press EXE, or by pressing C (Figure 7.2). Then press EXE again and press ◀ (the *left* arrow key) to lighten and ▶ (the *right* arrow key) to darken (Figure 7.3). If you wish to adjust the tint of the three colors, use the ▲ (the *up* arrow key) or ▼ (the *down* arrow key) to scroll the color you wish to adjust, and then use ◀ and ▶.

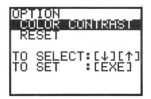

Figure 7.2: OPTION screen

Figure 7.3: CONTRAST

When you have finished with the calculator, turn it off to conserve battery power by pressing SHIFT and then OFF. Power is automatically switched off approximately six minutes after the last operation.

Technology Tip: To return to the main menu you can press the MENU key. In general, whenever you need to return to the main menu, you can press the MENU key. Note that to enter any mode from the main menu, you just need to select the appropriate number or letter. Thus, you can jump quickly to the contrast screen by pressing MENU C.

Check your Casio *CFX*-9800G's calculation mode settings by selecting COMP from the main menu (Menu 1). The following screen showing the COMP settings should appear (Figure 7.4).

Press SHIFT SET UP to change the G-type, Angle, Display and M-D/Cpy settings as shown (Figure 7.4). Use the up and down arrow keys to scroll through the three categories. When the cursor is at G-type press

F1 to select the rectangular coordinates mode and F5 for connected graphs. When the cursor is at Angle press F2 to select the radians mode. When the cursor is at Display press F3 for normal display (if the option changes to Nrm2 instead of Nrm1, just press F3 again). When the cursor is at M-D/Cpy press F1 to select M-DISP. (Note while both Figure 7.4 displays MONOCHR, this was in order to copy the image of the screen into this document.) Setting the option to MDisp allows you to see the current settings when you press the MDisp key. EXIT to return to the COMP screen and then press AC$^{/ON}$ to clear the screen.

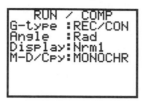

Figure 7.4: COMP settings

7.1.2 *Editing*: One advantage of the Casio *CFX*-9800G is that up to seven lines are visible at one time, so you can *see* a long calculation. For example, enter the calculation mode (MENU 1) and type this sum (Figure 7.5):

$$1 + 2 + 3 + 4 + 5 + 6 + 7 + 8 + 9 + 10 + 11 + 12 + 13 + 14 + 15 + 16 + 17 + 18 + 19 + 20$$

Then press EXE to see the answer too.

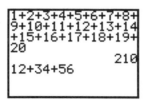

Figure 7.5: Calculation screen

Often we do not notice a mistake until we see how unreasonable an answer is. The Casio *CFX*-9800G permits you to redisplay an entire calculation, edit it easily, then execute the *corrected* calculation.

Suppose you had typed 12 + 34 + 56 as in Figure 7.5 but had *not yet* pressed EXE, when you realize that 34 should have been 74. Simply press ◄ (the *left* arrow key) as many times as necessary to move the blinking cursor left to 3, then type 7 to write over it. On the other hand, if 34 should have been 384, move the cursor back to 4, press SHIFT INS (the cursor changes to a blinking frame) and then type 8 (inserts at the cursor position and the other characters are pushed to the right). If the 34 should have been 3 only, move the cursor to 4, and press DEL to delete it.

Even if you had pressed EXE, you may still edit the previous expression. Press the *left* or *right* arrow key to *redisplay* the last expression that we entered. Now you can change it. If you press ◄, the cursor will be at the *end* of the previous expression; if you press ► the cursor will appear at the *beginning*. Even if you have already pressed some keys since the last EXE, but *not* EXE again, you can still recall the previous expression by first pressing AC$^{/ON}$ to clear the screen and then pressing ◄ or ►.

In fact the Casio *CFX*-9800G retains many prior entries. After pressing AC$^{/ON}$ to clear the screen, press ▲ repeatedly to cycle back through previous expressions. If you pass by an expression that you want, just press ▼ as many times as necessary to cycle forward.

Technology Tip: When you need to evaluate a formula for different values of a variable, use the editing feature to simplify the process. For example, suppose you want to find the balance in an investment account if there is now $5000 in the account and interest is compounded annually at the rate of 8.5%. The formula for the balance is $P = \left(1 + \frac{r}{n}\right)^{nt}$, where P = principal, r = rate of interest (expressed as a decimal), n = number of times interest is compounded each year, and t = number of years. In our example, this becomes $5000(1+.085)^t$. Here are the keystrokes for finding the balance after t = 3, 5, and 10 years. Figure 7.6 shows the first set of keystrokes and the result.

Years	Keystrokes	Balance
3	5000 (1 + .085) ∧ 3 EXE	$6386.45
5	◄ ◄ 5 EXE	$7518.28
10	◄ ◄ 10 EXE	$11,304.92

Figure 7.6: Editing expressions

Then to find the balance from the same initial investment but after 5 years when the annual interest rate is 7.5%, press the keys to change the last calculation above: ◄ ◄ DEL ◄ 5 ◄ ◄ ◄ ◄ ◄ 7 EXE.

7.1.3 Key Functions: Most keys on the Casio *CFX*-9800G offer access to more than one function, just as the keys on a computer keyboard can produce more than one letter ("g" and "G") or even quite different characters ("5" and "%"). The primary function of a key is indicated on the key itself, and you access that function by a simple press on the key.

To access the *second* function indicated to the *left* above a key, first press SHIFT (the cursor changes to a blinking **S** and a menu appears at the bottom of the screen) and *then* press the key. For example to calculate

$\sqrt{25}$, press SHIFT $\sqrt{}$ 25 EXE.

When you want to use a letter or other character printed to the *right* above a key, first press ALPHA (the cursor changes to a blinking **A** and a menu appears at the bottom of the screen) and then the key. For example, to use the letter K in a formula, press ALPHA K. If you need several letters in a row, press SHIFT A-LOCK, which is like the CAPS LOCK key on a computer keyboard, and then press all the letters you want. Remember to press ALPHA when you are finished and want to restore the keys to their primary functions.

7.1.4 Order of Operations: The Casio *CFX*-9800G performs calculations according to the standard algebraic rules. Working outwards from inner parentheses, calculations are performed from left to right. Powers and roots are evaluated first, followed by multiplications and divisions, and then additions and subtractions.

Enter these expressions to practice using your Casio *CFX*-9800G.

Expression	Keystrokes	Display
$7 - 5 \cdot 3$	7 – 5 × 3 EXE	–8
$(7-5) \cdot 3$	(7 – 5) × 3 EXE	6
$120 - 10^2$	120 – 10 x² EXE	20
$(120-10)^2$	(120 – 10) x² EXE	12100
$\dfrac{24}{2^3}$	24 ÷ 2 ∧ 3 EXE	3
$\left(\dfrac{24}{2}\right)^3$	(24 ÷ 2) ∧ 3 EXE	1728
$(7 - -5) \cdot -3$	(7 – –5) × – 3 EXE	–36

7.1.5 Algebraic Expressions and Memory: Your calculator can evaluate expressions such as $\dfrac{N(N+1)}{2}$ *after* you have entered a value for N. Suppose you want $N = 200$. Press 200 SHIFT → ALPHA N EXE to store the value 200 in memory location N. Whenever you use N in an expression, the calculator will substitute the value 200 until you make a change by storing *another* number in N. Next enter the expression $\dfrac{N(N+1)}{2}$ by typing ALPHA N (ALPHA N + 1) ÷ 2 EXE. For $N = 200$, you will find that $\dfrac{N(N+1)}{2} = 20100$.

The contents of any memory location may be revealed by typing just its letter name and then EXE. And the Casio *CFX*-9800G retains memorized values even when it is turned off, so long as its batteries are good.

7.1.6 Repeated Operations with Ans: The result of your *last* calculation is always stored in memory location Ans and replaces any previous result. This makes it easy to use the answer from one computation in another computation. For example, press 30 + 15 EXE so that 45 is the last result displayed. Then press SHIFT Ans ÷ 9 EXE and get 5 because 45 ÷ 9 = 5.

With a function like division, you press the ÷ *after* you enter an argument. For such functions, whenever you start a new calculation with the previous answer followed by pressing the function key, you may press just the function key. So instead of SHIFT Ans ÷ 9 in the previous example, you could have pressed simply ÷ 9 to achieve the same result. This technique also works for these functions: $+$ $-$ $\times$ $\wedge$ x^2 x^{-1}.

Here is a situation where this is especially useful. Suppose a person makes $5.85 per hour and you are asked to calculate earnings for a day, a week, and a year. Execute the given keystrokes to find the person's incomes during these periods (Figure 7.7).

Pay Period	Keystrokes	Earnings
8-hour day	5.85 × 8 EXE	$46.80
5-day week	SHIFT Ans × 5 EXE	$234
52-week year	× 52 EXE	$12,168

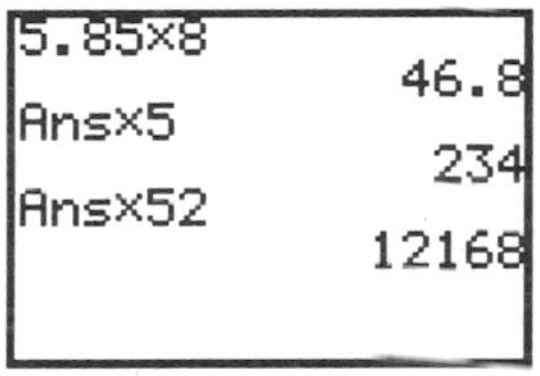

Figure 7.7: SHIFT Ans key

In general, the Casio *CFX*-9800G does not distinguish between the negative sign and the subtraction operator. But when you enter −4 as the *first* number in a calculation, you must use the negative key (−) rather than the − key. Press these keys for an illustration: 8 EXE − 5 EXE (−) 5 EXE.

7.1.7 The MATH Menu: Operators and functions associated with a scientific calculator are available either immediately from the keys of the Casio *CFX*-9800G or by the SHIFT keys. You have direct access to common arithmetic operations (x^2, SHIFT $\sqrt{\ }$, SHIFT x^{-1}, $\wedge$), trigonometric functions (sin, cos, tan), and their inverses (SHIFT sin^{-1}, SHIFT cos^{-1}, SHIFT tan^{-1}), exponential and logarithmic functions (log, SHIFT 10^X, ln, SHIFT e^X), and a famous constant (SHIFT π).

A significant difference between the Casio *CFX*-9800G graphing calculators and most scientific calculators is that the Casio *CFX*-9800G requires the argument of a function *after* the function, as you would see in a formula written in your textbook. For example, on the Casio *CFX*-9800G you calculate $\sqrt{16}$ by pressing the keys $\sqrt{\ }$ 16 in that order.

The Casio *CFX*-9800G has a special fraction key a^b/$_c$ for entering fractions and mixed numbers. To enter a fraction such as $\frac{2}{5}$, press 2 $a^b/_c$ 5 EXE. To enter a mixed number like $2\frac{3}{4}$, press 2 $a^b/_c$ 3 $a^b/_c$ 4 EXE. Press $a^b/_c$ to toggle between the mixed number and its decimal equivalent; press SHIFT $^d/_c$ and see $2\frac{3}{4}$ as an improper fraction, $\frac{11}{4}$.

Here are keystrokes for basic mathematical operations. Try them for practice on your Casio *CFX*-9800G.

Expression	Keystrokes	Display
$\sqrt{3^2 + 4^2}$	SHIFT $\sqrt{\ }$ (3 x^2 + 4 x^2) EXE	5
$2\frac{1}{3}$	2 $a^b/_c$ 1 $a^b/_c$ 3 EXE $a^b/_c$	2.333333333
log 200	LOG 200 EXE	2.301029996
$2.34 \cdot 10^5$	2.34 × SHIFT 10^x 5 EXE	234000

Additional mathematical operations and functions are available from the MATH menu. Press SHIFT MATH to see the six categories of mathematical functions. They are listed across the bottom of the Casio *CFX*-9800G's screen and correspond to the function keys, F1 to F6 (Figure 7.8).

You will learn in your mathematics textbook how to apply many of them. As an example, calculate $|-5|$ by pressing, with the MATH menu already on display, F3 *[NUM]* F1 *[Abs]* (−) 5 EXE (Figure 7.9). To clear a menu from the screen, press EXIT. In this case, you press EXIT twice, the first time to clear the NUM menu to get back to the MATH menu and then to clear the MATH menu.

Figure 7.8: MATH menu

Figure 7.9: MATH NUM menu

The *factorial* of a non-negative integer is the *product* of *all* the integers from 1 up to the given integer. The symbol for factorial is the exclamation point. So 4! (pronounced *four factorial*) is $1 \cdot 2 \cdot 3 \cdot 4 = 24$. You will learn more about applications of factorials in your textbook, but for now use the Casio *CFX*-9800G to calculate 4! Press these keystrokes: 4 SHIFT MATH F2 *[PRB]* F1 *[x!]* EXE

On the Casio *CFX*-9800G it is also possible to do calculations with complex numbers. Press SHIFT CMPLX to activate the complex number calculation menu at the bottom of the screen. For example, to divide $2 + 3i$ by $4 - 2i$, press (2 + 3 F1 *[i]*) ÷ (4 − 2 F1 *[i]*) EXE. The result is $0.1 + 0.8i$ (Figure 7.10).

To find the complex conjugate of $4 + 5i$ press F4 *[Conj]* (4 + 5 F1 *[i]*) EXE (Figure 7.10).

Casio *CFX*-9800G Color Power Graphic Calculator

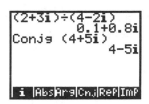

Figure 7.10: Complex number calculations

The Casio *CFX*-9800G can also solve for the real and complex roots of a quadratic or cubic function in the equation mode (MENU 9). The functions to be solved must be in the form $f(x) = ax^3 + bx^2 + cx + d$ or $f(x) = ax^2 + bx + c$ where $a \neq 0$. Enter the equation mode and press F2 *[PLY]*. You are then prompted for the degree (either 2 or 3) of your polynomial (Figure 7.11).

Figure 7.11: Prompt for degree

For example, to find all the zeros of $f(x) = x^3 - 4x^2 + 14x - 20$ select F2 *[3]* at the prompt and then enter the coefficients by pressing 1 EXE (–) 4 EXE 14 EXE (–) 20 EXE (Figure 7.12). If you had not pressed EXE yet, you can change a coefficient by pressing AC$^{/ON}$ and entering a new value. Otherwise, move your cursor over the coefficient with your arrow keys and enter a new value. Now press F1 *[SOL]* and the calculator will display the solutions (Figure 7.13).

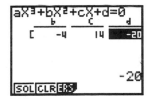

Figure 7.12: Entering the coefficients

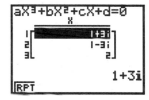

Figure 7.13: Zeros of $f(x) = x^3 - 4x^2 + 14x - 20$

To perform a new calculation press F1 *[RPT]*. If you are computing the roots of another cubic function then you can either edit the existing coefficients or press F2 *[CLR]* to reset all the coefficients to zero. If you are computing the roots of a quadratic function, press F3 *[ERS]* F1 *[YES]* F1 *[2]* and then enter in the coefficients of the quadratic and procced.

Note that it may take considerable time for the calculation result of a cubic equation to appear on the display. Failure of a result to appear immediately does not mean that the unit is not functioning properly.

7.2　Functions and Graphs

7.2.1 Evaluating Functions:　Suppose you receive a monthly salary of $1975 plus a commission of 10% of sales. Let $x =$ your sales in dollars; then your wages W in dollars are given by the equation $W = 1975 + .10x$. If your January sales were $2230 and your February sales were $1865, what was your income during those months?

Here's one method to use your Casio *CFX*-9800G to perform this task. First enter COMP mode (MENU 1) and press AC/ON to clear your screen. Then set $x = 2230$ by pressing 2230 SHIFT → X,θ,T. (The X,θ,T key lets you enter a variable x without having to use the ALPHA key.) Then press SHIFT ↵ to allow another expression to be input on a single command line. Finally, enter the expression $1975 + .10x$ by pressing these keys: 1975 + .10 X,θ,T. Now press EXE to calculate the answer (Figure 7.14).

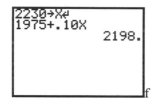

Figure 7.14: Evaluating a function

It is not necessary to repeat all these steps to find the February wages. Simply press ▶ to recall the entire previous line, change 2230 to 1865, and press EXE.

Another method is to use the function memory. The Casio *CFX*-9800G can store up to six functions. First be in the COMP mode and press AC/ON to clear your screen. Press SHIFT F-MEM to display the function memory menu at the bottom of the screen. Enter the expression $1975 + .10x$ by pressing these keys: 1975 + .10 X,θ,T. Then store this as function memory number 1 by pressing F1 *[STO]* F1 *[f1]* (Figure 7.15). Press AC/ON to clear the screen, leaving the function memory menu. Then set $x = 2230$ by pressing 2230 SHIFT → X,θ,T EXE. Recall the entire expression by pressing F2 *[RCL]* F1 *[f1]*, and then press EXE to calculate the answer (Figure 7.16). To find the February wages, set $x = 1865$ and then evaluate the function for the new value by pressing F2 *[RCL]* F1 *[f1]* EXE. By pressing F2 F1 you recall the entire expression. You can also press F3 F1 to get the variable f1.

In general, to *store* a function enter it first in the calculation screen, but do *not* press EXE. If the function memory menu is active, press F1 and the appropriate function key from F1 to F2, otherwise you must press SHIFT F-MEM before pressing F1. To *recall* a function from the calculation screen press F2 and the function key corresponding to the function you want. Press F5 to see the list of functions currently in the function memory.

　　　　　　　　　　　　　　　　　　　　Casio *CFX*-9800G Color Power Graphic Calculator

Figure 7.15: Storing a function

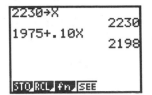

Figure 7.16: Evaluating a function

With the Casio *CFX*-9800G, you can also use your calculator's TABLE mode to create a table of values for a function. From the main menu select the TABLE icon (MENU 8) to get the TABLE&GRAPH screen (Figure 7.17).

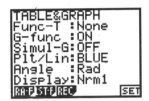

Figure 7.17: TABLE&GRAPH screen

Then press F1 *[RA-F]* to get the range function screen where you can store up to 30 functions. Move the highlight so Y1 is selected. and enter the function $1975 + .10x$ by pressing these keys: 1975 + .10 X,θ,T EXE (Figure 7.18). Press F5 *[RNG]* to set the conditions for the *x*-variable when generating a function table. Strt is the starting value of the *x*-variable, End is the ending value of the *x*-variable, and ptch is the change of the *x*-variable. In this case set Strt to 2230, End to 2230 and ptch to 0 (Figure 7.19).

Figure 7.18: RANGE FUNCTION screen

Figure 7.19: TABLE Range

Press EXIT F6 *[TBL]* to see the table (Figure 7.20). We will now add $x = 1865$ to the table by pressing F2 *[ROW]* F3 *[ADD]* 1865 EXE (Figure 7.21).

Press EXIT to get the previous menu at the bottom of the screen. To add more entries use the ROW and ADD commands as described above. If you wish to change the function press EXIT to return to the range function screen. You can also add other functions to the list and select which ones are to be evaluated by pressing F4 *[SEL]*.

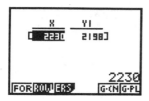

Figure 7.20: Initial table of values

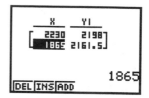

Figure 7.21: Evaluating a function in a table

Technology Tip: The Casio *CFX*-9800G does not require multiplication to be expressed between variables, so *xxx* means x^3. It is often easier to press two or three *x*'s together than to search for the square key or the powers key. Of course, expressed multiplication is also not required between a constant and a variable. Hence to enter $2x^3 + 3x^2 - 4x + 5$ in the Casio *CFX*-9800G, you might save keystrokes and press just these keys: 2 X,θ,T X,θ,T X,θ,T + 3 X,θ,T X,θ,T − 4 X,θ,T + 5

7.2.2 Functions in a Graph Window: On the Casio *CFX*-9800G, you can easily generate the graph of a function. The ability to draw a graph contributes substantially to our ability to solve problems.

For example, here is how to graph $y = -x^3 + 4x$ from the COMP mode. First press Graph and then (−) X,θ,T ∧ 3 + 4 X,θ,T to enter the function (as in Figure 7.22). Now press EXE and the Casio *CFX*-9800G changes to a window with the graph of $y = -x^3 + 4x$ (Figure 7.23).

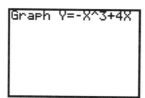

Figure 7.22: Graph command in COMP mode

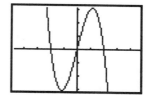

Figure 7.23: Graph of $y = -x^3 + 4x$

While the Casio *CFX*-9800G is busy calculating coordinates for a plot, it displays a solid square at the top right of the graph window. When you see this indicator, even though the screen does not change, you know that the calculator is working.

Switch back and forth between the graph window and the home screen by pressing G↔T.

The graph window on the Casio *CFX*-9800G may look like the one in Figure 7.23, Since the graph of $y = -x^3 + 4x$ extends infinitely far left and right and also infinitely far up and down, the Casio *CFX*-9800G can only display only a piece of the actual graph. This displayed rectangular part is called a *viewing rectangle*.

Casio *CFX*-9800G Color Power Graphic Calculator

You can easily change the viewing rectangle to enhance your investigation of a graph. For example, press any of the arrow keys to pan the graph window in the corresponding direction. If you press the down arrow, for example, the window will pan down so that you may look at points below the current window.

The viewing rectangle for the Casio *CFX*-9800G in Figure 7.23 shows the part of the graph that extends horizontally form –4.7 to 4.7 and vertically from –3.1 to 3.1. Press RANGE to see information about your viewing rectangle. Figures 7.24 shows the RANGE screen that corresponds to the viewing rectangle in Figure 7.23. This is the *standard* viewing rectangle for the Casio *CFX*-9800G.

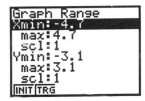

Figure 7.24: standard RANGE

The variables Xmin and Xmax are the minimum and maximum *x*-values of the viewing rectangles; Ymin and Ymax are the minimum and maximum *y*-values.

Xscl and Yscl set the spacing between tick marks on the axes.

Use the arrow keys ▲ and ▼ to move up and down from one line to another in this list; pressing the EXE key will move down the list. Enter a new value to over-write a previous value and then press EXE. Note that the DEL key does not work when you are in the RANGE menu. To leave the RANGE menu, press the EXIT key. Finally, press EXE to redraw the graph. The following figures show different RANGE screens and the corresponding viewing rectangle for each one.

To initialize the viewing rectangle quickly to the *standard* viewing rectangle, press RANGE F1 *[INIT]* EXIT. Then press EXE to redraw the graph.

As you pan over the graph by pressing the arrow keys, the RANGE to dimensions are updated automatically. More information about graph windows is presented later in Section 7.2.4.

Figure 7.25: –10 to 10 in both directions

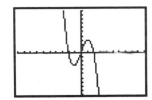

Figure 7.26: Graph of $y = -x^3 + 4x$

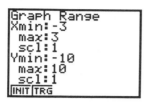

Figure 7.27: Custom window

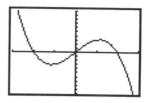

Figure 7.28: Graph of $y = -x^3 + 4x$

Technology Tip: Clear any graphs drawn COMP mode by pressing F5 when the Casio *CFX*-9800G is showing the graph screen or SHIFT F5 when it is displaying the calculation screen. (When you press SHIFT the graph commands are displayed at bottom, and you select the clear screen command Cls.)

If you are going to use a function later, or if you need to perform some calculations before returning to its graph, save it in the Casio *CFX*-9800G's function memory as described in Section 7.2.1.

Technology Tip: It's a good idea to reserve at least one function memory location, say f1, for temporary storage of functions, and use the remaining locations for longer-term storage.

Another procedure for graphing on the Casio *CFX*-9800G is to use the graph mode. To enter the graph mode select the GRAPH icon (MENU 6) from the main menu. The graph function (GRAPH FUNC) settings should appear as in Figure 7.29; press EXIT. If not, press F6 *[SET]* and then select the connected and rectangular mode for drawing by pressing F1 *[REC]* and F5 *[CON]*. Scroll down and select the appropriate choice for the other options. Press EXIT to return to the main GRAPH screen. Erase any existing functions by scrolling to the desired function and pressing F2 *[DEL]*. You can store up to 30 functions in memory. To enter $y = -x^3 + 4x$, move the highlight to Y1, and then press (–) X,θ,T ∧ 3 + 4 X,θ,T EXE. The screen in Figure 7.30 will appear. Now press F6 *[DRW]* to draw the graph.

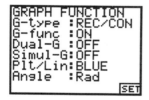

Figure 7.29: GRAPH FUNC settings

Figure 7.30: GRAPH FUNC screen

As before, you can switch back and forth between the graph window and the graph function screen by pressing G↔T.

For editing functions or graphing multiple functions it is preferable to use the GRAPH mode to draw functions. Unless otherwise noted, we will be using the GRAPH mode for our graphs.

To change a function in the GRAPH mode, recall it to the edit line by moving the highlight to the function

Casio *CFX*-9800G Color Power Graphic Calculator

you want to edit and press F1 *[EDIT]*. Edit the function as desired and press EXE when you are done.

Change the range in GRAPH mode just as you did in COMP mode.

7.2.3 Graphing Step and Piecewise Defined Functions: The greatest integer function, written $[[x]]$, gives the greatest *integer* less than or equal to a number x. On the Casio *CFX*-9800G, the greatest integer function is called Intg and is located as F5 under the NUM sub-menu of the MATH menu (see Figure 7.9). Calculate $[[6.78]] = 6$ in the COMP mode by pressing SHIFT MATH F3 *[NUM]* F5 *[Intg]* 6.78 EXE.

To graph $y = [[x]]$, enter the GRAPH mode (MENU 6 EXIT) and press SHIFT MATH F3 *[NUM]* F5 *[Intg]* X,θ,T EXE SHIFT QUIT F6 *[DRW]*. Figure 7.31 shows this graph in a viewing rectangle from −5 to 5 in both directions.

The true graph of the greatest integer function is a step graph, like the one in Figure 7.32. For the graph of $y = [[x]]$, a segment should *not* be drawn between every pair of successive points. You can change from CON (connected) to PLT (plotted) by pressing SHIFT SET UP F6 *[PLT]*. Press EXIT and then F6 to draw the new graph (Figure 7.32).

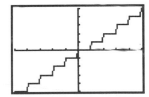

Figure 7.31: Connected graph of $y = [[x]]$

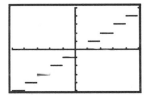

Figure 7.32: Plotted graph of $y = [[x]]$

In general, you'll want the graph to be connected, so do that by returning to the set up screen and pressing F5 *[CON]*.

It is not possible to graph piecewise defined functions on the Casio *CFX*-9800G.

7.2.4 Graphing a Circle: Here is a useful technique for graphs that are not functions but can be "split" into a top part and a bottom part, or into multiple parts. Suppose you wish to graph the circle of radius 6 whose equation is $x^2 + y^2 = 36$. First solve for y and get an equation for the top semicircle, $y = \sqrt{36 - x^2}$, and for the bottom semicircle, $y = -\sqrt{36 - x^2}$. Then graph the two semicircles simultaneously.

Use the following keystrokes to draw this circle's graph in the GRAPH mode. Store $\sqrt{36 - x^2}$ as Y1 by scrolling to Y1 and pressing SHIFT $\sqrt{\ }$ (36 − X,θ,T x^2) EXE. Then store $-\sqrt{36 - x^2}$ as Y2 by scrolling to Y2 and pressing (−) SHIFT $\sqrt{\ }$ (36 − X,θ,T x^2) EXE (Figure 7.33). Next press F6 *[DRW]* to graph both halves (Figure 7.34). Make sure that the RANGE is set large enough to display a circle of radius 6.

Figure 7.33: Two semicircles

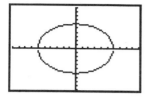

Figure 7.34: One view of circle's graph

Instead of entering $-\sqrt{36-x^2}$ as Y2, you could have entered $-$Y1 as Y2 and saved some keystrokes. Try this by going into the GRAPH screen, scrolling to Y2 pressing $(-)$ SHIFT VAR F3 *[GPH]* F1 *[Y]* 1 EXE. The graph should be as before. The VAR menu (displayed along the bottom of the screen in Figure 7.35) enables you to recall graphic functions and other information from memory.

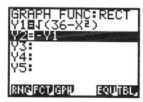

Figure 7.35: Using the VAR menu

If your range were set to a viewing rectangle extending from -10 to 10 in both directions, your graph would look like Figure 7.34. Now this does *not* look a circle, because the units along the axes are not the same. You need what is called a "square" viewing rectangle.

The Casio *CFX*-9800G's standard viewing rectangle is square but too small to display a circle of radius 6, so you should double the dimensions of the standard viewing window. Change the range to extend horizontally from -9.4 to 9.4 and vertically from -6.2 to 6.2 (Figure 7.36). The graph for the better circle is shown in Figure 7.37.

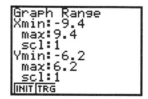

Figure 7.36: Twice standard range

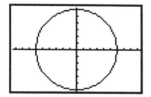

Figure 7.37: Better circle

Technology Tip: Another way to get an approximately square graph on the Casio *CFX*-9800G is to change the range variables so that the value of Ymax − Ymin is $\frac{2}{3}$ times Xmax − Xmin. For example, use the RANGE values in Figure 7.38 to get the corresponding graph in Figure 7.39. This method works because the dimensions of the Casio *CFX*-9800G's display are such that the ratio of vertical to horizontal is approximately $\frac{2}{3}$.

Figure 7.38: $\frac{\text{vertical}}{\text{horizontal}} = \frac{16}{24} = \frac{2}{3}$

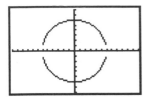

Figure 7.39: "Square" circle

The two semicircles in Figure 7.39 do not meet because of an idiosyncrasy in the way the Casio *CFX*-9800G plots a graph.

Technology Tip: The square viewing rectangle is also important when you want to judge whether two lines are perpendicular. The intersection of perpendicular lines will always *look* like a right angle in a square viewing rectangle.

7.2.5 TRACE: In the graph mode (MENU 6) graph the function $y = -x^3 + 4x$ from Section 7.2.2 using the standard viewing rectangle. (Remember to clear or cancel any other functions in the graph function screen.) When the graph window is displayed, press F1 to enable ◀ (the *left* arrow key) and ▶ (the *right* arrow key) to trace along the function. The coordinates that are displayed belong to points on the function's graph, so the *y*-coordinate is the calculated value of the function at the corresponding *x*-coordinate (Figure 7.40). (Note that on the Casio *CFX*-9800G the *x*-coordinate of the trace begins at the left-most value of the rectangle so the cursor will not appear until it is traced onto the screen.)

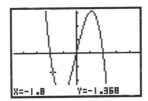

Figure 7.40: Trace

Press F6 to cycle between the *x*-coordinate alone, the *y*-coordinate alone, and both coordinates.

Now plot a second function, $y = -.25x$, along with $y = -x^3 + 4x$. From the above graph window, return to the GRAPH screen by pressing G↔T, and enter the second function as Y2 by pressing (−) .25 X,θ,T EXE

(Figure 7.41). Finally, press F6 to draw both functions (Figure 7.42).

Figure 7.41: Two functions in GRAPH mode

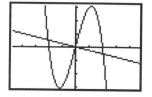

Figure 7.42: Graph of $y = -.25x$ and $y = -x^3 + 4x$

Technology Tip: The Casio *CFX*-9800G can display a graph in one of three colors: blue, orange, and green. So when you are plotting more than one function, it's helpful to color their graphs distinctly. In the graph function screen, move the highlight onto a function and press F4 *[COLR]* to activate the color menu, and then select F1 *[BLU]* for blue, F2 *[ORN]* for orange, and F3 *[GRN]* for green. Notice that each function's formula is colored to match its graph. Press EXIT to remove the color menu.

Technology Tip: Since the Casio *CFX*-9800G always draws axes in green, you may wish to color graphs in blue and orange for contrast. Also, the trace cursor will always be orange.

Notice that in Figure 7.41 the equal signs next to Y1 and Y2 are *both* highlighted. This means that *both* functions will be graphed. In the GRAPH screen, move the cursor onto Y1 and press F5 *[SEL]*. The equal sign next to Y1 should no longer be highlighted (Figure 7.43). Now press F6 and see that only Y2 is plotted (Figure 7.44).

Figure 7.43: Only Y2 active

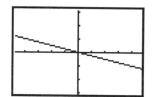

Figure 7.44: Graph of $y = -.25x$

Up to thirty different functions can be stored in list of functions in the GRAPH screen and any combination of them may be graphed simultaneously. You can make a function active or inactive for graphing by scrolling to the function pressing the F5 *[SEL]*. Now go back to the GRAPH screen and do what is need in order to graph Y1 but not Y2.

Now activate both functions so that both graphs are plotted. Press F1 to trace and the cursor will be on the graph of $y = -x^3 + 4x$ because it is higher up on the list of active functions in the GRAPH FUNC screen. Press the up ▲ or down ▼ arrow key to move the cursor vertically to the graph of $y = -.25x$. Next press the left and right arrow keys to trace along the graph of $y = -.25x$. When more than one function is plotted, you can move the trace cursor vertically from one graph to another in this way.

Technology Tip: On the Casio *CFX*-9800G, when there is more than one function plotted, the function of the graph being graphed or traced can be displayed. In SET UP, move your cursor down to G-func and press F1 *[ON]* to turn the display on or F2 *[OFF]* to turn it off.

Technology Tip: By the way, trace the graph of $y = -.25x$ and press and hold either ◀ or ▶. Eventually you will reach the left or right edge of the window. Keep pressing the arrow key and the Casio *CFX*-9800G will allow you to continue the trace by panning the viewing rectangle. Check the RANGE screen to see that the Xmin and Xmax are automatically updated.

The Casio *CFX*-9800G has a display of 95 horizontal columns of pixels and 63 vertical rows, so when you trace a curve across a graph window, you are actually moving from Xmin to Xmax in 94 equal jumps, each called Δx. You would calculate the size of each jump to be $\Delta x = \dfrac{\text{Xmax} - \text{Xmin}}{94}$. Sometimes you may want the jumps to be friendly numbers like 0.1 or 0.25 so that, when you trace along the curve, the x-coordinates will be incremented by such a convenient amount. Just set your viewing rectangle for a particular increment Δx by making Xmax = Xmin + 94 · Δx. For example, if you want Xmin = −5 and $\Delta x = 0.3$, set Xmax = −5 + 94 · 0.3 = 23.2.

On the Casio *CFX*-9800G, to center your window around a particular point, say (h, k), and also have a certain Δx, set Xmin = $h - 47 \cdot \Delta x$ and make Xmax = $h + 47 \cdot \Delta x$. Likewise, make Ymin = $k - 31 \cdot \Delta y$ and make Ymax = $h + 31 \cdot \Delta x$. For example, to center a window around the origin (0, 0), with both horizontal and vertical increments of 0.25, set the range so that Xmin = 0 − 47 · 0.25 = −11.75, Xmax = 0 + 47 · 0.25 = 11.75, Ymin = 0 − 31 · 0.25 = −7.75 and Ymax = $h + 31 \cdot 0.25$ = 7.75.

The Casio *CFX*-9800G's standard viewing window is already a friendly viewing rectangle, centered at the origin (0, 0) with $\Delta x = \Delta y = 0.1$.

See the benefit by first plotting $y = x^2 + 2x + 1$ in a window that extends from −10 to 10 in both directions. Trace near its y-intercept, which is (0, 1), and move towards its x-intercept, which is (−1, 0). Then initialize the range to the standard window and trace again near the intercepts.

7.2.6 **ZOOM:** Plot again the two graphs, for $y = -x^3 + 4x$ and $y = -.25x$. There appears to be an intersection near $x = 2$. The Casio *CFX*-9800G provides several ways to enlarge the view around this point. You can change the viewing rectangle directly by pressing RANGE and editing the values of Xmin, Xmax, Ymin, and Ymax. Figure 7.45 shows a new viewing rectangle for the range extending from 1.5 to 2.5 horizontally and from −2.5 to 2.5 vertically.

Trace along the graphs until coordinates of a point that is close to the intersection are displayed.

A more efficient method for enlarging the view is to draw a new viewing rectangle with the cursor. Start again with a graph of the two functions $y = -x^3 + 4x$ and $y = -.25x$ in a standard viewing rectangle.

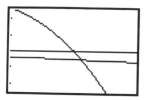

Figure 7.45: Closer view

Now imagine a small rectangular box around the intersection point, near $x = 2$. Press F2 *[Zoom]* to activate the Zoom menu at the bottom of the screen (Figure 7.46).

Now press F1 *[BOX]* to draw a box to define this new viewing rectangle. Use the arrow keys to move the cursor, which is now free-moving and whose coordinates are displayed at the bottom of the window, to one corner of the new viewing rectangle you imagine (Figure 7.47).

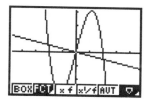

Figure 7.46: Zoom menu

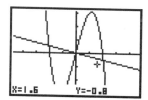

Figure 7.47: One corner selected

Press EXE to fix the corner where you moved the cursor. Use the arrow keys again to move the cursor to the diagonally opposite corner of the new rectangle (Figure 7.48). If this box looks all right to you, press EXE. The rectangular area you have enclosed will now enlarge to fill the graph window (Figure 7.49).

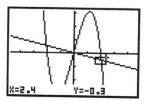

Figure 7.48: Box drawn

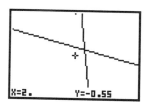

Figure 7.49: New viewing rectangle

You may cancel the zoom any time *before* you press this last EXE. Press another function key such as F1 to cancel the zoom and initiate a trace instead, or press F2 to zoom again and start over. Even if you did execute the zoom, you may still return to the original viewing rectangle and start over by pressing F2 *[Zoom]* F6 *[▾]* F1 *[ORG]*.

The Casio *CFX*-9800G has a split screen feature that enables you to see two views of a graph simultaneously.

In SET UP, move the cursor down to Dual-G and toggle it on (press F1). Now when you zoom, the left window displays the original graph and the right window displays the result of the zoom (Figure 7.50).

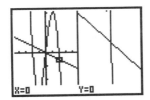

Figure 7.50: DUAL GRAPH

In the Casio *CFX*-9800G's DUAL GRAPH mode, only the left side can be acted on. So to achieve another zoom, first press F6 F2 *[CHNG]* to exchange the left and right windows. When you press RANGE, you will find *two* ranges that can be changed independently. The F6 key toggles between the left side range and the right side range.

Technology Tip: Use the G↔T key to toggle the Casio *CFX*-9800G from the dual graph to full-screen left side to full-screen right side to the GRAPH FUNC screen.

The Casio *CFX*-9800G can quickly magnify a graph around the cursor's location. Return once more to the standard range for the graph of the two functions $y = -x^3 + 4x$ and $y = -.25x$. Trace along the graphs to move the cursor as close as you can to the point of intersection near $x = 2$ (Figure 7.51). Then press F2 F3 *[×f]* and the calculator draws a magnified graph, centered at the cursor's position (Figure 7.52). The range values are changed to reflect this new viewing rectangle. Look in the RANGE menu to check.

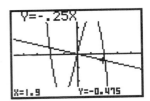

Figure 7.51: Before a zoom in

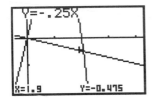

Figure 7.52: After a zoom in

As you see in the Zoom menu (press F2), the Casio *CFX*-9800G can zoom in (press F2 F3 *[× f]*) or zoom out (press F2 F4 *[× 1/f]*). Zoom out to see a larger view of the graph, centered at the cursor position. You can change the horizontal and vertical scale of the magnification by pressing F2 F2 and editing Xfct and Yfct, the horizontal and vertical magnification factors.

Technology Tip: An advantage of zooming in from the default viewing rectangle is that subsequent windows will also be square. Likewise, if you zoom in from a friendly viewing rectangle, the zoomed windows will also be friendly.

The default zoom factor is 2 in both directions (press F1 *[INIT]* in the Zoom Factor menu). It is not necessary for Xfct and Yfct to be equal. Sometimes, you may prefer to zoom in one direction only, so the other factor should be set to 1, Press EXIT to leave the Zoom Factor menu.

Technology Tip: If you should zoom in too much and lose the curve, zoom back to the original viewing rectangle and start over. Or use the arrow keys to pan over if you think the curve is not too far away. You can also just initialize the range to the Casio *CFX-9800G*'s standard window.

Technology Tip: The Casio *CFX-9800G* can automatically select the necessary *vertical* range for a function. For auto scaling, press F2 *[Zoom]* F5 *[AUTO]*. Take care, because sometimes when you are graphing two functions together, the calculator will auto scale for one function in such a way that the other function will no longer be visible. For example, plot the two functions $y = -x^3 + 4x$ and $y = -.25x$ in the Casio *CFX-9800G*'s standard viewing rectangle, then auto scale and trace along both functions.

7.2.7 Relative Minimums and Maximums: Graph $y = -x^3 + 4x$ once again in the standard viewing rectangle. This function appears to have a relative minimum near $x = -1$ and a relative maximum near $x = 1$. You may zoom and trace to approximate these extreme values.

First trace along the curve near the local minimum. Notice by how much the *x*-values and *y*-values change as you move from point to point Trace along the curve until the *y*-coordinate is as *small* as you can get it, so that you are as close as possible to the local minimum, and zoom in (press F2 F3 or use a zoom box). Now trace again along the curve and, as you move from point to point, see that the coordinates change by smaller amounts than before. Keep zooming and tracing until you find the coordinates of the local minimum point as accurately as you need them, approximately (−1.15, −3.08).

Follow a similar procedure to find the local maximum. Trace along the curve until the *y*-coordinate is as *great* as you can get it, so that you are as close as possible to the local maximum, and zoom in. The local maximum point on the graph of $y = -x^3 + 4x$ is approximately (1.15, 3.08).

The Casio *CFX-9800G* can automatically find the maximums and minimums for functions drawn in the GRAPH mode. After graphing $y = -x^3 + 4x$, press SHIFT G-SOLV to activate the graph solve menu at the bottom of the screen (Figure 7.53).

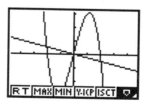

Figure 7.53: G-SOLVE menu

After activating the graph solve menu, press F3 *[MIN]* to calculate the minimum (Figure 7.54). Then find the maximum by pressing SHIFT G-SOLV F2 *[MAX]* (Figure 7.55).

Casio *CFX-9800G* Color Power Graphic Calculator

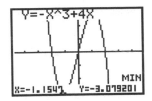

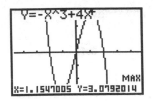

Figure 7.54: Minimum of $y = -x^3 + 4x$

Figure 7.55: Maximum of $y = -x^3 + 4x$

Note that if you have more than one graph on the screen, the calculator will pause until you specify the graph whose maximum or minimum you want to calculate. As you use the up and down arrow keys to move between graphs, press EXE when the equation you want to evaluate appears on the screen.

If your graph has more than one maximum or minimum, you can use the left and right arrows to move between them.

7.3   Solving Equations and Inequalities

7.3.1 Intercepts and Intersections: Tracing and zooming are also used to locate an x-intercept of a graph, where a curve crosses the x-axis. For example, the graph of $y = x^3 - 8x$ crosses the x-axis three times (Figure 7.56). After tracing over to the x-intercept point that is farthest to the left, zoom in (Figure 7.57). Continue this process until you have located all three intercepts with as much accuracy as you need. The three x-intercepts of $y = x^3 - 8x$ are approximately -2.828, 0, and 2.828.

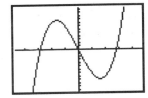

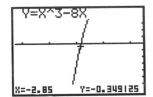

Figure 7.56: Graph of $y = x^3 - 8x$

Figure 7.57: Near an x-intercept of $y = x^3 - 8x$

Technology Tip: As you zoom in, you may also wish to change the spacing between tick marks on the x-axis so that the viewing rectangle shows scale marks near the intercept point. Then the accuracy of your approximation will be such that the error is less than the distance between two tick marks. Change the x-scale from the RANGE menu. Move the cursor down to Xscale and enter an appropriate value.

The x-intercept of a function's graph is a *root* of the equation $f(x) = 0$, and the Casio *CFX*-9800G can automatically search for the roots. First plot the function in GRAPH mode and then activate the graph solve menu by pressing SHIFT G-SOLV. (Refer back to figure 7.53.) Then press F1 *[ROOT]* to locate an x-intercept on the graph in the current window (Figure 7.58). The calculator searches from left to right to find an x-intercept in the current window; press the right arrow key to search for the next *x*-intercept to the right.

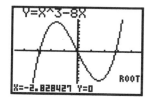

Figure 7.58: A root of $y = x^3 - 8x$

TRACE and ZOOM are especially important for locating the intersection points of two graphs, say the graphs of $y = -x^3 + 4x$ and $y = -.25x$. Trace along one of the graphs until you arrive close to an intersection point. Then press ▲ or ▼ to jump to the other graph. Notice that the *x*-coordinate does not change, but the *y*-coordinate is likely to be different (Figures 7.59 and 7.60).

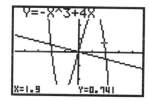

Figure 7.59: Trace on $y = -x^3 + 4x$

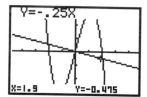

Figure 7.60: Trace on $y = -.25x$

When two *y*-coordinates are as close as they can get, you have come as close as you now can to the point of intersection. So zoom in around the intersection point, then trace again until the two *y*-coordinates are as close as possible. Continue this process until you have located the point of intersection with as much accuracy as necessary.

Technology Tip: Press F6 a couple of times to display only the *y*-coordinate. Then while tracing towards an intersection, it's easier to see where the *y*-coordinates are closest.

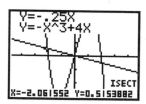

Figure 7.61: An intersection of $y = -x^3 + 4x$ and $y = -.25x$

While the graphs are displayed, automate the Casio *CFX*-9800G to search for points of intersection by press-

Casio *CFX*-9800G Color Power Graphic Calculator

ing SHIFT G-SOLV F5 *[ISCT]*. If more than two functions are being plotted, the calculator will ask you to specify the two whose intersection you seek. The calculator searches from left to right to find an intersection point in the current window; press the right arrow to continue the search for the *next* intersection point. Figure 7.61 shows one of the intersection points.

7.3.2 Solving Equations by Graphing: Suppose you need to solve the equation $24x^3 - 36x + 17 = 0$. First graph $y = 24x^3 - 36x + 17$ in a window large enough to exhibit *all* its x-intercepts, corresponding to all the equation's roots. Then use trace and zoom or the ROOT command, to locate each one. In fact, this equation has just one solution, approximately $x = -1.414$.

Remember that when an equation has more than one root, it may be necessary to change the viewing rectangle a few times to locate all of them.

Technology Tip: To solve an equation like $24x^3 + 17 = 36x$, you may first transform it into standard form, $24x^3 - 36x + 17 = 0$, and proceed as above. However, you may also graph the *two* functions $y = 24x^3 + 17$ and $y = 36x$, then zoom and trace to locate their point of intersection or use the ISCT command.

7.3.3 Solving Systems by Graphing: The solutions to a system of equations correspond to the points of intersection of their graphs (Figure 7.62). For example, to solve the system $y = x^3 + 3x^2 - 2x - 1$ and $y = x^2 - 3x - 4$, first graph them together. Then use zoom and trace or the ISCT command, to locate their point of intersection, approximately $(-2.17, 7.25)$.

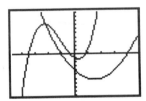

Figure 7.62: Graph of $y = x^3 + 3x^2 - 2x - 1$ and $y = x^2 - 3x - 4$

If you zoom and trace, you must judge whether the two current y-coordinates are sufficiently close for $x = -2.17$ or whether you should continue to zoom and trace to improve the approximation.

The solutions of the system of two equations $y = x^3 + 3x^2 - 2x - 1$ and $y = x^2 - 3x - 4$ correspond to the solutions of the single equation $x^3 + 3x^2 - 2x - 1 = x^2 - 3x - 4$, which simplifies to $x^3 + 2x^2 + x + 3 = 0$. So you may also graph $y = x^3 + 2x^2 + x + 3$ and find its x-intercepts to solve the system.

7.3.4 Solving Inequalities by Graphing: Consider the inequality $1 - \dfrac{3x}{2} \geq x - 4$. To solve it with your Casio

CFX-9800G, graph the two functions $y = 1 - \dfrac{3x}{2}$ and $y = x - 4$ (Figure 7.63). First locate their point of inter-section, at $x = 2$. The inequality is true when the graph of $y = 1 - \dfrac{3x}{2}$ lies *above* the graph of $y = x - 4$, and that occurs when $x < 2$. So the solution is the half-line $x \le 2$, or $(-\infty, 2]$.

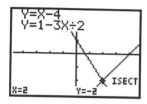

Figure 7.63: Solving $1 - \dfrac{3x}{2} \ge x - 4$

The Casio *CFX*-9800G is capable of graphing inequalities of the form $y \le f(x)$, $y < f(x)$, $y \ge f(x)$, or $y > f(x)$. For example, to graph $y \ge x^2 - 1$ in GRAPH mode, press F3 *[TYPE]* F4 *[INEQ]*. Scroll to Y3, input $x^2 - 1$, press F3 *[Y≥]*, and press EXE (Figure 7.64). (Be sure the other graphs are not selected.) Now press F6 to draw the graph (Figure 7.65).

Figure 7.64: Inequality options

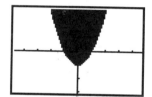

Figure 7.65: Graph of $y \ge x^2 - 1$

Next press F1 to trace along the boundaries of the inequality. Notice that the Casio *CFX*-9800G displays co-ordinates appropriately as inequalities. Zooming is also available for inequality graphs.

Solve a system of inequalities, such as $1 - \dfrac{3x}{2} \ge y$ and $y > x - 4$, by plotting the two inequality graphs simul-taneously. First, clear the graph window and reset the range to a convenient window. Input $1 - \dfrac{3x}{2}$ as an ine-quality type in Y1 and store it by pressing F4 *[Y≤]* EXE. Likewise, input $x - 4$ as an inequality type in Y2 and press F1 *[Y>]* EXE. Now press F6 to draw the two inequalities as in Figure 7.66.

Technology Tip: Since you can change the mode of the Casio *CFX*-9800G at any time, you can graph ine-qualities and equations together at the same time. Simply change to inequality type before entering an ine-

Casio *CFX*-9800G Color Power Graphic Calculator

quality, and change to rectangular type before entering an equation.

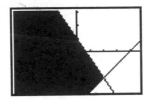

Figure 7.66: Graphs of $1 - \dfrac{3x}{2} \geq y$ and $y > x - 4$

7.4   Trigonometry

7.4.1 Degrees and Radians: The trigonometric functions can be applied to angles measured either in radians or degrees, but you should take care that the Casio *CFX*-9800G is configured for whichever measure you need. From the main menu, enter the calculation mode (**MENU 1**) to see the current settings. To change your angle setting while in the calculation mode, press **SHIFT DRG** to activate the angular unit menu at the bottom of the screen. Press **F1** for degree measure or **F2** for radian measure, then press **EXE** to make it so.

It's a good idea to check the angle measure setting before executing a calculation that depends on a particular measure. You may change a mode setting at any time and not interfere with pending calculations. Try the following keystrokes to see this in action.

Expression	*Keystrokes*	*Display*
sin 45°	SHIFT DRG F1 EXE	
	sin 45 EXE	0.7071067812
sin π°	sin SHIFT π EXE	0.05480366515
sin π	SHIFT DRG F2 EXE	
	sin SHIFT π EXE	0
sin 45	sin 45 EXE	0.8509035245
$\sin\dfrac{\pi}{6}$	sin (SHIFT π ÷ 6) EXE	0.5

The first line of keystrokes sets the Casio *CFX*-9800G in degree mode and calculates the sine of 45 *degrees*. While the calculator is still in degree mode, the second line keystrokes calculates the sine of π *degrees*, approximately 3.1415°. The third line changes to radian mode just before calculating the sine of π *radians*. The fourth line calculates the sine of 45 *radians* (the calculator remains in radian mode).

The Casio *CFX*-9800G makes it possible to mix degrees and radians in a calculation. Execute these keystrokes to calculate $\tan 45° + \sin\frac{\pi}{6}$ as shown in Figure 7.67: tan 45 SHIFT DRG F4 + sin (SHIFT π ÷ 6) F5 EXE. Do you get 1.5 whether your calculator is in set *either* in degree mode *or* in radian mode?

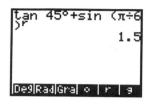

Figure 7.67: Angle measure

7.4.2 Graphs of Trigonometric Functions: When you graph a trigonometric function, you need to pay careful attention to the choice of graph window. For example, graph $y = \dfrac{\sin 30x}{30}$ in the standard viewing rectangle. Trace along the curve to see where it is. Zoom in to a better window, or use the period and amplitude to establish a better window.

Technology Tip: In the RANGE menu of the Casio *CFX*-9800G the viewing rectangle can be set by pressing F2 to a special window for trigonometric functions so that horizontal range is from -2π to 2π in radian mode or from $-360°$ to $360°$ in degree mode and the vertical range is from -1.6 to 1.6.

7.5   Scatter Plots

7.5.1 Entering Data: The table shows the total prize money (in millions of dollars) awarded at the Indianapolis 500 race from 1981 to 1989. (*Source*: Indianapolis Motor Speedway Hall of Fame.)

Year	1981	1982	1983	1984	1985	1986	1987	1988	1989
Prize ($million)	$1.61	$2.07	$2.41	$2.80	$3.27	$4.00	$4.49	$5.03	$5.72

We'll now use the Casio *CFX*-9800G to construct a scatter plot that represents these points and to find a linear model that approximates the given data.

In the main menu select the REG icon (MENU 4) to enter the regression mode. You want to set the mode to the settings shown in Figure 7.68. In particular you want S-data set to STO and S-graph set to DRAW for data storage and graphing. If you need to change the set up, press SHIFT SET UP. For example, to set STAT GRAPH to DRAW press ▼ ▼ F1 *[DRW]* (Figure 7.69). Press EXIT AC$^{/ON}$ to enter the regression screen.

Instead of entering the full year 198x we will enter only x. Since the data will be plotted as it is entered, choose an appropriate viewing rectangle, say from 0 to 10 horizontally and vertically. First clear any existing statistical data by pressing F2 *[EDIT]* and then F1 *[DEL]* as many times as necessary to delete any existing entries. Press EXIT when you are done. Here are the keystrokes to enter the data for the first three years: 1 F3 1.61 F1 2 F3 2.07 F1 3 F3 2.41 F1. Continue to enter all the given data .

Figure 7.68: REG settings

Figure 7.69: Setting up REG

You can go back and forth from your graph to the REG screen by pressing the G↔T. You may edit statistical data in the same way you edit expressions in the home screen. In the REG screen press F2 to display the data table (Figure 7.70). Move the cursor to the x and y value for any data point you wish to change, then type the correction and press EXE. To insert or delete data, move the cursor to the x or y value for any data point you wish to add or delete. Press F2 and a new data point is created; press F1 and the data point is deleted. Press EXIT to return to the regression screen.

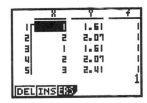

Figure 7.70: Editing data points

7.5.2 Plotting Data: Once all the data points have been entered, press SHIFT CLR F2 *[SCL]* EXE to clear statistical memory. Then press F6 *[CAL]* to calculate the statistics associated with the data. Recalculation is necessary whenever you edit data or change the graph window. Figure 7.71 shows the scatter plot in a viewing rectangle extending form 0 to 10 along the horizontal and vertical axes.

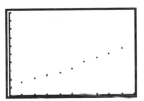

Figure 7.71: Scatter plot

7.5.3 Regression Line: The Casio *CFX*-9800G calculates and the slope and y-intercept for the line that best fits all the data. From the plot of the data, return to the regression screen by pressing G↔T. Press F6 *[REG]* to calculate a linear regression model. As you can see in Figure 7.72, the Casio *CFX*-9800G displays the regression menu at the bottom of the screen and names the y-intercept A and calls the slope B. The number r (between −1 and 1) is called the *correlation coefficient* and measures the goodness of fit of the linear regression with the data. The closer the absolute value of r is to 1, the better the fit; the closer the absolute value of r is to 0, the worse the fit. Press the function key F1 EXE for A, F2 EXE for B, and F3 EXE for r.

Graph the line $y = A + Bx$ by pressing GRAPH SHIFT F4 1 EXE. See how well this line fits with your data points (Figure 7.73).

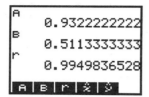

Figure 7.72: Linear regression model

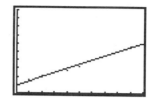

Figure 7.73: Linear regression line

7.5.4 Exponential Growth Model: The table shows the world population (in millions) from 1980 to 1992.

Year	1980	1985	1986	1987	1988	1989	1990	1991	1992
Population (millions)	4453	4850	4936	5024	5112	5202	5294	5384	5478

In the regression mode, press SHIFT SET UP. At REG press F3 *[EXP]*, so the set up screen looks like Figure 7.74. Then EXIT the screen and follow the procedure described above and enter the data to find an exponential model that approximates the given data. Use 0 for 1980, 5 for 1985, and so on. You may find it easier to EDIT the existing data rather than entering new data.

Now EXIT the data screen and press F6 *[CALC]* to process the data. Then press F6 *[REG]* to compute the exponential growth model $y = ae^{bx}$. Press F1 *[A]* EXE F2 *[B]* EXE to find the values of A and B (Figure 7.75). In this case, the exponential growth model is $y = 4451e^{0.017303t}$.

Figure 7.74: Setting up REG

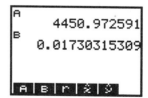

Figure 7.75: Exponential growth model

If you wish to plot and graph the data, follow the method for linear regression. Set an appropriate range for the data and then press SHIFT CLR F2 *[SCL]* EXE F6 *[CAL]*. The data will now be plotted in the range. As in the linear regression model, press GRAPH SHIFT F4 1 EXE to graph the exponential growth model.

Casio *CFX*-9800G Color Power Graphic Calculator

7.6 Matrices

7.6.1 Making a Matrix: The Casio *CFX*-9800G can work with 26 different matrices (Mat A through Mat Z).

Here's how to create this 3×4 matrix $\begin{bmatrix} 1 & -4 & 3 & 5 \\ -1 & 3 & -1 & -3 \\ 2 & 0 & -4 & 6 \end{bmatrix}$ in your calculator as Mat A.

Enter the matrix mode by selecting the MAT icon from the main menu (MENU 5). Then press EXIT F4 *[EDIT]* to see the matrix list. The display will show the dimension of each matrix if the matrix exists; otherwise, it will display None (Figure 7.76). Move the cursor to Mat A and press F2 *[DIM]* 3 EXE 4 EXE to enter its dimensions of 3 rows by 4 columns. Return to the matrix list by pressing EXIT once, then press F1 *[EDIT]* to edit Mat A.

Use the arrow keys or press EXE repeatedly to move the cursor to a matrix element you want to change. If you press EXE, you will move right across a row and then back to the first column of the next row. The element in the second row and first column in Figure 7.77 is highlighted, so that the element's current value is displayed at the bottom right corner of the screen. Continue to enter all the elements of Mat A; press EXE after inputting each value.

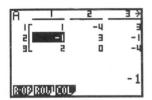

Figure 7.76: Matrix list   Figure 7.77: Editing a matrix

When you are finished, leave the editing screen by pressing EXIT once to return to the matrix list.

7.6.2 Matrix Math: You can perform many calculations with matrices in the matrix mode. To calculate the scalar multiplication 2 Mat A, enter the matrix calculation screen. This is the primary screen in the matrix mode. If you are in the matrix list, press EXIT. Now press 2 F1 *[Mat]* ALPHA A EXE. The resulting matrix is displayed on the screen and is stored in matrix memory as Mat Ans (Figure 7.78). If you would rather have the matrix stored as specific matrix, say Mat C, you should press 2 F1 *[Mat]* ALPHA A SHIFT → F1 *[Mat]* ALPHA C EXE. The calculator will display the matrix. Press EXIT to return to the matrix calculation screen (Figure 7.79).

To add two matrices, say Mat A and Mat B, create Mat B (with the same dimensions as Mat A) and then press F1 *[Mat]* ALPHA A + F1 *[Mat]* ALPHA B EXE. Again, if you want to store the answer as a specific matrix, say Mat C, then press → F1 *[Mat]* ALPHA C before executing the above command. Subtraction is performed in a similar manner.

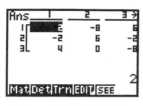

Figure 7.78: MAT A

Figure 7.79: Matrix calculation

Now set the dimensions of **Mat C** as 2×3 and enter the matrix $\begin{bmatrix} 2 & 0 & 3 \\ 1 & -5 & -1 \end{bmatrix}$ as **Mat C**. For matrix multiplication of **Mat C** by **Mat A**, press **F1** *[Mat]* ALPHA C × **F1** *[Mat]* ALPHA A EXE. If you tried to multiply **Mat A** by **Mat C**, your calculator would signal an error because the dimensions of the two matrices do not permit multiplication in this way.

The *transpose* of a matrix is another matrix with the rows and columns interchanged. To calculate the transpose of **Mat A** , press **F3** *[Trn]* **F1** *[Mat]* ALPHA A EXE.

7.6.3 Row Operations: Here are the keystrokes necessary to perform elementary row operations on a matrix. Your textbook provides a more careful explanation of the elementary row operations and their uses.

Enter the editing screen for **Mat A**. Press **F1** to activate the row operations menu at the bottom of the calculator screen (Figure 7.80).

After you select a row operation, your calculator will prompt you through it. For example, to interchange the second and third rows of **Mat A** defined above, press **F1** 2 EXE 3 EXE, while the calculator prompts for the row numbers (Figure 7.81). The format of this command is **Swap Row m ↔ Row n**.

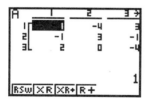

Figure 7.80: Row operations menu

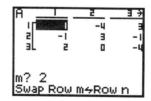

Figure 7.81: Swap rows 2 and 3

To add row 2 and row 3 and *store* the results in row 3, press **F4** 2 EXE 3 EXE. The format of this command is **Row m + Row n → Row n**.

To multiply row 2 by –4 and *store* the results in row 2, thereby replacing row 2 with new values, press **F2** (–) 4 EXE 2 EXE. The format of this command is **k × Row m → Row m**.

To multiply row 2 by –4 and *add* the results to row 3, thereby replacing row 3 with new values, press **F3** (–) 4 EXE 2 EXE 3 EXE. The format of this command is **k × Row m + Row n → Row n**.

Note that as you perform row operations on the Casio *CFX*-9800G, your old matrix is replaced by the new matrix, so if you want to keep the original matrix in case you need it, you should save it under another name.

For example, use row operations to solve this system of linear equations: $\begin{cases} x - 2y + 3z = 9 \\ -x + 3y = -4 \\ 2x - 5y + 5z = 17 \end{cases}$.

First enter this *augmented matrix* as Mat A in your Casio *CFX*-9800G: $\begin{bmatrix} 1 & -2 & 3 & 9 \\ -1 & 3 & 0 & -4 \\ 2 & -5 & 5 & 17 \end{bmatrix}$. Next store this

matrix as Mat E (press EXIT a couple of times to go back to the matrix home screen, then press F1 *[Mat]* ALPHA A SHIFT → F1 *[Mat]* ALPHA E EXE, as in Figure 7.82), so you may keep the original in case you need to recall it.

Figure 7.82: Storing as Mat E

We now edit Mat E. Here are the row operations and their associated keystrokes. At each step, the result is stored as Mat E and replaces the previous Mat E. The completion of the row operations is shown in Figure 7.83. First press F1 to begin performing row operations.

Row Operations	Keystrokes
add row 1 to row 2	F4 1 EXE 2 EXE
add −2 times row 1 to row 3	F3 (−) 2 EXE 1 EXE 3 EXE
add row 2 to row 3	F4 2 EXE 3 EXE
multiply row 3 by $\frac{1}{2}$	F2 1 ÷ 2 EXE 3 EXE

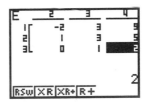

Figure 7.83: Final matrix after row operations

Thus $z = 2$, so $x = 1$ and $y = -1$.

7.6.4 Determinants and Inverses: Enter this 3×3 square matrix as Mat A: $\begin{bmatrix} 1 & -2 & 3 \\ -1 & 3 & 0 \\ 2 & -5 & 5 \end{bmatrix}$. To calculate its

determinant $\begin{vmatrix} 1 & -2 & 3 \\ -1 & 3 & 0 \\ 2 & -5 & 5 \end{vmatrix}$, go to the matrix home screen and press F2 *[Det]* F1 *[Mat]* ALPHA A EXE. You

should find that the determinant is 2 as shown in Figure 7.84.

Since the determinant of the matrix is not zero, it has an inverse matrix. Press F1 *[Mat]* ALPHA A SHIFT x^{-1} EXE to calculate the inverse. The result is shown in Figure 7.85.

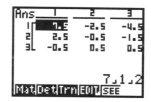

Figure 7.84: Determinant of Mat A   Figure 7.85: Inverse of Mat A

Now let's solve a system of linear equations by matrix inversion. Once again, consider $\begin{cases} x - 2y + 3z = 9 \\ -x + 3y = -4 \\ 2x - 5y + 5z = 17 \end{cases}$.

The coefficient matrix for this system is the matrix $\begin{bmatrix} 1 & -2 & 3 \\ -1 & 3 & 0 \\ 2 & -5 & 5 \end{bmatrix}$ which was entered in the previous example.

Now enter the matrix $\begin{bmatrix} 9 \\ -4 \\ 17 \end{bmatrix}$ as **Mat B**. Then in matrix mode, press F1 *[Mat]* ALPHA A SHIFT x^{-1} × F1 *[Mat]*

ALPHA B EXE to get the answer as shown in Figure 7.86.

Casio *CFX*-9800G Color Power Graphic Calculator

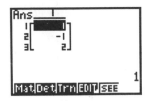

Figure 7.86: Solution matrix

The solution is still $x = 1$, $y = -1$, and $z = 2$.

7.7   Sequences

7.7.1 Iteration with the Ans key: Compute the following in the COMP mode (MENU 1). The Ans feature enables you to perform iterations to evaluate a function repeatedly. As an example, calculate $\dfrac{n-1}{3}$ for $n = 27$.

Then calculate $\dfrac{n-1}{3}$ for $n =$ the answer to the previous calculation. Continue to use each answer as n in the *next* calculation. here are keystrokes to accomplish this iteration on the Casio *CFX*-9800G calculator. (See the results in Figure 7.87.) Notice that when you use Ans in place of n in a formula, it is sufficient to press EXE to continue an iteration.

Iteration	Keystrokes	Display
1	27 EXE	27
2	(SHIFT Ans − 1) ÷ 3 EXE	8.666666667
3	EXE	2.555555556
4	EXE	0.5185185185
5	EXE	−0.1604938272

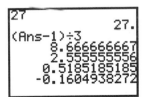

Figure 7.87: Iteration

Press EXE several more times and see what happens with this iteration. You may wish to try it again with a different starting value.

7.7.2 Arithmetic and Geometric Sequences:

7.7.2 Arithmetic and Geometric Sequences: Use iteration with the Ans variable to determine the *n*-th term of a sequence. For example, find the 18th term of an *arithmetic* sequence whose first term is 7 and whose common difference is 4. Enter the first term by pressing 7 EXE, then start the progression with the recursion formula, SHIFT Ans + 4 EXE. This yields the 2nd term, so press EXE sixteen more times to find the 18th term. For a *geometric* sequence whose common ratio is 4, start the progression with SHIFT Ans × 4 EXE.

Of course, you could also use the *explicit* formula for the *n*-th term of an arithmetic sequence $t_n = a + (n-1)d$. First enter values for the variables *a*, *d*, and *n*, then evaluate the formula by pressing ALPHA A + (ALPHA N − 1) ALPHA D EXE. For a geometric sequence whose *n*-th term is given by $t_n = a \cdot r^{n-1}$, enter values for the variables *a*, *d*, and *n*, then evaluate the formula by pressing ALPHA A ALPHA R ∧ (ALPHA N − 1) EXE.

You can also define the sequence recursively with the Casio *CFX*-9800G in the table mode. From the main menu enter the table mode by selecting the TABLE icon (MENU 8) and then press F3 *[REC]*. Next press F4 *[TYPE]* F2 *[a$_{n+1}$]* to select the recursion type. Once again, let's find the 18th term of an *arithmetic* sequence whose first term is 7 and whose common difference is 4. Input the recursion formula $a_{n+1} = a_n + 4$ by pressing F2 *[a$_n$]* + 4 (Figure 7.88). Now make $a_1 = 7$ (because the first term is a_1 where $n = 1$) and display a table that contains the 16th term a_{16} to the 20th term a_{20} by pressing F5 *[RNG]* F2 *[a$_1$]* 16 EXE 20 EXE 7 EXE EXIT F6 *[TBL]* (Figures 7.89 and 7.90).

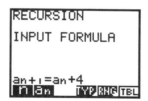

Figure 7.88: Recursion formula

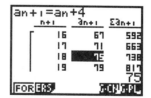

Figure 7.89: TABLE Range

Figure 7.90: $a_{18} = 75$

To use the explicit formula in a Casio *CFX*-9800G recursion table, start a new table by pressing F2 *[ERS]* F1 *[YES]*. Now make $a_n = 7 + (n-1) \cdot 4$ by press F4 *[TYPE]* F1 *[a$_n$]* 7 + (F1 *[n]* − 1) × 4. Once more, calculate term a_{18} by pressing F5 *[RNG]* 18 EXE 18 EXE EXIT F6 *[TABL]* (Figures 7.91 – 7.93).

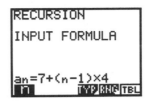

Figure 7.91: Explicit formula

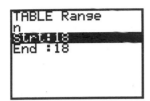

Figure 7.92: TABLE Range

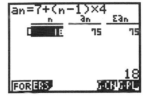

Figure 7.93: $a_{16} = 75$

Technology Tip: A table whose starting and ending range values are the same has just one entry. So to display a single *n*-th term series, set both the starting and ending range values to *n*.

There are more detailed instructions for using the table recursion mode in the Casio *CFX*-9800G manual.

7.7.3 Finding Sums of Sequences: You can use recursion option in the table mode of the Casio *CFX*-9800G to find the sum of a sequence. For example, suppose you want to find the sum $\sum\limits_{n=1}^{12} 4(0.3)^n$. Erase any existing formula by pressing F2 *[ERS]* F1 *[YES]*. Now press F4 *[TYP]* F1 *[a$_n$]* 4 × (.3) ∧ F1 *[n]*. Now you must set the range from 1 to 12 by pressing F5 *[RANG]* 1 EXE 12 EXE EXIT. Press F6 *[TABL]*. The last entry in the table is the sum (Figure 7.94).

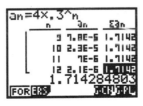

Figure 7.94: $\sum\limits_{n=1}^{12} 4(0.3)^n$

Now calculate the sum starting at $n = 0$ by editing the range. You should obtain a sum of approximately 5.712848.

7.8   Parametric and Polar Graphs

7.8.1 Graphing Parametric Equations: The Casio *CFX-9800G* plots parametric equations as easily as it plots functions. Enter the graph mode by selecting the GRAPH icon from the main menu (MENU 6). Press EXIT and then change to parametric mode by pressing F3 *[TYP]* F3 *[PRM]*. Be sure, if the independent parameter is an angle measure, that the angle measure has been set to whichever you need, Rad or Deg.

You can now enter the parametric functions. For example, here are the keystrokes need to graph the parametric equations $x = \cos^3 t$ and $y = \sin^3 t$. First check that angle measure is in radians. Then in the graph screen press (cos X,θ,T) ∧ 3 F1 *[,]* (sin X,θ,T) ∧ 3 EXE (Figure 7.95).

Figure 7.95: $x = \cos^3 t$ and $y = \sin^3 t$

Press RANGE F1 *[INIT]* to set the standard graphing window and to initialize the values of T. Press RANGE again to see that the values of T go from 0 to 2π in steps of $\frac{2\pi}{100} \approx 0.062832$. In order to provide a better viewing rectangle press RANGE twice and set the rectangle to go from −2 to 2 horizontally and vertically. EXIT back to the graph screen and press F6 *[DRW]* to draw the graph (Figure 7.96).

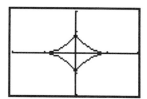

Figure 7.96: Parametric graph of $x = \cos^3 t$ and $y = \sin^3 t$

You may ZOOM and TRACE along parametric graphs just as you did with function graphs. As you trace along this graph, notice that the cursor moves in the counterclockwise direction as T increases.

Note that you can also graph parametric equations in the COMP mode (MENU 1) by using the GRAPH key.

Casio *CFX-9800G* Color Power Graphic Calculator

First clear any existing graph(s) from the graph window by pressing SHIFT F5 *[CLS]* EXE. After setting the graph type as parametric and choosing an appropriate RANGE, press GRAPH (cos X,θ,T) ∧ 3 , (sin X,θ,T) ∧ 3) EXE.

7.8.2 Rectangular-Polar Coordinate Conversion: Conversion between rectangular and polar coordinate systems is accomplished directly through keystrokes on the Casio *CFX*-9800G. These functions use the current angle measure setting, so it is a good idea to check the default angle measure before any conversion. Of course, you may override the current angle measure setting, as explained in Section 7.4.1. For the following examples, the Casio *CFX*-9800G is set to radian measure.

The Casio *CFX*-9800G uses the variables I and J to store the results of a conversion. So going from rectangular to polar coordinates, you get $(r, \theta) = (I, J)$. Going from polar to rectangular, you get $(x, y) = (I, J)$.

We perform these calculations in the COMP mode. To convert between rectangular and polar coordinates, activate the coordinate menu at the bottom of the screen by pressing SHIFT MATH F5 *[COR]*.

Given the rectangular coordinates $(x, y) = (4, -3)$, convert to polar coordinates (r, θ) by pressing F1 *[Pol]* 4 , – 3) EXE. The value of r is displayed; now press ALPHA J EXE to display the value of θ (Figure 7.97). The polar coordinates are approximately $(5, -0.6435)$.

Suppose $(r, \theta) = (3, \pi)$. Convert to rectangular coordinates (x, y) by pressing F2 *[Rec]* 3 , SHIFT π) EXE. The x-coordinate is displayed; press ALPHA J EXE to display the y-coordinate (Figure 7.98). The rectangular coordinates are $(-3, 0)$.

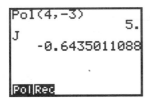

Figure 7.97: Rectangular to polar coordinates

Figure 7.98: Polar to rectangular coordinates

7.8.3 Graphing Polar Equations: The Casio *CFX*-9800G graphs polar functions in the form $r = f(\theta)$. Enter the graph mode by selecting the GRAPH icon from the main menu (MENU 6). Press EXIT and change to polar mode by pressing F3 *[TYP]* F2 *[POL]*. Be sure that the angle measure has been set to whichever you need, Rad or Deg. Here we will use radian measure. Press RANGE F1 *[INIT]* to initialize the graph window so θ goes from 0 to 2π.

For example, to graph $r = 4\sin\theta$, select a location for the function and press 4 sin X,θ,T EXE in the graph screen. Now, when you press the variable key X,θ,T, you get a θ because the calculator is in polar mode..

Choose a good viewing rectangle and an appropriate interval and increment for θ. In Figure 7.99, the viewing rectangle is roughly "square" and extends from –6 to 6 horizontally and from –4 to 4 vertically. (Refer back to the Technology Tip in Section 7.2.4.)

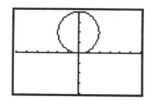

Figure 7.99: Polar graph of $r = 4 \sin \theta$

Trace along this graph to see the polar coordinates of the cursor's location displayed at the bottom of the window. Zooming works just the same as before.

Note that you can also graph polar equations in the COMP mode (MENU 1) by using the GRAPH key. First clear any existing graph(s) from the graph window by pressing SHIFT F5 *[CLS]* EXE. After setting the graph type as polar and choosing an appropriate RANGE, press GRAPH 4 sin X,θ,T EXE.

7.9   Probability

7.9.1 Random Numbers: The command Rn# generates a number between 0 and 1. You will find this command in the PRB (probability) sub-menu of the MATH menu. In the COMP mode (MENU 1) press SHIFT MATH F2 *[PRB]* to activate the probability menu at the bottom of the screen. We will assume that this menu is active for the rest of this section. Then press F4 *[Rn#]* EXE to generate a random number. Press EXE to generate another number; keep pressing EXE to generate more of them.

If you need a random number between, say, 0 and 10, then press 10 F4 *[Rn#]* EXE. To get a random number between 5 and 15, press 5 + 10 F4 *[Rn#]* EXE.

7.9.2 Permutations and Combinations: To calculate the number of permutations of 12 objects taken 7 at a time, $_{12}P_7$, press 12 F2 *[nPr]* 7 EXE (Figure 7.100). Thus $_{12}P_7 = 3{,}991{,}680$.

For the number of combinations of 12 objects taken 7 at a time, $_{12}C_7$, press 12 F3 *[nCr]* 7 EXE (Figure 7.100). Thus $_{12}C_7 = 792$.

Figure 7.100: $_{12}P_7$ and $_{12}C_7$

7.9.3 Probability of Winning: A state lottery is configured so that each player chooses six different numbers from 1 to 40. If these six numbers match the six numbers drawn by the State Lottery Commission, the player

Casio *CFX-9800G Color Power Graphic Calculator*

wins the top prize. There are $_{40}C_6$ ways for the six numbers to be drawn. If you purchase a single lottery ticket, your probability of winning is 1 in $_{40}C_6$. Press 1 ÷ 40 F3 [nCr] 6 EXE to calculate your chances, but don't be disappointed.

7.10   Programming

7.10.1 Entering a Program: The Casio *CFX*-9800G is a programmable calculator that can store sequences of commands for later replay. Here's an example to show you how to enter a useful program that solves quadratic equations by the quadratic formula.

Select the PRGM icon from the main menu (MENU A). Press F1 *[PRGM]*. You should now have a program list on your calculator (Figure 7.101). The Casio *CFX*-9800G has space for up to 38 programs, each named by a number or letter. If a program location is not used, the word empty appears to the right of its name in the list.

Figure 7.101: Program list

Press the up or down arrow keys to move the cursor to an empty program area; you may also press the key corresponding to a program's name and jump directly there. For example, to go to program 5, press 5; to edit program B, press ALPHA B.

With the program area you've chosen highlighted, press EXE to write a new program in that area or to edit a program that is already there.

Now enter a descriptive title, so press SHIFT A-LOCK and name this program QUADRATIC. Press ALPHA to cancel the alpha lock. Then press EXE to begin writing the actual program. If you do not enter a title, the first line of the program appears in the program list.

Any command you could enter directly in the Casio *CFX*-9800G's computation screen can be entered as a line in a program. There are also special programming commands.

Each time you press EXE while writing a program, the Casio *CFX*-9800G *automatically* inserts the ⏎ character at the end of the previous line. For simplicity, since this happens every time you press EXE, the ⏎ character is not shown in the program listing below.

Note that while entering a program the program menu can be activated by pressing SHIFT PRGM (Figure 7.102). This menu is used to access a variety of commands that are needed for writing a program.

Figure 7.102: Program menu

The instruction manual for your Casio *CFX*-9800G gives detailed information about programming. Refer to it to learn more about programming and how to use other features of your calculator.

Enter the program QUADRATIC by pressing the given keystrokes.

Program Line	Keystrokes
"ENTER A"? → A	SHIFT A-LOCK F2 E N T E R SPACE A F2 SHIFT PRGM F4 SHIFT → ALPHA A EXE

 displays the words ENTER A on the Casio *CFX*-9800G screen and waits
 for you to input a value that will be assigned to the variable A

"ENTER B"? → B	SHIFT A-LOCK F2 E N T E R SPACE B F2 SHIFT PRGM F4 SHIFT → ALPHA B EXE
"ENTER C"? → C	SHIFT A-LOCK F2 E N T E R SPACE C F2 SHIFT PRGM F4 SHIFT → ALPHA C EXE
$B^2 - 4AC → D$	ALPHA B x^2 – 4 ALPHA A ALPHA C SHIFT → ALPHA D EXE

 calculates the discriminant and stores its value as D

$(-B + \sqrt{D}) ÷ (2A) → M$	((–) ALPHA B + SHIFT √ ALPHA D) ÷ (2 ALPHA A) SHIFT → ALPHA M EXE

 calculates one root and stores it as M

$(-B - \sqrt{D}) ÷ (2A) → N$	((–) ALPHA B – SHIFT √ ALPHA D) ÷ (2 ALPHA A) SHIFT → ALPHA N EXE
$D < 0 ⇒$ Goto 1	ALPHA D F2 F4 0 EXIT F1 F1 F2 1 EXE

 tests to see if the discriminant is negative;

 if the discriminant is negative, jumps to the line Lbl 1 below; if the discriminant is not negative, continues on to the next line

                Casio *CFX*-9800G Color Power Graphic Calculator

$D = 0 \Rightarrow$ Goto 2	ALPHA D EXIT F2 F1 0 EXIT F1 F1 F2 2 EXE

tests to see if the discriminant is zero;

if the discriminant is zero, jumps to the line Lbl 2 below; if the discriminant is not zero, continues on to the next line

"TWO REAL ROOTS"	SHIFT A-LOCK F2 T W O SPACE R E A L SPACE R O O T S F2 ALPHA EXE
M⌐	ALPHA M EXIT F5

displays one root and pauses

N	ALPHA N EXE
Goto 3	F1 F2 3 EXE

jumps to the end of the program

Lbl 1	F3 1 EXE

jumping point for the Goto command above

"COMPLEX ROOTS"	SHIFT A-LOCK F2 C O M P L E X SPACE R O O T S F2 ALPHA EXE

displays a message in case the roots are complex numbers

M⌐	ALPHA M EXIT F5
N	ALPHA N EXE
Goto 3	F1 F2 3 EXE
Lbl 2	F3 2 EXE
"DOUBLE ROOT"	SHIFT A-LOCK F2 D O U B L E SPACE R O O T F2 ALPHA EXE

displays a message in case there is a double root

M	ALPHA M EXE
Lbl 3	F3 3

When you have finished, press MENU to leave the program editor and move on.

If you want to clear a program, enter the program editor again. Move to the program you want to delete, and when the cursor is blinking next to its name, press F2 to remove it from the calculator's memory.

7.10.2 Executing a Program: To run the program you have entered, enter the COMP mode (MENU 1) and then press SHIFT PRGM F3 and the number or letter that it was named; finally, press EXE to run it. If you have forgotten its name, you must go back to the program editor to find the program, then press F1 to run it.

The program has been written to prompt you for values of the coefficients a, b, and c in a quadratic equation $ax^2 + bx + c = 0$. Input a value, then press EXE to continue the program.

If you need to interrupt a program during execution, press AC/ON.

7.11   Differentiation

7.11.1 Limits: Suppose you need to find this limit: $\lim\limits_{x \to 0} \dfrac{\sin 4x}{x}$. Plot the graph of $f(x) = \dfrac{\sin 4x}{x}$ in a convenient viewing rectangle that contains the point where the function appears to intersect the line $x = 0$ (because you want the limit as $x \to 0$). Your graph should support the conclusion that $\lim\limits_{x \to 0} \dfrac{\sin 4x}{x} = 4$ (Figure 7.103).

To test whether the conclusion that $\lim\limits_{x \to \infty} \dfrac{2x - 1}{x + 1} = 2$ is reasonable, evaluate $f(x) = \dfrac{2x - 1}{x + 1}$ for several large positive values of x (since you want the limit as $x \to \infty$). For example, evaluate $f(100)$, $f(1000)$, and $f(10,000)$. Another way to test the conclusion is to examine the graph of $f(x) = \dfrac{2x - 1}{x + 1}$ in a viewing rectangle that extends over large values of x. See, as in Figure 7.104 (where the viewing rectangle extends horizontally from 0 to 100), whether the graph is asymptotic to the horizontal line $y = 2$. Enter $\dfrac{2x - 1}{x + 1}$ for Y1 and 2 for Y2.

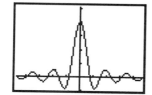

Figure 7.103: Checking $\lim\limits_{x \to 0} \dfrac{\sin 4x}{x} = 4$

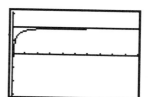

Figure 7.104: Checking $\lim\limits_{x \to \infty} \dfrac{2x - 1}{x + 1} = 2$

7.11.2 Numerical Derivatives: The derivative of a function f at x can be defined as the limit of the slopes of secant lines, so $f'(x) = \lim\limits_{\Delta x \to 0} \dfrac{f(x + \Delta x) - f(x - \Delta x)}{2\Delta x}$. And for small values of Δx, the expression $\dfrac{f(x + \Delta x) - f(x - \Delta x)}{2\Delta x}$ gives a good approximation to the limit.

Casio *CFX-9800G* Color Power Graphic Calculator

The Casio *CFX*-9800G has a function which is accessed by pressing SHIFT *d/dx* that will calculate the *symmetric difference*, $\dfrac{f(x+\Delta x)-f(x-\Delta x)}{2\Delta x}$. So to find a numerical approximation to $f'(2.5)$ when $f(x)=x^3$ and with $\Delta x = 0.001$, enter the calculation mode (MENU 1) and press SHIFT *d/dx* X,θ,T $\wedge$ 3, 2.5 , .001) EXE as shown in Figure 7.105. The format of this command is d/dx(*expression, variable, value,* Δx). The same derivative is also approximated in Figure 7.105 using $\Delta x = 0.0001$. For most purposes, $\Delta x = 0.001$ gives a very good approximation to the derivative.

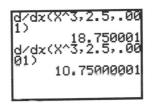

Figure 7.105: Using *d/dx*

Technology Tip: It is sometimes helpful to plot both a function and its derivative together. In Figure 7.107, the function $f(x)=\dfrac{5x-2}{x^2+1}$ and its numerical derivative (actually, an approximation to the derivative given by the symmetric difference) are graphed on viewing window that extends from −6 to 6 vertically and horizontally. In the graph mode (MENU 6), enter $\dfrac{5x-2}{x^2+1}$ for Y1. Move the highlight to Y2 and press SHIFT *d/dx* SHIFT VAR F3 *[GPH]* F1 *[Y]* 1, X,θ,T , .001) EXE (Figure 7.106).

Figure 7.106: Entering $f(x)$ and $f'(x)$

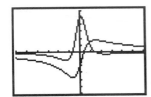

Figure 7.107: Graphs of $f(x)$ and $f'(x)$

The Casio *CFX*-9800G can compute the derivative of a point on a graph drawn in the graph mode. For example, enter the graph mode (MENU 6) and graph the function $f(x)=\dfrac{5x-2}{x^2+1}$ in a viewing rectangle so that the horizontal values are "friendly"; for example let the rectangle extend horizontally from −4.7 to 4.7. Then press SHIFT G-SOLV F6 *[◕]* F3 *[d/dx]*. The coordinates for the left-most *x*-value of the range will appear, along with the derivative at that point above the *y*-coordinate. You can use the left and right arrow keys to move to a specific point, say $x = -2.3$. Figure 7.108 shows the derivative at that point to be about −0.774692.

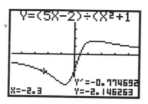

Figure 7.108: Derivative of $f(x) = \dfrac{5x-2}{x^2+1}$ at $x = -2.3$

If more than one function is graphed you can use ▲ and ▼ to scroll between the functions.

7.11.3 Newton's Method: With the Casio *CFX*-9800G, you may iterate using Newton's method to find the zeros of a function. Recall that Newton's Method determines each successive approximation by the formula $x_{n+1} = x_n - \dfrac{f(x_n)}{f'(x_n)}$.

As an example of the technique, consider $f(x) = 2x^3 + x^2 - x + 1$. In the calculation mode (MENU 1) enter this function in the function memory as f1 (refer back to Section 7.2.1). Set the range to the standard viewing window and clear any graphs from the window (SHIFT F5 *[CLS]* EXE). Graph the function in the calculation mode by pressing GRAPH SHIFT F-MEM F3 *[fn]* F1 EXE. A look at the graph suggests that it has a zero near $x = -1$, so start the iteration by storing -1 as x. Then, with the function choices still active at the bottom of the screen, press these keystrokes: X,θ,T − F1 *[f1]* ÷ SHIFT *d/dx* F1 *[f1]* , X,θ,T , .001) SHIFT → X,θ,T EXE EXE (Figure 7.109) to calculate the first two iterations of Newton's method. Press EXE repeatedly until two successive approximations differ by less than some predetermined value, say 0.0001. Note that each time you press EXE, the Casio will use the *current* value of x, and that value is changing as you continue the iteration.

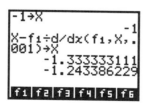

Figure 7.109: Newton's method

Technology Tip: Newton's Method is sensitive to your initial value for x, so look carefully at the function's graph to make a good first estimate. Also, remember that the method sometimes fails to converge!

You may want to write a short program for Newton's Method. See your calculator's manual for further information.

Casio *CFX*-9800G Color Power Graphic Calculator

7.12   Integration

7.12.1 Approximating Definite Integrals: The Casio *CFX*-9800G has a function which is accessed by pressing SHIFT ∫*dx* that will approximate a definite integral. For example, to find a numerical approximation to $\int_0^1 \cos x^2 dx$ first enter the calculation mode (MENU 1) and then press SHIFT ∫*dx* cos X,θ,T x^2 , 0 , 1) EXE (first two lines in Figure 7.110). The Casio *CFX*-9800G uses a method known as Simpson's Rule to perform the calculation. The format of this command is ∫(*expression, lower limit, upper limit, n*), where the last optional entry *n* is an integer from 1 to 9 which gives the number of intervals 2^n used in computing the integral. If *n* is not specified the calculator will automatically assign a value. The last two lines in Figure 7.110 shows the calculation when *n* is 7.

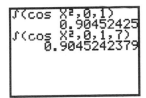

Figure 7.110: Using ∫*dx*

7.12.2 Areas: You may approximate the area under the graph of a function $y = f(x)$ between $x = A$ and $x = B$ with your Casio *CFX*-9800G. To do this you must be in the calculation mode (MENU 1) and access the command by pressing SHIFT G-∫*dx*. For example, here are the keystrokes for finding the area under the graph of the function $y = \cos x^2$ between $x = 0$ and $x = 1$. The area is represented by the definite integral $\int_0^1 \cos x^2 dx$. First clear any existing graphs and then press SHIFT G-∫*dx* cos X,θ,T x^2 , 0 , 1 (Figure 7.111) followed by EXE to draw the graph. Notice that these keystrokes are nearly identical to the keystrokes used above in computing the integral; the difference is from using the command G-∫*dx* instead of ∫*dx*. You can also add a third value *n* to your command to specify the number of intervals used in computing the integral. The range in Figure 7.112 extends from –5 to 5 horizontally and from –2 to 2 vertically. If you need to change your range press the RANGE key, enter your range, press EXIT to enter the calculation screen, and then press EXE to redraw.

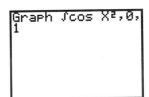

Figure 7.111: Using G-∫*dx*

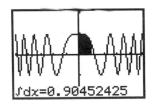

Figure 7.112: Graph and area

Technology Tip: Suppose that you want to find the area between two functions, $y = f(x)$ and $y = g(x)$ from $x = A$ and $x = B$. If $f(x) \geq g(x)$ for $A \leq x \leq B$, then graph the expression $f(x) - g(x)$ in the manner described above to find the required area.

Chapter 8

Casio CFX-9850G

8.1 Getting started with the Casio CFX-9850G

8.1.1 Basics: Press the AC/ON key to begin using your Casio CFX-9850G. The main menu screen will appear on your calculator (Figure 8.1).

If you need to adjust the display contrast, enter the contrast screen by selecting the CONT icon from the main menu using the arrow keys and pressing EXE, or by pressing D (Figure 8.2). Then press ◄ (the *left* arrow key) to lighten and ► (the *right* arrow key) to darken. If you wish to adjust the tint of the three colors, use the ▲ (the *up* arrow key) or ▼ (the *down* arrow key) to scroll the color you wish to adjust, and then use ◄ and ►.

Figure 8.1: Main menu

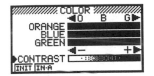

Figure 8.2: Contrast screen

When you have finished with the calculator, turn it off to conserve battery power by pressing SHIFT and then OFF. Power is automatically switched off approximately six minutes after the last operation.

Technology Tip: To return to the main menu you can press the MENU key. In general, whenever you need to return to the main menu, you can press the MENU key. Note that to enter any mode from the main menu, you just need to select the appropriate number or letter. Thus, you can jump quickly to the contrast screen by pressing MENU D.

Enter the Casio CFX-9850G's calculation mode by selecting the RUN icon from the main menu (Menu 1). You can check the settings for this mode (and for any mode you are using) by pressing SHIFT SET UP. Figure 8.3 displays an example for the calculation mode.

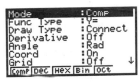

Figure 8.3: Calculation set up screen

To change the settings, use ▼ and ▲ to highlight the item you wish to change and then press the function key that corresponds to the option you want. For example, when the highlight is at Mode press F1 to select the Comp option. To start select the following options: press F1 *[Y =]* for Func Type, F1 *[Con]* for Draw Type F2 *[Off]* for Derivative, F2 *[Rad]* for Angle, F1 *[On]* for Coord, F2 *[Off]* for Grid, F1 *[On]* for Axes, F2 *[Off]* for Label, and F3 *[Norm]* for Display. (Note that you want Norm1 not Norm2 for Display; if Norm2 appears, just press F3 *[Norm]* again.) Press EXIT to leave the set up screen.

8.1.2 Editing: One advantage of the Casio *CFX*-9850G is that up to seven lines are visible at one time, so you can *see* a long calculation. For example, enter the calculation mode (MENU 1) and type this sum (Figure 8.4):

$$1 + 2 + 3 + 4 + 5 + 6 + 7 + 8 + 9 + 10 + 11 + 12 + 13 + 14 + 15 + 16 + 17 + 18 + 19 + 20$$

Then press EXE to see the answer.

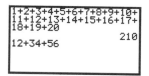

Figure 8.4: Calculation screen

Often we do not notice a mistake until we see how unreasonable an answer is. The Casio *CFX*-9850G permits you to redisplay an entire calculation, edit it easily, then execute the *corrected* calculation.

Suppose you had typed 12 + 34 + 56 as in Figure 8.4 but had *not yet* pressed EXE, when you realize that 34 should have been 74. Simply press ◄ (the *left* arrow key) as many times as necessary to move the blinking cursor left to 3, then type 7 to write over it. On the other hand, if 34 should have been 384, move the cursor back to 4, press SHIFT INS (the cursor changes to a blinking frame) and then type 8 (inserts at the cursor position and the other characters are pushed to the right). If the 34 should have been 3 only, move the cursor to 4, and press DEL to delete it.

Even if you had pressed EXE, you may still edit the previous expression. Press the *left* or *right* arrow key to *redisplay* the last expression that we entered. Now you can change it. If you press ◄, the cursor will be at the *end* of the previous expression; if you press ► the cursor will appear at the *beginning*. Even if you have already pressed some keys since the last EXE, but *not* EXE again, you can still recall the previous expression by first pressing AC$^{/ON}$ to clear the screen and then pressing ◄ or ►.

In fact the Casio *CFX*-9850G retains many prior entries. After pressing AC$^{/ON}$ to clear the screen, press ▲ repeatedly to cycle back through previous expressions. If you pass by an expression that you want, just press ▼ as many times as necessary to cycle forward.

Technology Tip: When you need to evaluate a formula for different values of a variable, use the editing feature to simplify the process. For example, suppose you want to find the balance in an investment account if there is now $5000 in the account and interest is compounded annually at the rate of 8.5%. The formula for the balance is $P = \left(1 + \frac{r}{n}\right)^{nt}$, where P = principal, r = rate of interest (expressed as a decimal), n = number of times interest is compounded each year, and t = number of years. In our example, this becomes

$5000(1+.085)^t$. Here are the keystrokes for finding the balance after $t = 3, 5,$ and 10 years. Figure 8.5 shows the first set of keystrokes and the result.

Years	Keystrokes	Balance
3	5000 (1 + .085) ∧ 3 EXE	$6386.45
5	◄ ◄ 5 EXE	$7518.28
10	◄ ◄ 10 EXE	$11,304.92

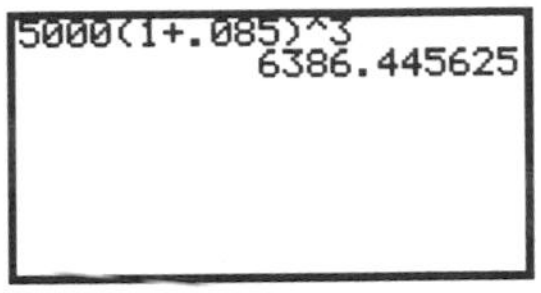

```
5000(1+.085)^3
            6386.445625
```

Figure 8.5: Editing expressions

Then to find the balance from the same initial investment but after 5 years when the annual interest rate is 7.5%, press the keys to change the last calculation above: ◄ ◄ DEL ◄ 5 ◄ ◄ ◄ ◄ ◄ 7 EXE.

8.1.3 Key Functions: Most keys on the Casio *CFX*-9850G offer access to more than one function, just as the keys on a computer keyboard can produce more than one letter ("g" and "G") or even quite different characters ("5" and "%"). The primary function of a key is indicated on the key itself, and you access that function by a simple press on the key.

To access the *second* function indicated to the *left* above a key, first press SHIFT (the cursor changes to a blinking **S** and a menu appears at the bottom of the screen) and *then* press the key. For example to calculate $\sqrt{25}$, press SHIFT $\sqrt{}$ 25 EXE.

When you want to use a letter or other character printed to the *right* above a key, first press ALPHA (the cursor changes to a blinking **A** and a menu appears at the bottom of the screen) and then the key. For example, to use the letter K in a formula, press ALPHA K. If you need several letters in a row, press SHIFT A-LOCK, which is like the CAPS LOCK key on a computer keyboard, and then press all the letters you want. Remember to press ALPHA when you are finished and want to restore the keys to their primary functions.

8.1.4 Order of Operations: The Casio *CFX*-9850G performs calculations according to the standard algebraic rules. Working outwards from inner parentheses, calculations are performed from left to right. Powers and roots are evaluated first, followed by multiplications and divisions, and then additions and subtractions.

Enter these expressions to practice using your Casio *CFX*-9850G.

Expression	Keystrokes	Display
$7 - 5 \cdot 3$	$7 - 5 \times 3$ EXE	-8
$(7 - 5) \cdot 3$	$(7 - 5) \times 3$ EXE	6
$120 - 10^2$	$120 - 10 \; x^2$ EXE	20
$(120 - 10)^2$	$(120 - 10) \; x^2$ EXE	12100
$\dfrac{24}{2^3}$	$24 \div 2 \wedge 3$ EXE	3
$\left(\dfrac{24}{2}\right)^3$	$(24 \div 2) \wedge 3$ EXE	1728
$(7 - -5) \cdot -3$	$(7 - -5) \times - 3$ EXE	-36

8.1.5 Algebraic Expressions and Memory: Your calculator can evaluate expressions such as $\dfrac{N(N+1)}{2}$ after you have entered a value for N. Suppose you want $N = 200$. Press 200 → ALPHA N EXE to store the value 200 in memory location N. Whenever you use N in an expression, the calculator will substitute the value 200 until you make a change by storing *another* number in N. Next enter the expression $\dfrac{N(N+1)}{2}$ by typing

ALPHA N (ALPHA N + 1) ÷ 2 EXE. For $N = 200$, you will find that $\dfrac{N(N+1)}{2} = 20100$.

The contents of any memory location may be revealed by typing just its letter name and then EXE. And the Casio *CFX*-9850G retains memorized values even when it is turned off, so long as its batteries are good.

8.1.6 Repeated Operations with Ans: The result of your *last* calculation is always stored in memory location Ans and replaces any previous result. This makes it easy to use the answer from one computation in another computation. For example, press 30 + 15 EXE so that 45 is the last result displayed. Then press SHIFT Ans ÷ 9 EXE and get 5 because $45 \div 9 = 5$.

With a function like division, you press the ÷ *after* you enter an argument. For such functions, whenever you would start a new calculation with the previous answer followed by pressing the function key, you may press just the function key. So instead of SHIFT Ans ÷ 9 in the previous example, you could have pressed simply ÷ 9 to achieve the same result. This technique also works for these functions: $+ \quad - \quad \times \quad \wedge \quad x^2 \quad x^{-1}$.

Here is a situation where this is especially useful. Suppose a person makes \$5.85 per hour and you are asked to calculate earnings for a day, a week, and a year. Execute the given keystrokes to find the person's incomes during these periods (Figure 8.6).

Pay Period	Keystrokes	Earnings
8-hour day	5.85 × 8 EXE	$46.80
5-day week	SHIFT Ans × 5 EXE	$234
52-week year	× 52 EXE	$12,168

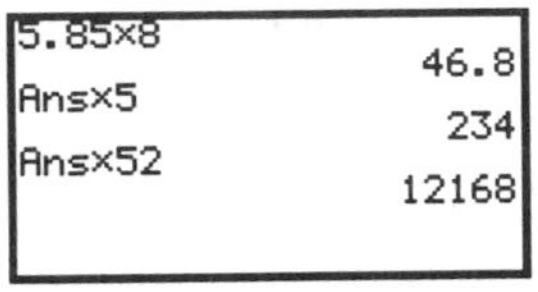

```
5.85×8
              46.8
Ans×5
               234
Ans×52
             12168
```

Figure 8.6: SHIFT Ans key

In general, the Casio *CFX*-9850G does not distinguish between the negative sign and the subtraction operator. But when you enter −4 as the *first* number in a calculation, you must use the negative key (−) rather than the − key. Press these keys for an illustration: 8 EXE − 5 EXE (−) 5 EXE.

8.1.7 The **OPTION** *Menu:*   Operators and functions associated with a scientific calculator are available either immediately from the keys of the Casio *CFX*-9850G or by the SHIFT keys. You have direct access to common arithmetic operations (x^2, SHIFT $\sqrt{}$, SHIFT x^{-1}, ∧), trigonometric functions (sin, cos, tan), and their inverses (SHIFT $\sin^{-1}$, SHIFT $\cos^{-1}$, SHIFT $\tan^{-1}$), exponential and logarithmic functions (log, SHIFT 10^x, ln, SHIFT e^x), and a famous constant (SHIFT π).

A significant difference between the Casio *CFX*-9850G graphing calculators and most scientific calculators is that the Casio *CFX*-9850G requires the argument of a function *after* the function, as you would see in a formula written in your textbook. For example, on the Casio *CFX*-9850G you calculate $\sqrt{16}$ by pressing the keys $\sqrt{}$ 16 in that order.

The Casio *CFX*-9850G has a special fraction key $a^b/_c$ for entering fractions and mixed numbers. To enter a fraction such as $\frac{2}{5}$, press 2 $a^b/_c$ 5 EXE. To enter a mixed number like $2\frac{3}{4}$, press 2 $a^b/_c$ 3 $a^b/_c$ 4 EXE. Press F↔D to toggle between the mixed number and its decimal equivalent. Press SHIFT $^d/_c$ and see $2\frac{3}{4}$ as an improper fraction, $\frac{11}{4}$.

Here are keystrokes for basic mathematical operations. Try them for practice on your Casio *CFX*-9850G.

Expression	Keystrokes	Display
$\sqrt{3^2 + 4^2}$	SHIFT $\sqrt{\ }$ (3 x^2 + 4 x^2) EXE	5
$2\frac{1}{3}$	2 $a^b\!/_c$ 1 $a^b\!/_c$ 3 EXE F↔D	2.333333333
log 200	LOG 200 EXE	2.301029996
$2.34 \cdot 10^5$	2.34 × SHIFT 10^x 5 EXE	234000

Additional mathematical operations and functions are available from the option (OPTN) menu. Press OPTN to see the first five options. The first five options in the calculation mode are listed across the bottom of the Casio *CFX*-9850G's screen and correspond to the function keys, F1 to F5 (Figure 8.7). Press F6 *[▷]* to see five more options, and press F6 *[▷]* once again to see four more options. The options available differ according to the mode you are in.

You will learn in your mathematics textbook how to apply many of them. As an example, calculate $|-5|$ by pressing OPTN F6 *[▷]* F4 *[NUM]* (for access to the numeric calculations (NUM) menu) and then F1 *[Abs]* (−) 5 EXE (Figure 8.8). To return to a previous menu, press EXIT. In this case, you press EXIT to clear the NUM menu to get back to the OPTN menu; if you press EXIT again, you clear the OPTN menu since there is no previous menu.

Figure 8.7: Option menu

Figure 8.8: Numeric calculations menu

The *factorial* of a non-negative integer is the *product* of *all* the integers from 1 up to the given integer. The symbol for factorial is the exclamation point. So 4! (pronounced *four factorial*) is $1 \cdot 2 \cdot 3 \cdot 4 = 24$. You will learn more about applications of factorials in your textbook, but for now use the Casio *CFX*-9850G to calculate 4! Press these keystrokes: 4 OPTN F6 *[▷]* F3 *[PROB]* F1 *[x!]* EXE.

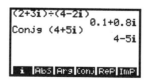

Figure 8.9: Complex number calculations

On the Casio *CFX*-9850G it is also possible to do calculations with complex numbers. Press OPTN F3 *[CPLX]* to activate the complex number calculation menu at the bottom of the screen. For example, to divide

         Casio *CFX*-9850G Color Power Graphic Calculator

$2 + 3i$ by $4 - 2i$, press (2 + 3 F1 *[i]*) ÷ (4 – 2 F1 *[i]*) EXE. The result is $0.1 + 0.8i$ (Figure 8.9).

To find the complex conjugate of $4 + 5i$ press F4 *[Conj]* (4 + 5 F1 *[i]*) EXE (Figure 8.9).

The Casio *CFX*-9850G can also solve for the real and complex roots of a quadratic or cubic function in the equation mode (MENU A). The functions to be solved must be in the form $f(x) = ax^3 + bx^2 + cx + d$ or $f(x) = ax^2 + bx + c$ where $a \neq 0$. Enter the equation mode and press F2 *[POLY]*. You are then prompted to for the degree (either 2 or 3) of your polynomial (Figure 8.10).

Figure 8.10: Prompt for degree

For example, to find all the zeros of $f(x) = x^3 - 4x^2 + 14x - 20$ select F2 *[3]* at the degree prompt and then enter the coefficients by pressing 1 EXE (–) 4 EXE 14 EXE (–) 20 EXE (Figure 8.11). If you had not pressed EXE yet, you can change a coefficient by pressing AC$^{/ON}$ and entering a new value. Else, move your cursor over the coefficient with your arrow keys and enter a new value. Now press F1 *[SOLV]* and the calculator will display the solutions (Figure 8.12).

Figure 8.11: Entering the coefficients

Figure 8.12: Zeros of $f(x) = x^3 - 4x^2 + 14x - 20$

To perform a new calculation press F1 *[RPT]*. If you are computing the roots of another cubic function then you can either edit the existing coefficients or press F3 *[CLR]* to reset all the coefficients to zero. If you are computing the roots of a quadratic function, press F2 *[DEL]* F1 *[YES]* F1 *[2]* and then enter in the coefficients of the quadratic and proceed.

Note that it may take considerable time for the calculation result of a cubic equation to appear on the display. Failure of a result to appear immediately does not mean that the unit is not functioning properly.

8.2   Functions and Graphs

8.2.1 Evaluating Functions: Suppose you receive a monthly salary of $1975 plus a commission of 10% of sales. Let x = your sales in dollars; then your wages W in dollars are given by the equation $W = 1975 + .10x$. If your January sales were \$2230 and your February sales were \$1865, what was your income during those months?

Here's one method to use your Casio *CFX*-9850G to perform this task. First enter the calculation (RUN) mode (MENU 1) and press AC/ON to clear your screen. Then set $x = 2230$ by pressing 2230 → X,θ,T. (The X,θ,T key lets you enter a variable x without having to use the ALPHA key.) Then press SHIFT ↵ to allow another expression to be input on a single command line. Finally, enter the expression $1975 + .10x$ by pressing these keys: 1975 + .10 X,θ,T. Now press EXE to calculate the answer (Figure 8.13).

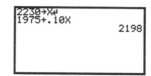

Figure 8.13: Evaluating a function

It is not necessary to repeat all these steps to find the February wages. Simply press ▶ to recall the entire previous line, change 2230 to 1865, and press EXE.

Another method is to use the function memory. The Casio *CFX*-9850G can store up to six functions. First be in the RUN mode and press AC/ON to clear your screen. Press OPTN F6 [▷] F6 [▷] F3 *[FMEM]* to display the function memory menu at the bottom of the screen. Enter the expression $1975 + .10x$ by pressing these keys: 1975 + .10 X,θ,T. Then store this as function memory number 1 by pressing F1 *[STO]* F1 *[f1]* (Figure 8.14). Press AC/ON to clear the screen, leaving the function memory menu. Then set $x = 2230$ by pressing 2230 → X,θ,T EXE. Recall the entire expression by pressing F2 *[RCL]* F1 *[f1]*, and then press EXE to calculate the answer (Figure 8.15). To find the February wages, set $x = 1865$ and then evaluate the function for the new value by pressing F2 *[RCL]* F1 *[f1]* EXE. By pressing F2 F1 you recall the entire expression. You can also press F3 *[RCL]* F1 *[f1]* to get the variable f1.

Figure 8.14: Storing a function

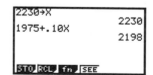

Figure 8.15: Evaluating a function

In general, to *store* a function enter it first in the calculation screen, but do *not* press EXE. If the function memory menu is active, press F1 *[STO]* and the appropriate function key from F1 to F6, otherwise you must activate the function memory menu before pressing F1 *[STO]*. To *recall* a function from the calculation screen press F2 *[RCL]* and the function key corresponding to the function you want. Press F5 *[SEE]* to see the list of functions currently in the function memory.

With the Casio *CFX*-9850G, you can also use your calculator's table mode to create a table of values for a

Casio CFX-9850G Color Power Graphic Calculator

function. From the main menu select the TABLE icon (MENU 7). Press SHIFT SET UP to get the settings for the table mode. For this section, use the setting shown in Figures 8.16 and 8.17. Press EXIT to leave the set up.

Figure 8.16: Table settings

Figure 8.17: Table settings

Move the highlight so Y1 is selected. and enter the function $1975 + .10x$ by pressing these keys: 1975 + .10 X,θ,T EXE (Figure 8.18). Press F5 *[RANG]* to set the conditions for the x-variable when generating a function table. Start is the starting value of the x-variable, End is the ending value of the x-variable, and pitch is the change of the x-variable. In this case set Start to 2230, End to 2230 and pitch to 0 (Figure 8.19).

Figure 8.18: Table function screen

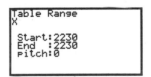

Figure 8.19: Table Range

Press EXIT F6 *[TBL]* to see the table (Figure 8.20). We will now add $x = 1865$ to the table by pressing F3 *[ROW]* F3 *[ADD]* 1865 EXE (Figure 8.21).

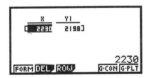

Figure 8.20: Initial table of values

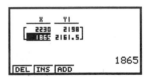

Figure 8.21: Evaluating a function in a table

Press EXIT to get the previous menu at the bottom of the screen. To add more entries use the ROW and ADD commands as described above. To delete a row use the DEL command. If you wish to change the function press EXIT to return to the range function screen. You can also add other functions to the list and select which ones are to be evaluated by pressing F4 *[SEL]*.

Technology Tip: The Casio *CFX*-9850G does not require multiplication to be expressed between variables, so *xxx* means x^3. It is often easier to press two or three *x*'s together than to search for the square key or the powers key. Of course, expressed multiplication is also not required between a constant and a variable. Hence to enter $2x^3 + 3x^2 - 4x + 5$ in the Casio *CFX*-9850G, you might save keystrokes and press just these keys: 2

X,θ,T X,θ,T X,θ,T + 3 X,θ,T X,θ,T − 4 X,θ,T + 5

8.2.2 Functions in a Graph Window: On the Casio *CFX*-9850G, you can easily generate the graph of a function. The ability to draw a graph contributes substantially to our ability to solve problems.

To graph a function on the Casio *CFX*-9850G, we use the graph mode. To enter the graph mode select the GRAPH icon (MENU 5) from the main menu. The graph function (Graph Func) screen will appear. Observe that any previous function entered in the table function screen will appear.

Press SHIFT SET UP to get the settings for the graph mode. For this section, use the setting shown in Figures 8.22 and 8.23. Press EXIT to leave the set up.

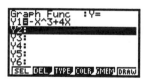

Figure 8.22: Graph settings

Figure 8.23: Graph settings

Erase any existing functions by scrolling to the desired function and pressing F2 *[DEL]* F1 *[YES]*. You can store up to twenty functions in memory. If the graph function type (Graph Func) is not set to Y= (see top of Figure 8.24), press F3 *[TYPE]* F1 *[Y=]*. To enter $y = -x^3 + 4x$, move the highlight to Y1, and then press (−) X,θ,T $\wedge$ 3 + 4 X,θ,T EXE (Figure 8.24). Now press F6 *[DRAW]* to draw the graph (Figure 8.25).

While the Casio *CFX*-9850G is busy calculating coordinates for a plot, it displays a solid square at the top right of the graph window. When you see this indicator, even though the screen does not change, you know that the calculator is working.

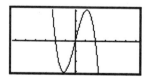

Figure 8.24: Graph function screen

Figure 8.25: Graph of $y = -x^3 + 4x$

Switch back and forth between the graph window and the graph function screen by pressing F6 *[G↔T]* in the graph window and SHIFT G↔T in the graph function screen.

To change a function in the GRAPH mode, move the highlight to the function you want to edit and use ◀ and ▶ to move the cursor to where you want to edit. Press EXE when you are done.

Technology Tip: A useful feature on the Casio *CFX*-9850G is the graph memory. The graph memory lets you store up to six sets of graph function data and recall it later when you need. For example, we can save our current screen as GM1 by pressing F5 *[GMEM]* F1 *[STO]* F1 *[GM1]* EXIT. Now you can delete and enter new functions. If you wish to recall your graph function data, press F5 *[GMEM]* F2 *[RCL]* F1 *[GM1]* EXIT. We will later use $y = -x^3 + 4x$ in Section 8.2.5.

The graph window on the Casio *CFX*-9850G may look like the one in Figure 8.25. Since the graph of $y = -x^3 + 4x$ extends infinitely far left and right and also infinitely far up and down, the Casio *CFX*-9850G can only display only a piece of the actual graph. This displayed rectangular part is called a *viewing window*.

You can easily change the viewing window to enhance your investigation of a graph. For example, press any of the arrow keys to pan the graph window in the corresponding direction. If you press the down arrow, for example, the window will pan down so that you may look at points below the current window.

The viewing window for the Casio *CFX*-9850G in Figure 8.25 shows the part of the graph that extends horizontally form –6.3 to 6.3 and vertically from –3.1 to 3.1. Press SHIFT V-Window (located at the F3 key) to see information about your viewing window. Figure 8.26 shows the View Window screen that corresponds to the viewing window in Figure 8.25. This is the *initialized* viewing window for the Casio *CFX*-9850G.

Figure 8.26: Initialized viewing window

The variables Xmin and Xmax are the minimum and maximum *x*-values of the viewing windows; Ymin and Ymax are the minimum and maximum *y*-values.

Xscale and Yscale set the spacing between tick marks on the axes.

Use the arrow keys ▲ and ▼ to move up and down from one line to another in this list. Enter a new value to over-write a previous value and then press EXE. To leave the View Window screen, press the EXIT key. Finally, press F6 *[DRAW]* to redraw the graph. The following figures show different View Window screens and the corresponding viewing window for each one. To get the window that extends horizontally and vertically from –10 to 10, press F3 *[STD]* while in the View Window screen. This is the *standard* viewing window.

Figure 8.27: Standard viewing window

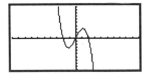

Figure 8.28: Graph of $y = -x^3 + 4x$

Figure 8.29: Custom window

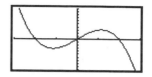

Figure 8.30: Graph of $y = -x^3 + 4x$

To return the viewing window quickly to the *initialized* viewing window, press SHIFT V-Window F1 EXIT. Then press F6 *[DRAW]* to redraw the graph.

As you pan over the graph by pressing the arrow keys, the dimensions in the View Window are updated automatically. More information about windows is presented later in Section 8.2.4.

Technology Tip: A useful feature on the Casio *CFX-9850G* is that you can save custom viewing windows to memory. You can store up to six sets of viewing windows and recall it later when needed. For example, we can save the custom window in Figure 8.29 as V·W1 by pressing F4 *[STO]* F1 *[V·W1]*. To recall it in the View Window, just press F5 *[RCL]* F1 *[V·W1]*.

Sometimes you may wish to display grid points corresponding to tick marks on the axes. Press SHIFT SET UP, scroll down to Grid, and press F1 *[On]*. You can now redraw your graph with the grid points. In general, you'll want the grid turned *off*, so do that by returning to the graph settings, scrolling down to Grid, and selecting off.

Technology Tip: It is also possible to graph a function manually from the calculation mode (MENU 1). First press SHIFT Sketch F1 *[Cls]* EXE to clear any existing graphs. Then press F5 *[GRPH]* to activate the graph command menu. Suppose you want to graph $y = -x^3 + 4x$. Press F1 *[Y=]* (−) X,θ,T ∧ 3 + 4 X,θ,T EXE. As before, you can press SHIFT V-Window to set the viewing window and F6 *[G↔T]* to switch from the graph to the calculation screen and SHIFT G↔T to switch from the calculation screen to the graph. Unless otherwise noted, we will use the graph mode for our graphs.

8.2.3 Graphing Step and Piecewise–Defined Functions:

The greatest integer function, written $[[x]]$, gives the greatest *integer* less than or equal to a number x. On the Casio *CFX-9850G*, the greatest integer function is called Intg and is located as under the numeric calculations (NUM) sub-menu of the options menu. Calculate $[[6.78]] = 6$ in the calculation mode (MENU 1) by pressing OPTN F6 *[▷]* F4 *[NUM]* F5 *[Intg]* 6.78 EXE.

To graph $y = [[x]]$, enter the graph mode (MENU 5) and enter the function in Y1 by pressing OPTN F5 *[NUM]* F5 *[Intg]* X,θ,T EXE F6. (Note that the numeric calculations menu is located in a different location in the graph mode). Figure 8.31 shows this graph in a viewing window from −5 to 5 in both directions.

The true graph of the greatest integer function is a step graph, like the one in Figure 8.32. For the graph of $y = [[x]]$, a segment should *not* be drawn between every pair of successive points. You can change the Draw Type from the connected setting to the plotted setting by pressing SHIFT SET UP F2 *[Plot]*. Press EXIT and then F6 to draw the new graph (Figure 8.32).

Casio *CFX-9850G* Color Power Graphic Calculator

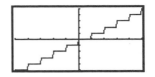

Figure 8.31: Connected graph of $y = [[x]]$

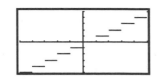

Figure 8.32: Plotted graph of $y = [[x]]$

It is not possible to graph piecewise–defined functions on the Casio *CFX*-9850G.

In general, you'll want the graph to be connected, so do that by returning to the set up screen and pressing F1 *[CON]*.

8.2.4 Graphing a Circle: Here is a useful technique for graphs that are not functions but can be "split" into a top part and a bottom part, or into multiple parts. Suppose you wish to graph the circle of radius 6 whose equation is $x^2 + y^2 = 36$. First solve for y and get an equation for the top semicircle, $y = \sqrt{36 - x^2}$, and for the bottom semicircle, $y = -\sqrt{36 - x^2}$. Then graph the two semicircles simultaneously.

Use the following keystrokes to draw this circle's graph in the graph mode (MENU 5). Store $\sqrt{36 - x^2}$ as Y1 by scrolling to Y1 and pressing SHIFT $\sqrt{}$ (36 – X,θ,T x²) EXE. Then store $-\sqrt{36 - x^2}$ as Y2 by pressing (–) SHIFT $\sqrt{}$ (36 – X,θ,T x²) EXE (Figure 8.33). Next press F6 *[DRW]* to graph both halves (Figure 8.34). Make sure that the RANGE is set large enough to display a circle of radius 6.

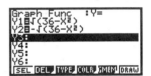

Figure 8.33: Two semicircles

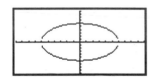

Figure 8.34: One view of the circle's graph

Instead of entering $\sqrt{36 - x^2}$ as Y2, you could have entered –Y1 as Y2 and saved some keystrokes. Try this by going into the graph function screen, scrolling to Y2 pressing (–) VARS F4 *[GRPH]* F1 *[Y]* 1 EXE. The graph should be as before. The VARS key enables you to recall graphic functions and other information from memory.

If your viewing window was set to the standard viewing window, your graph would look like Figure 8.34. Now this does *not* look a circle, because the units along the axes are not the same. You need what is called a "square" viewing window. With the graph on the screen press SHIFT F2 *[ZOOM]* F6 *[▷]* F2 *[SQR]*. The graph should now be "square" (Figure 8.35).

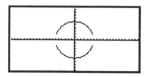

Figure 8.35: Changing to a "square" circle

Note that the Casio *CFX*-9850G's initialized viewing window is square but too small to display a circle of radius 6, so you should double the dimensions of the initialized window. Change the range to extend horizontally from −12.6 to 12.6 and vertically from −6.2 to 6.2 (Figure 8.36). The graph for the better circle is shown in Figure 8.37.

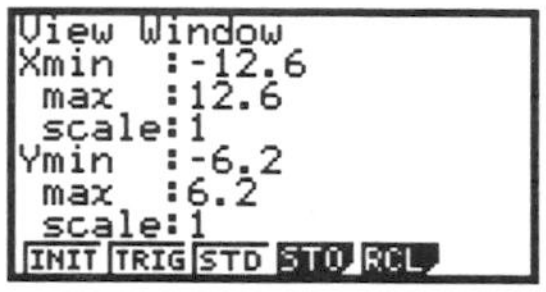

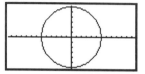

Figure 8.36: Twice standard range

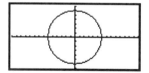

Figure 8.37: Better circle

Technology Tip: Another way to get an approximately square graph on the Casio *CFX*-9850G is to change the viewing window values so Ymax − Ymin is $\frac{1}{2}$ (Xmax − Xmin). For example, use the values in Figure 8.38 to get the corresponding graph in Figure 8.39. This method works because the dimensions of the Casio *CFX*-9850G's display are such that the ratio of vertical to horizontal is approximately $\frac{1}{2}$.

Figure 8.38: $\frac{\text{vertical}}{\text{horizontal}} = \frac{14}{28} = \frac{1}{2}$

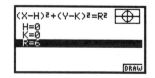

Figure 8.39: "Square" circle

Technology Tip: The square viewing window is also important when you want to judge whether two lines are perpendicular. The intersection of perpendicular lines will always *look* like a right angle in a square viewing window.

Figure 8.40: General equation of circle

Figure 8.41: Setting the circle

Casio *CFX*-9850G **Color Power Graphic Calculator**

Technology Tip: The conics mode allows you to directly graph circles on the Casio *CFX*-9850G From the main menu select the CONICS icon (MENU 9). Scroll down until the general equation of the circle is highlighted (Figure 8.40). Press EXE. To graph $x^2 + y^2 = 36$, set H = 0, K = 0, and R = 6 (Figure 8.41). Set an appropriate viewing window and press F6 *[DRAW]* or EXE.

8.2.5 TRACE: In the graph mode (MENU 5) graph the function $y = -x^3 + 4x$ from Section 8.2.2 using the initialized window. (If you had saved this as GM1, recall it now by pressing F5 *[GMEM]* F2 *[RCL]* F1 *[GM1]* EXIT.) Press F6 *[DRAW]*. When the graph window is displayed, press F1 to enable ◄ (the *left* arrow key) and ► (the *right* arrow key) to trace along the function. The coordinates that are displayed belong to points on the function's graph, so the *y*-coordinate is the calculated value of the function at the corresponding *x*-coordinate (Figure 8.42). (Note that on the Casio *CFX*-9850G the *x*-coordinate of the trace begins at the leftmost value of the window so the cursor will not appear until it is traced onto the screen.)

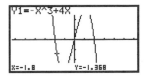

Figure 8.42: Using trace

Now plot a second function, $y = -.25x$, along with $y = -x^3 + 4x$. From the above graph window, return to the GRAPH screen by pressing EXIT, and enter the second function as Y2 by highlighting Y2 and pressing (–) .25 X,θ,T EXE (Figure 8.43). Finally, press F6 to draw both functions (Figure 8.44).

Figure 8.43: Two functions

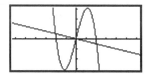

Figure 8.44: Graph of $y = -.25x$ and $y = -x^3 + 4x$

Technology Tip: The Casio *CFX*-9850G can display a graph in one of three colors: blue, orange, and green. So when you are plotting more than one function, it's helpful to color their graphs distinctly. In the graph function screen, move the highlight onto a function and press F4 *[COLR]* to activate the color menu, and then select F1 *[Blue]* for blue, F2 *[Orng]* for orange, and F3 *[Grn]* for green. Notice that each function's formula is colored to match its graph. Press EXIT to remove the color menu.

Technology Tip: Since the Casio *CFX*-9850G always draws axes in green, you may wish to color graphs in blue and orange for contrast. Also, the trace cursor will always be orange.

Notice that in Figure 8.43 the equal signs next to Y1 and Y2 are *both* highlighted. This means that *both* functions will be graphed. In the GRAPH screen, move the cursor onto Y1 and press F1 *[SEL]*. The equal sign next to Y1 should no longer be highlighted (Figure 8.45). Now press F6 and see that only Y2 is plotted (Figure 8.46).

Figure 8.45: Only Y2 active

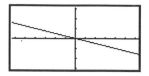

Figure 8.46: Graph of $y = -.25x$

Up to twenty different functions can be stored in list of functions in the graph function screen and any combination of them may be graphed simultaneously. You can make a function active or inactive for graphing by scrolling to the function pressing the F1 *[SEL]*. Now go back to the graph function screen and do what is needed in order to graph Y1 but not Y2.

Now activate both functions so that both graphs are plotted. Press F1 to trace and the cursor will be on the graph of $y = -x^3 + 4x$ because it is higher up on the list of active functions in the graph function screen. Press the up ▲ or down ▼ arrow key to move the cursor vertically to the graph of $y = -.25x$. Next press the left and right arrow keys to trace along the graph of $y = -.25x$. When more than one function is plotted, you can move the trace cursor vertically from one graph to another in this way.

Technology Tip: On the Casio *CFX*-9850G, when there is more than one function plotted, the function of the graph being graphed or traced can be displayed. In SET UP, move your cursor down to Graph Func and press F1 *[ON]* to turn the display on or F2 *[OFF]* to turn it off.

Technology Tip: By the way, trace the graph of $y = -.25x$ and press and hold either ◄ or ►. Eventually you will reach the left or right edge of the window. Keep pressing the arrow key and the Casio *CFX*-9850G will allow you to continue the trace by panning the viewing window. Check the View Window screen to see that the Xmin and Xmax are automatically updated.

The Casio *CFX*-9850G has a display of 127 horizontal columns of pixels and 63 vertical rows, so when you trace a curve across a graph window, you are actually moving from Xmin to Xmax in 126 equal jumps, each called Δx. You would calculate the size of each jump to be $\Delta x = \dfrac{\text{Xmax} - \text{Xmin}}{126}$. Sometimes you may want the jumps to be friendly numbers like 0.1 or 0.25 so that, when you trace along the curve, the x-coordinates will be incremented by such a convenient amount. Just set your viewing window for a particular increment Δx by making Xmax = Xmin + 126 · Δx. For example, if you want Xmin = −5 and Δx = 0.3, set Xmax = −5 + 126 · 0.3 = 32.8.

On the Casio *CFX*-9850G, to center your window around a particular point, say (h, k), and also have a certain Δx, set Xmin = $h - 63 \cdot \Delta x$ and make Xmax = $h + 63 \cdot \Delta x$. Likewise, make Ymin = $k - 31 \cdot \Delta y$ and make Ymax = $h + 31 \cdot \Delta x$. For example, to center a window around the origin (0, 0), with both horizontal and vertical increments of 0.25, set the range so that $\mathsf{Xmin} = 0 - 63 \cdot 0.25 = -15.75$, $\mathsf{Xmax} = 0 + 63 \cdot 0.25 = 15.75$, $\mathsf{Ymin} = 0 - 31 \cdot 0.25 = -7.75$ and $\mathsf{Ymax} = 0 + 31 \cdot 0.25 = 7.75$.

The Casio *CFX*-9850G's initialized window is already a friendly viewing window, centered at the origin (0, 0) with $\Delta x = \Delta y = 0.1$.

See the benefit by first plotting $y = x^2 + 2x + 1$ in a window that extends from -10 to 10 in both directions. Trace near its y-intercept, which is (0, 1), and move towards its x-intercept, which is $(-1, 0)$. Then initialize the range to the standard window and trace again near the intercepts.

8.2.6 ZOOM: Plot again the two graphs, for $y = -x^3 + 4x$ and $y = -.25x$. There appears to be an intersection near $x = 2$. The Casio *CFX*-9850G provides several ways to enlarge the view around this point. You can change the viewing window directly by pressing SHIFT V-Window and editing the values of Xmin, Xmax, Ymin, and Ymax. Figure 8.47 shows a new viewing window for the range extending from 1.5 to 2.5 horizontally and from -2.5 to 2.5 vertically.

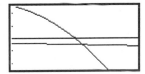

Figure 8.47: Closer view

Trace along the graphs until coordinates of a point that is close to the intersection are displayed.

A more efficient method for enlarging the view is to draw a new viewing window with the cursor. Start again with a graph of the two functions $y = -x^3 + 4x$ and $y = -.25x$ in the initialized window.

Now imagine a small rectangular box around the intersection point, near $x = 2$. Press F2 *[Zoom]* to activate the Zoom menu at the bottom of the screen (Figure 8.48).

Now press F1 *[BOX]* to draw a box to define this new viewing window. Use the arrow keys to move the cursor, which is now free-moving and whose coordinates are displayed at the bottom of the window, to one corner of the new viewing window you imagine (Figure 8.49).

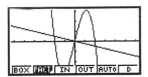

Figure 8.48: Zoom menu

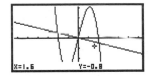

Figure 8.49: One corner selected

Press EXE to fix the corner where you moved the cursor. Use the arrow keys again to move the cursor to the diagonally opposite corner of the new rectangle (Figure 8.50). If this box looks all right to you, press EXE. The rectangular area you have enclosed will now enlarge to fill the graph window (Figure 8.51).

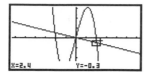

Figure 8.50: Box drawn

Figure 8.51: New viewing window

You may cancel the zoom any time *before* you press this last EXE. Press another function key such as F1 to cancel the zoom and initiate a trace instead, or press F2 to zoom again and start over. Even if you did execute the zoom, you may still return to the original viewing window and start over by pressing F2 *[Zoom]* F6 *[▷]* F1 *[ORIG]*.

The Casio *CFX*-9850G has a split screen feature that enables you to see two views of a graph simultaneously. In SET UP, move the cursor down to Dual Screen and toggle it to Graph (press F1). Now when you zoom, the left window displays the original graph and the right window displays the result of the zoom (Figure 8.52).

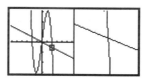

Figure 8.52: Dual graph

In the Casio *CFX*-9850G's dual graph mode, only the left side can be acted on. So to achieve another zoom, first press OPTN F2 *[SWAP]* to exchange the left and right windows. When you press SHIFT V-Window, you will find *two* ranges that can be changed independently. The F6 key toggles between the left side range and the right side range.

Technology Tip: Use the F6 key to toggle the Casio *CFX*-9850G from the dual graph to full-screen left side to full-screen right side to the graph function screen.

The Casio *CFX*-9850G can quickly magnify a graph around the cursor's location. Return once more to the initialized window for the graph of the two functions $y = -x^3 + 4x$ and $y = -.25x$. Trace along the graphs to move the cursor as close as you can to the point of intersection near $x = 2$ (Figure 8.53). Then press F2 F3 *[IN]* and the calculator draws a magnified graph, centered at the cursor's position (Figure 8.54). The range values are changed to reflect this new viewing window. Look in the View Window screen to check.

As you see in the Zoom menu (press F2), the Casio *CFX*-9850G can zoom in (press F3 *[IN]*) or zoom out (F4 *[OUT]*). Zoom out to see a larger view of the graph, centered at the cursor position. You can change the horizontal and vertical scale of the magnification by pressing F2 *[FACT]* and editing Xfact and Yfact, the hori-

Casio *CFX*-9850G Color Power Graphic Calculator

zontal and vertical magnification factors.

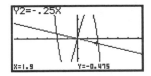

Figure 8.53: Before a zoom in

Figure 8.54: After a zoom in

Technology Tip: An advantage of zooming in from the default viewing window is that subsequent windows will also be square. Likewise, if you zoom in from a friendly viewing window, the zoomed windows will also be friendly.

The default zoom factor is 2 in both directions (press F1 *[INIT]* in the zoom factor screen). It is not necessary for Xfact and Yfact to be equal. Sometimes, you may prefer to zoom in one direction only, so the other factor should be set to 1, Press EXIT to leave the zoom factor screen.

Technology Tip: If you should zoom in too much and lose the curve, zoom back to the original viewing window and start over. Or use the arrow keys to pan over if you think the curve is not too far away. You can also just initialize the range to the Casio *CFX*-9850G's standard window.

Technology Tip: The Casio *CFX*-9850G can automatically select the necessary *vertical* range for a function. For auto scaling, activate the zoom menu (F2 *[Zoom]*) and then press F5 *[AUTO]*. Take care, because sometimes when you are graphing two functions together, the calculator will auto scale for one function in such a way that the other function will no longer be visible. For example, plot the two functions $y = -x^3 + 4x$ and $y = -.25x$ in the Casio *CFX*-9850G's initialized window, then auto scale and trace along both functions.

8.2.7 Relative Minimums and Maximums: Graph $y = -x^3 + 4x$ once again in the initialized window. This function appears to have a relative minimum near $x = -1$ and a relative maximum near $x = 1$. You may zoom and trace to approximate these extreme values.

First trace along the curve near the local minimum. Notice by how much the *x*-values and *y*-values change as you move from point to point Trace along the curve until the *y*-coordinate is as *small* as you can get it, so that you are as close as possible to the local minimum, and zoom in (press F2 *[Zoom]* F3 *[IN]* or use a zoom box). Now trace again along the curve and, as you move from point to point, see that the coordinates change by smaller amounts than before. Keep zooming and tracing until you find the coordinates of the local minimum point as accurately as you need them, approximately $(-1.15, -3.08)$.

Follow a similar procedure to find the local maximum. Trace along the curve until the *y*-coordinate is as *great* as you can get it, so that you are as close as possible to the local maximum, and zoom in. The local maximum point on the graph of $y = -x^3 + 4x$ is approximately $(1.15, 3.08)$.

The Casio *CFX*-9850G can automatically find the maximums and minimums for functions drawn in the

GRAPH mode. After graphing $y = -x^3 + 4x$, press F5 *[G-Solv]* to activate the graph solve menu at the bottom of the screen (Figure 8.55).

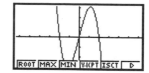

Figure 8.55: Graph solve menu

After activating the graph solve menu, press F3 *[MIN]* to calculate the minimum (Figure 8.56). Then find the maximum by pressing SHIFT G-SOLV F2 *[MAX]* (Figure 8.57).

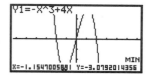

Figure 8.56: Minimum of $y = -x^3 + 4x$

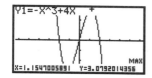

Figure 8.57: Maximum of $y = -x^3 + 4x$

Note that if you have more than one graph on the screen, the calculator will pause until you specify the graph whose maximum or minimum you want to calculate. As you use the up and down arrow keys to move between graphs, press EXE when the equation you want to evaluate appears on the screen.

If your graph has more than one maximum or minimum, you can use the left and right arrows to move between them.

8.3   Solving Equations and Inequalities

8.3.1 Intercepts and Intersections: Tracing and zooming are also used to locate an *x*-intercept of a graph, where a curve crosses the *x*-axis. For example, the graph of $y = x^3 - 8x$ crosses the *x*-axis three times (Figure 8.58). After tracing over to the *x*-intercept point that is farthest to the left, zoom in (Figure 8.59). Continue this process until you have located all three intercepts with as much accuracy as you need. The three *x*-intercepts of $y = x^3 - 8x$ are approximately −2.828, 0, and 2.828.

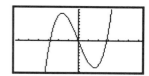

Figure 8.58: Graph of $y = x^3 - 8x$

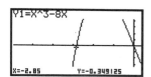

Figure 8.59: Near an *x*-intercept of $y = x^3 - 8x$

Casio *CFX*-9850G Color Power Graphic Calculator

Technology Tip: As you zoom in, you may also wish to change the spacing between tick marks on the *x*-axis so that the viewing window shows scale marks near the intercept point. Then the accuracy of your approximation will be such that the error is less than the distance between two tick marks. Change the *x*-scale from the View Window screen. Move the cursor down to Xscale and enter an appropriate value.

The *x*-intercept of a function's graph is a *root* of the equation $f(x) = 0$, and the Casio *CFX*-9850G can automatically search for the roots. First plot the function in the graph mode and then activate the graph solve menu by pressing F5 *[G-Solv]*. (Refer back to Figure 8.55.) Then press F1 *[ROOT]* to locate an *x*-intercept on the graph in the current window (Figure 8.58). The calculator searches from left to right to find an *x*-intercept in the current window; press the right arrow key to search for the next *x*-intercept to the right.

Figure 8.60: A root of $y = x^3 - 8x$

TRACE and ZOOM are especially important for locating the intersection points of two graphs, say the graphs of $y = -x^3 + 4x$ and $y = -.25x$. Trace along one of the graphs until you arrive close to an intersection point. Then press ▲ or ▼ to jump to the other graph. Notice that the *x*-coordinate does not change, but the *y*-coordinate is likely to be different (Figures 8.61 and 8.62).

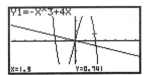

Figure 8.61: Trace on $y = -x^3 + 4x$

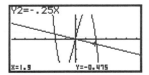

Figure 8.62: Trace on $y = -.25x$

When two *y*-coordinates are as close as they can get, you have come as close as you now can to the point of intersection. So zoom in around the intersection point, then trace again until the two *y*-coordinates are as close as possible. Continue this process until you have located the point of intersection with as much accuracy as necessary.

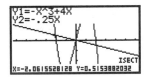

Figure 8.63: An intersection of $y = -x^3 + 4x$ and $y = -.25x$

While the graphs are displayed, automate the Casio *CFX*-9850G to search for points of intersection by pressing F5 *[G-Solv]* F5 *[ISCT]*. If more than two functions are being plotted, the calculator will ask you to specify the two whose intersection you seek. The calculator searches from left to right to find an intersection point in the current window; press the right arrow to continue the search for the *next* intersection point. Figure 8.63 shows one of the intersection points.

8.3.2 Solving Equations by Graphing: Suppose you need to solve the equation $24x^3 - 36x + 17 = 0$. First graph $y = 24x^3 - 36x + 17$ in a window large enough to exhibit *all* its x-intercepts, corresponding to all the equation's roots. Then use trace and zoom or the ROOT command, to locate each one. In fact, this equation has just one solution, approximately $x = -1.414$.

Remember that when an equation has more than one root, it may be necessary to change the viewing window a few times to locate all of them.

Technology Tip: To solve an equation like $24x^3 + 17 = 36x$, you may first transform it into standard form, $24x^3 - 36x + 17 = 0$, and proceed as above. However, you may also graph the *two* functions $y = 24x^3 + 17$ and $y = 36x$, then zoom and trace to locate their point of intersection or use the ISCT command.

8.3.3 Solving Systems by Graphing: The solutions to a system of equations correspond to the points of intersection of their graphs (Figure 8.64). For example, to solve the system $y = x^3 + 3x^2 - 2x - 1$ and $y = x^2 - 3x - 4$, first graph them together. Then use zoom and trace or the ISCT command, to locate their point of intersection, approximately $(-2.17, 7.25)$.

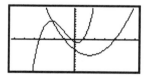

Figure 8.64: Graph of $y = x^3 + 3x^2 - 2x - 1$ and $y = x^2 - 3x - 4$

If you zoom and trace, you must judge whether the two current y-coordinates are sufficiently close for $x = -2.17$ or whether you should continue to zoom and trace to improve the approximation.

The solutions of the system of two equations $y = x^3 + 3x^2 - 2x - 1$ and $y = x^2 - 3x - 4$ correspond to the solutions of the single equation $x^3 + 3x^2 - 2x - 1 = x^2 - 3x - 4$, which simplifies to $x^3 + 2x^2 + x + 3 = 0$. So you may also graph $y = x^3 + 2x^2 + x + 3$ and find its x-intercepts to solve the system.

8.3.4 Solving Inequalities by Graphing: Consider the inequality $1 - \dfrac{3x}{2} \geq x - 4$. To solve it with your Casio

Casio *CFX*-9850G Color Power Graphic Calculator

CFX-9850G, graph the two functions $y = 1 - \dfrac{3x}{2}$ and $y = x - 4$ and locate their point of intersection, at $x = 2$ (Figure 8.65). The inequality is true when the graph of $y = 1 - \dfrac{3x}{2}$ lies *above* the graph of $y = x - 4$, and that occurs when $x < 2$. So the solution is the half-line $x \leq 2$, or $(-\infty,\ 2]$.

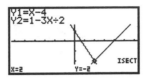

Figure 8.65. Solving $1 - \dfrac{3x}{2} \geq x - 4$

The Casio *CFX*-9850G is capable of graphing inequalities of the form $y \leq f(x)$, $y < f(x)$, $y \geq f(x)$, or $y > f(x)$. For example, to graph $y \geq x^2 - 1$ in the graph mode, scroll to **Y3**, press **F3** *[TYPE]* **F6** *[▷]* **F3** *[Y≥]*, input $x^2 - 1$, and press **EXE** (Figure 8.66). (Be sure the other graphs are not selected.) Now press **F6** to draw the graph (Figure 8.67).

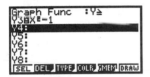

Figure 8.66: Inequality options

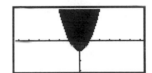

Figure 8.67: Graph of $y \geq x^2 - 1$

Next press **F1** to trace along the boundaries of the inequality. Notice that the Casio *CFX*-9850G displays co-ordinates appropriately as inequalities. Zooming is also available for inequality graphs.

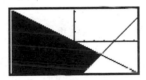

Figure 8.68: Graphs of $1 - \dfrac{3x}{2} \geq y$ and $y > x - 4$

Solve a system of inequalities, such as $1 - \dfrac{3x}{2} \geq y$ and $y > x - 4$, by plotting the two inequality graphs simultaneously. First, clear the graph window and reset the range to a convenient window. Input $1 - \dfrac{3x}{2}$ as an inequality type in Y1 by scrolling to Y1 and pressing F3 *[TYPE]* F6 *[▷]* F4 *[Y≤]* 1 – 3 X,θ,T ÷ 2 EXE. Likewise, input $x - 4$ in Y2 by scrolling to Y2 and pressing F3 *[TYPE]* F6 *[▷]* F1 *[Y>]* X,θ,T – 4 EXE. Now press F6 to draw the two inequalities as in Figure 8.68

Technology Tip: Since you can change the mode of the Casio *CFX*-9850G at any time, you can graph inequalities and equations together at the same time. Simply change to inequality type before entering an inequality, and change to rectangular type (Y=) before entering an equation.

8.4   Trigonometry

8.4.1 Degrees and Radians: The trigonometric functions can be applied to angles measured either in radians or degrees, but you should take care that the Casio *CFX*-9850G is configured for whichever measure you need. From the main menu, enter the calculation mode (MENU 1) and press SHIFT SET UP to see the current settings. To change your angle setting while in the calculation mode, press SHIFT SET UP, scroll to Angle and select the appropriate angle setting.

It's a good idea to check the angle measure setting before executing a calculation that depends on a particular measure. You may change a mode setting at any time and not interfere with pending calculations. Try the following keystrokes to see this in action.

Expression	Keystrokes	Display
sin 45°	SHIFT SET UP ▼ ▼ ▼ ▼ F1	
	EXIT sin 45 EXE	0.7071067812
sin π°	sin SHIFT π EXE	0.05480366515
sin π	SHIFT SET UP ▼ ▼ ▼ ▼ F2	
	EXIT sin SHIFT π EXE	0
sin 45	sin 45 EXE	0.8509035245
$\sin \dfrac{\pi}{6}$	sin (SHIFT π ÷ 6) EXE	0.5

The first line of keystrokes sets the Casio *CFX*-9850G in degree mode and calculates the sine of 45 *degrees*. While the calculator is still in degree mode, the second line keystrokes calculates the sine of π *degrees*, approximately 3.1415°. The third line changes to radian mode just before calculating the sine of π *radians*. The fourth line calculates the sine of 45 *radians* (the calculator remains in radian mode).

The Casio *CFX*-9850G makes it possible to mix degrees and radians in a calculation. Execute these keystrokes to calculate $\tan 45° + \sin \frac{\pi}{6}$ as shown in Figure 8.69: OPTN F6 *[▷]* F5 *[ANGL]* tan 45 F1 + sin (SHIFT π ÷ 6) F2 EXE. Pressing OPTN F6 *[▷]* F5 *[ANGL]* activates the angular unit menu at the bottom of the screen. Do you get 1.5 whether your calculator is in set *either* in degree mode *or* in radian mode?

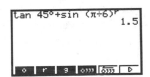

Figure 8.69: Angle measure

8.4.2 Graphs of Trigonometric Functions: When you graph a trigonometric function, you need to pay careful attention to the choice of graph window. For example, graph $y = \dfrac{\sin 30x}{30}$ in the standard viewing window. Trace along the curve to see where it is. Zoom in to a better window, or use the period and amplitude to establish a better window.

Technology Tip: In the View Window screen of the Casio *CFX*-9850G the viewing window can be set by pressing F2 to a special window for trigonometric functions so that horizontal range is from -3π to 3π in radian mode or from $-540°$ to $540°$ in degree mode and the vertical range is from -1.6 to 1.6.

8.5   Scatter Plots

8.5.1 Entering Data: The table shows the total prize money (in millions of dollars) awarded at the Indianapolis 500 race from 1981 to 1989. (*Source*: Indianapolis Motor Speedway Hall of Fame.)

Year	1981	1982	1983	1984	1985	1986	1987	1988	1989
Prize ($million)	$1.61	$2.07	$2.41	$2.80	$3.27	$4.00	$4.49	$5.03	$5.72

We'll now use the Casio *CFX*-9850G to construct a scatter plot that represents these points and to find a linear model that approximates the given data.

In the main menu select the STAT icon (MENU 2) to enter the statistics mode. Press SHIFT SET UP. You want to set the mode to the following settings: Auto for Stat Wind, On for Graph Func, None for Background, Blue for Plot/Line, Rad for Angle, On for Coord, Off for Grid, On for Axes, Off for Label, and Norm1 for Display. The first screen of the set up is shown in Figure 8.70.

Press EXIT to get back to the list screen. First clear any lists by moving the highlight to column you want to delete and pressing F6 *[▷]* F4 *[DEL-A]* F1 *[YES]*. Instead of entering the full year 198x we will enter only x. Now enter the x's in List 1 by moving the cursor to the first entry in List 1 and press 1 EXE. Continue to enter all the other years (pressing EXE after each entry). Then move to the first entry of List 2 by pressing ▶ and enter the corresponding prize data (Figure 8.71).

You may edit the lists in the same way you edit expressions in the home screen. Move the cursor to the value for any entry you wish to change, then type the correction and press **EXE**. To insert or delete data, move the cursor to the value for any entry you wish to add or delete. Press **F6** *[▷]* **F5** *[INS]* and a new entry in the list is created; press **F3** *[DEL]* and the entry is deleted. (You may have to scroll the bottom menu by pressing **F6** *[▷]*.)

Figure 8.70: Statistics mode set up

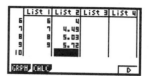

Figure 8.71: Entering the data

8.5.2 Plotting Data: Once all the data points have been entered, press **F1** *[GRPH]* to activate the statistics graph menu. (You may have to scroll the bottom menu by pressing **F6** *[▷]*.) Press **F4** *[SEL]* to see if Stat-Graph1 is set on. Set this screen as in Figure 8.72. Press **EXIT** and then **F6** *[SET]* **F1** *[GPH1]* to check the set up for StatGraph1. Set this screen as in Figure 8.73. Press **EXIT**.

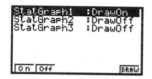

Figure 8.72: Select StatGraph 1

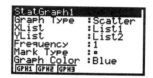

Figure 8.73: Setting up StatGraph 1

To get the scatter plot of the data press **F1** *[GPH1]*. Figure 8.74 shows the scatter plot in a viewing window automatically determined by the calculator.

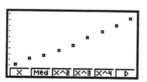

Figure 8.74: Scatter plot

8.5.3 Regression Line: The Casio *CFX-9850G* calculates the slope and y-intercept for the line that best fits all the data. From the plot of the data, press **F1** *[x]* to calculate the linear regression. As you can see in Figure 8.75 the result of the calculation names the y-intercept **b** and calls the slope **a**. The number **r** (between -1 and 1) is called the *correlation coefficient* and measures the goodness of fit of the linear regression with the data. The closer the absolute value of **r** is to 1, the better the fit; the closer the absolute value of **r** is to 0, the worse the fit.

Casio *CFX-9850G* Color Power Graphic Calculator

To graph the line $y = ax + b$ by pressing F6 *[DRAW]*. See how well this line fits with your data points (Figure 8.76).

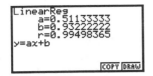

Figure 8.75: Linear regression model

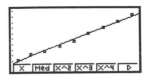

Figure 8.76: Linear regression line

8.5.4 Exponential Growth Model: The table shows the world population (in millions) from 1980 to 1992.

Year	1980	1985	1986	1987	1988	1989	1990	1991	1992
Population (millions)	4453	4850	4936	5024	5112	5202	5294	5384	5478

In the statistics mode (MENU 2), follow the procedure described above and enter the year into the List 1 and the corresponding population into List 2. Use 0 for 1980, 5 for 1985, and so on.

Now press F1 *[GRPH]* F1 *[GPH1]* as before to make a scatter plot. The calculator will automatically choose a suitable viewing window. To compute the exponential growth model $y = ae^{bx}$, press F6 *[▷]* F2 *[Exp]* (Figure 8.77). In this case, the exponential growth model is $y = 4451e^{0.017303t}$.

Press F6 *[DRAW]* to plot the exponential growth model with the scatter plot (Figure 8.78).

Figure 8.77: Exponential growth model

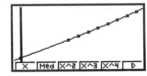

Figure 8.78: Graph of exponential growth model

8.6   Matrices

8.6.1 Making a Matrix: The Casio *CFX*-9850G can work with 26 different matrices (Mat A through Mat Z).

Here's how to create this 3×4 matrix $\begin{bmatrix} 1 & -4 & 3 & 5 \\ -1 & 3 & -1 & -3 \\ 2 & 0 & -4 & 6 \end{bmatrix}$ in your calculator as Mat A.

Enter the matrix mode by selecting the MAT icon from the main menu (MENU 3). The display will show the dimension of each matrix if the matrix exists; otherwise, it will display None (Figure 8.79). Move the highlight to Mat A and press 3 EXE 4 EXE to enter its dimensions of 3 rows by 4 columns. You will now be in the screen for Mat A.

Use the arrow keys or press EXE repeatedly to move the cursor to a matrix element you want to change. If you press EXE, you will move right across a row and then back to the first column of the next row. The element in the second row and first column in Figure 8.80 is highlighted, so that the element's current value is displayed at the bottom right corner of the screen. Continue to enter all the elements of Mat A; press EXE after inputting each value.

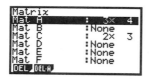

Figure 8.79: Matrix list

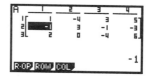

Figure 8.80: Editing a matrix

When you are finished, leave the editing screen by pressing EXIT once to return to the matrix list. To edit an existing matrix, scroll to the matrix and press EXE.

8.6.2 Matrix Math: To perform matrix calculations you must enter the calculation mode (MENU 1) and then press OPTN F2 *[MAT]* to activate the matrix calculation menu. We will assume that this menu will be active throughout our matrix calculations. To calculate the scalar multiplication 2 Mat A, press 2 F1 *[Mat]* ALPHA A EXE. The resulting matrix is displayed on the screen and is stored in matrix memory as Mat Ans (Figure 8.81). If you would rather have the matrix stored as specific matrix, say Mat C, you should press 2 F1 *[Mat]* ALPHA A → F1 *[Mat]* ALPHA C EXE. The calculator will display the matrix.

Figure 8.81: 2 Mat A

To add two matrices, say Mat A and Mat B, create Mat B (with the same dimensions as Mat A) in the matrix mode (MENU 3). Then enter the calculation mode (MENU 1), activate the matrix calculation menu by pressing OPTN F2 *[MAT]*, and then press F1 *[Mat]* ALPHA A + F1 *[Mat]* ALPHA B EXE. Again, if you want to store the answer as a specific matrix, say Mat C, then press → F1 *[Mat]* ALPHA C before executing the above command. Subtraction is performed in a similar manner.

Casio *CFX-9850G* Color Power Graphic Calculator

Now set the dimensions of Mat C as 2×3 and enter the matrix $\begin{bmatrix} 2 & 0 & 3 \\ 1 & -5 & -1 \end{bmatrix}$ as Mat C. Again you must switch between matrix and calculation mode. For matrix multiplication of Mat C by Mat A, press (assuming the matrix calculation menu is active) F1 *[Mat]* ALPHA C × F1 *[Mat]* ALPHA A EXE. If you tried to multiply Mat A by Mat C, your calculator would signal an error because the dimensions of the two matrices do not permit multiplication in this way.

The *transpose* of a matrix is another matrix with the rows and columns interchanged. To calculate the transpose of Mat A , press (assuming the matrix calculation menu is active) F4 *[Trn]* F1 *[Mat]* ALPHA A EXE.

8.6.3 Row Operations: Here are the keystrokes necessary to perform elementary row operations on a matrix. Your textbook provides a more careful explanation of the elementary row operations and their uses.

In the matrix mode enter the editing screen for Mat A. Press F1 to activate the row operations menu at the bottom of the calculator screen (Figure 8.82).

After you select a row operation, your calculator will prompt you through it. For example, to interchange the second and third rows of Mat A defined above, press F1 2 EXE 3 EXE, while the calculator prompts for the row numbers (Figure 8.83). The format of this command is Swap Row m ↔ Row n.

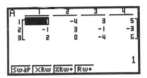

Figure 8.82: Row operations menu

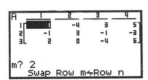

Figure 8.83: Swap rows 2 and 3

To add row 2 and row 3 and *store* the results in row 3, press F4 2 EXE 3 EXE. The format of this command is Row m + Row n → Row n.

To multiply row 2 by –4 and *store* the results in row 2, thereby replacing row 2 with new values, press F2 (–) 4 EXE 2 EXE. The format of this command is k × Row m → Row m.

To multiply row 2 by –4 and *add* the results to row 3, thereby replacing row 3 with new values, press F3 (–) 4 EXE 2 EXE 3 EXE. The format of this command is k × Row m + Row n → Row n.

Note that as you perform row operations on the Casio *CFX*-9850G, your old matrix is replaced by the new matrix, so if you want to keep the original matrix in case you need it, you should save it under another name.

For example, use row operations to solve this system of linear equations: $\begin{cases} x - 2y + 3z = 9 \\ -x + 3y = -4 \\ 2x - 5y + 5z = 17 \end{cases}$.

First enter this *augmented matrix* as Mat A in your Casio *CFX*-9850G: $\begin{bmatrix} 1 & -2 & 3 & 9 \\ -1 & 3 & 0 & -4 \\ 2 & -5 & 5 & 17 \end{bmatrix}$. Next store this

matrix as Mat E, so you may keep the original in case you need to recall it. Press MENU 1 to enter the calculation mode and then press OPTN F2 *[MAT]* F1 *[Mat]* ALPHA A → F1 *[Mat]* ALPHA E EXE.

Now return to the matrix mode and edit Mat E. Here are the row operations and their associated keystrokes. At each step, the result is stored as Mat E and replaces the previous Mat E. The completion of the row operations is shown in Figure 8.84. First press F1 to begin performing row operations.

Row Operations	*Keystrokes*
add row 1 to row 2	F4 1 EXE 2 EXE
add −2 times row 1 to row 3	F3 (−) 2 EXE 1 EXE 3 EXE
add row 2 to row 3	F4 2 EXE 3 EXE
multiply row 3 by $\frac{1}{2}$	F2 1 ÷ 2 EXE 3 EXE

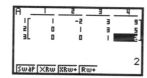

Figure 8.84: Final matrix after row operations

Thus $z = 2$, so $x = 1$ and $y = -1$.

8.6.4 Determinants and Inverses: Enter this 3×3 square matrix as Mat A: $\begin{bmatrix} 1 & -2 & 3 \\ -1 & 3 & 0 \\ 2 & -5 & 5 \end{bmatrix}$. To calculate its

determinant $\begin{vmatrix} 1 & -2 & 3 \\ -1 & 3 & 0 \\ 2 & -5 & 5 \end{vmatrix}$, enter the calculation mode and press OPTN F2 *[MAT]* F3 *[Det]* F1 *[Mat]*

ALPHA A EXE. You should find that the determinant is 2 as shown in Figure 8.85.

Figure 8.85: Determinant of Mat A

Figure 8.86: Inverse of Mat A

Since the determinant of the matrix is not zero, it has an inverse matrix. Press F1 *[Mat]* ALPHA A SHIFT x^{-1} EXE to calculate the inverse. The result is shown in Figure 8.86.

Now let's solve a system of linear equations by matrix inversion. Once again, consider $\begin{cases} x - 2y + 3z = 9 \\ -x + 3y = -4 \\ 2x - 5y + 5z = 17 \end{cases}$.

The coefficient matrix for this system is the matrix $\begin{bmatrix} 1 & -2 & 3 \\ -1 & 3 & 0 \\ 2 & -5 & 5 \end{bmatrix}$ which was entered in the previous example.

In the matrix mode enter the matrix $\begin{bmatrix} 9 \\ -4 \\ 17 \end{bmatrix}$ as Mat B. Then in the calculation mode, press OPTN F2 *[MAT]* F1 *[Mat]* ALPHA A SHIFT x^{-1} × F1 *[Mat]* ALPHA B EXE to get the answer as shown in Figure 8.87.

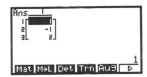

Figure 8.87: Solution matrix

The solution is still $x = 1$, $y = -1$, and $z = 2$.

8.7   Sequences

8.7.1 Iteration with the Ans key: Compute the following in the calculation mode (MENU 1). The Ans feature enables you to perform iterations to evaluate a function repeatedly. As an example, calculate $\dfrac{n-1}{3}$ for $n = 27$. Then calculate $\dfrac{n-1}{3}$ for $n =$ the answer to the previous calculation. Continue to use each answer as n in the *next* calculation. here are keystrokes to accomplish this iteration on the Casio *CFX-9850G* calculator. (See the results in Figure 8.88.) Notice that when you use Ans in place of n in a formula, it is sufficient to press EXE to continue an iteration.

Iteration	Keystrokes	Display
1	27 EXE	27
2	(SHIFT Ans − 1) ÷ 3 EXE	8.666666667
3	EXE	2.555555556
4	EXE	0.5185185185
5	EXE	−0.1604938272

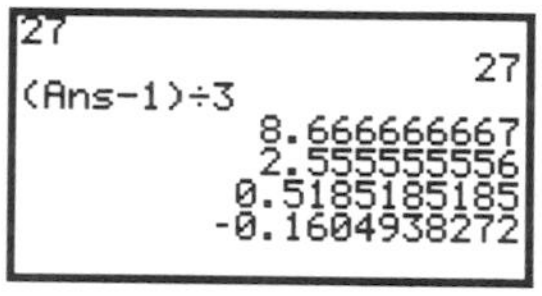

Figure 8.88: Iteration

Press EXE several more times and see what happens with this iteration. You may wish to try it again with a different starting value.

8.7.2 Arithmetic and Geometric Sequences: Use iteration with the Ans variable to determine the *n*-th term of a sequence. For example, find the 18th term of an *arithmetic* sequence whose first term is 7 and whose common difference is 4. Enter the first term 7, then start the progression with the recursion formula, SHIFT Ans + 4 EXE. This yields the 2nd term, so press EXE sixteen more times to find the 18th term. For a *geometric* sequence whose common ratio is 4, start the progression with SHIFT Ans × 4 EXE.

Of course, you could also use the *explicit* formula for the *n*-th term of an arithmetic sequence $t_n = a + (n-1)d$. First enter values for the variables a, d, and n, then evaluate the formula by pressing ALPHA A + (ALPHA N − 1) ALPHA D EXE. For a geometric sequence whose *n*-th term is given by $t_n = a \cdot r^{n-1}$, enter values for the variables a, d, and n, then evaluate the formula by pressing ALPHA A ALPHA R ∧ (ALPHA N − 1) EXE.

Figure 8.89: Recursion formula

You can also define the sequence recursively with the Casio *CFX*-9850G in the recursion mode. From the main menu enter the recursion mode by selecting the RECUR icon (MENU 8). If the recursion type a_{n+1} is not selected, press F3 *[TYPE]* F2 *[a_{n + 1}]* to select this type. Once again, let's find the 18th term of an *arithmetic* sequence whose first term is 7 and whose common difference is 4. Input the recursion formula $a_{n+1} = a_n + 4$ by pressing F4 *[n,a_n]* F2 *[a_n]* + 4 EXE (Figure 8.89). Now make $a_1 = 7$ (because the first term is a_1 where $n = 1$) and display a table that contains the 16th term a_{16} to the 20th term a_{20} by pressing F5 *[RANG]* F2 *[a₁]* 16 EXE 20 EXE 7 EXE EXIT F6 *[TABL]* (Figures 8.90 and 8.91).

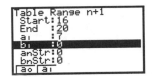

Figure 8.90: TABLE Range

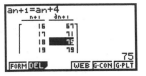

Figure 8.91: $a_{18} = 75$

To use the explicit formula in a Casio *CFX*-9850G recursion table, start a new table by pressing EXIT to return to the recursion table, moving the highlight onto the formula, and pressing F2 *[DEL]* F1 *[YES]*. Now make $a_n = 7 + (n-1) \cdot 4$ by press F3 *[TYPE]* F1 *[a_n]* 7 + (F4 *[n]* – 1) × 4 EXE. Once more, calculate term a_{18} by pressing F5 *[RANG]* 18 EXE 18 EXE EXIT F6 *[TABL]* (Figures 8.92 – 8.94).

Figure 8.92: Explicit formula

Figure 8.93: TABLE Range

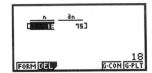

Figure 8.94: $a_{18} = 75$

Technology Tip: A table whose starting and ending range values are the same has just one entry. So to display a single *n*-th term series, set both the starting and ending range values to *n*.

There are more detailed instructions for using the recursion mode in the Casio *CFX*-9850G manual.

8.7.3 Finding Sums of Sequences: You can use recursion mode of the Casio *CFX*-9850G to find the sum of a sequence. For example, suppose you want to find the sum $\sum_{n=1}^{12} 4(0.3)^n$. First check the set up by pressing

SHIFT SET UP. If Σ Display is set to Off, press F1 *[On]*. Press EXIT and erase any existing formula by pressing F2 *[DEL]* F1 *[YES]*. Now press F3 *[TYP]* F1 *[a_n]* 4 × (.3) ∧ F4 *[n]* EXE. Now you must set the range from 1 to 12 by pressing F5 *[RANG]* 1 EXE 12 EXE EXIT. Press F6 *[TABL]*. The last entry in the table is the sum (Figure 8.95).

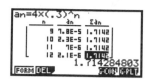

Figure 8.95: $\displaystyle\sum_{n=1}^{12} 4(0.3)^n$

Now calculate the sum starting at $n = 0$ by editing the range. You should obtain a sum of approximately 5.712848.

8.8   Parametric and Polar Graphs

8.8.1 Graphing Parametric Equations: The Casio *CFX-9850G* plots parametric equations as easily as it plots functions. Enter the graph mode by selecting the GRAPH icon from the main menu (MENU 5). Change to parametric mode by pressing F3 *[TYP]* F3 *[Parm]*. Be sure, if the independent parameter is an angle measure, that the angle measure has been set to whichever you need, Rad or Deg.

You can now enter the parametric functions. For example, here are the keystrokes need to graph the parametric equations $x = \cos^3 t$ and $y = \sin^3 t$. First check that angle measure is in radians and unselect any existing functions. In the graph function screen select a location for the equation of x and press (cos X,θ,T) ∧ 3 EXE (sin X,θ,T) ∧ 3 EXE (Figure 8.96).

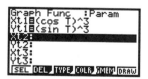

Figure 8.96: $x = \cos^3 t$ and $y = \sin^3 t$

Press SHIFT V-Window F1 *[INIT]* to set the standard graphing window and to initialize the values of T. Scroll down the screen to see that the values of T go from 0 to 2π in steps of $\frac{2\pi}{100} \approx 0.062832$. In order to provide a better viewing window scroll back up the screen and set the rectangle to go from −2 to 2 horizontally and vertically. EXIT back to the graph screen and press F6 *[DRAW]* to draw the graph (Figure 8.97).

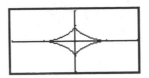

Figure 8.97: Parametric graph of $x = \cos^3 t$ and $y = \sin^3 t$

You may zoom and trace along parametric graphs just as you did with function graphs. As you trace along this graph, notice that the cursor moves in the counterclockwise direction as T increases.

Note that you can also graph parametric equations in the calculation mode (MENU 1). First in the calculation mode set up screen, set Func Type as Param. Exit the set up and clear any exiting graph(s) from the graph window by pressing SHIFT Sketch F1 *[Cls]* EXE. After setting an appropriate viewing window, press F5 *[GRPH]* F3 *[Parm]* (cos X,θ,T) ∧ 3 , (sin X,θ,T) ∧ 3) EXE.

8.8.2 Rectangular-Polar Coordinate Conversion: Conversion between rectangular and polar coordinate systems is accomplished directly through keystrokes on the Casio *CFX*-9850G. These functions use the current angle measure setting, so it is a good idea to check the default angle measure before any conversion. Of course, you may override the current angle measure setting, as explained in Section 8.4.1. For the following examples, the Casio *CFX*-9850G is set to radian measure.

We perform these calculations in the calculation mode (MENU 1). To convert between rectangular and polar coordinates, activate the coordinate conversion menu at the bottom of the screen by pressing OPTN F6 *[▷]* F5 *[ANGL]* F6 *[▷]*.

Given the rectangular coordinates $(x, y) = (4, -3)$, convert to polar coordinates (r, θ) by pressing F1 *[Pol(]* 4 , − 3) EXE. The value of r is displayed in the first row of the column and the value of θ is displayed in the second row (Figure 8.98). The polar coordinates are approximately $(5, -0.6435)$.

Suppose $(r, \theta) = (3, \pi)$. Convert to rectangular coordinates (x, y) by pressing F2 *[Rec(]* 3 , SHIFT π) EXE. The x-coordinate is displayed in the first row and the y-coordinate is displayed in the second row (Figure 8.99). The rectangular coordinates are $(-3, 0)$.

Figure 8.98: Rectangular to polar coordinates

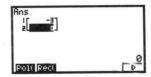

Figure 8.99: Polar to rectangular coordinates

8.8.3 Graphing Polar Equations: The Casio *CFX*-9850G graphs polar functions in the form $r = f(\theta)$. Enter the graph mode by selecting the GRAPH icon from the main menu (MENU 5). Clear or unselect any existing graphs. Change to polar mode by pressing F3 *[TYP]* F2 *[r =]*. Be sure that the angle measure has been set to whichever you need, **Rad** or **Deg**. Here we will use radian measure. Press **SHIFT V-WINDOW F1** *[INIT]* to initialize the graph window so θ goes from 0 to 2π.

For example, to graph $r = 4\sin\theta$, select a location for the function, and press **4 sin** X,θ,T **EXE** in the graph screen. Now, when you press the variable key X,θ,T, you get a θ because the calculator is in polar mode.

Choose a good viewing window and an appropriate interval and increment for θ. In Figure 8.100, the viewing window is roughly "square" and extends from −8 to 8 horizontally and from −4 to 4 vertically. (Refer back to the Technology Tip in Section 8.2.4.)

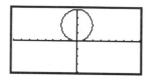

Figure 8.100: Polar graph of $r = 4\sin\theta$

Trace along this graph to see the polar coordinates of the cursor's location displayed at the bottom of the window. Zooming works just the same as before.

Note that you can also graph polar equations in the calculation mode (**MENU 1**). First in the calculation mode set up screen, set **Func Type** as **r =**. Exit the set up and clear any exiting graph(s) from the graph window by pressing **SHIFT Sketch F1** *[Cls]* **EXE**. After setting an appropriate viewing window, press **F5** *[GRPH]* **F2** *[r =]* **4 sin** X,θ,T **EXE**.

8.9   Probability

8.9.1 Random Numbers: The command **Ran#** generates a number between 0 and 1. You will find this command in the **PROB** (probability) sub-menu of the **MATH** menu. In the calculation mode (**MENU 1**) press **OPTN F6** *[▷]* **F3** *[PROB]* to activate the probability menu at the bottom of the screen. We will assume that this menu is active for the rest of this section. Then press **F4** *[Ran#]* **EXE** to generate a random number. Press **EXE** to generate another number; keep pressing **EXE** to generate more of them.

If you need a random number between, say, 0 and 10, then press **10 F4** *[Ran#]* **EXE**. To get a random number between 5 and 15, press **5 + 10 F4** *[Ran#]* **EXE**.

8.9.2 Permutations and Combinations: To calculate the number of permutations of 12 objects taken 7 at a time, $_{12}P_7$, press **12 F2** *[nPr]* **7 EXE** (Figure 8.101). Thus $_{12}P_7 = 3{,}991{,}680$.

For the number of combinations of 12 objects taken 7 at a time, $_{12}C_7$, press **12 F3** *[nCr]* **7 EXE** (Figure 8.101). Thus $_{12}C_7 = 792$.

Casio *CFX*-9850G Color Power Graphic Calculator

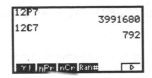

Figure 8.101: $_{12}P_7$ and $_{12}C_7$

8.9.3 Probability of Winning: A state lottery is configured so that each player chooses six different numbers from 1 to 40. If these six numbers match the six numbers drawn by the State Lottery Commission, the player wins the top prize. There are $_{40}C_6$ ways for the six numbers to be drawn. If you purchase a single lottery ticket, your probability of winning is 1 in $_{40}C_6$. Press 1 ÷ 40 F3 *[nCr]* 6 EXE to calculate your chances, but don't be disappointed.

8.10   Programming

8.10.1 Entering a Program: The Casio *CFX*-9850G is a programmable calculator that can store sequences of commands for later replay. Here's an example to show you how to enter a useful program that solves quadratic equations by the quadratic formula.

Select the PRGM icon from the main menu (MENU B). You will get the Program List screen. If you have not entered a program yet, the message No Programs will appear (Figure 8.102). Programs will be listed by the file name of the program.

Figure 8.102: Program list

To write a program press F3 *[NEW]*. Now enter a descriptive title, say QUADRATI (note only eight characters can be used in the title). The alpha lock is automatically turned on, so just type name. Then press EXE to begin writing the actual program.

Any command you could enter directly in the Casio *CFX*-9850G's computation screen can be entered as a line in a program. There are also special programming commands.

Each time you press EXE while writing a program, the Casio *CFX*-9850G *automatically* inserts the ⌐ character at the end of the previous line. For simplicity, since this happens every time you press EXE, the ⌐ character is not shown in the program listing below.

Note that while entering a program the program menu can be activated at the bottom of the screen by pressing

SHIFT PRGM (Figure 8.103). This menu is used to access a variety of commands that are need for writing a program.

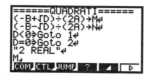

Figure 1.103: Program menu

The instruction manual for your Casio *CFX*-9850G gives detailed information about programming. Refer to it to learn more about programming and how to use other features of your calculator.

Enter the program QUADRATIC by pressing the given keystrokes.

Program Line	*Keystrokes*
"ENTER A"? → A	SHIFT A-LOCK F2 *["]* E N T E R SPACE A F2 *["]* SHIFT PRGM F4 *[?]* → ALPHA A EXE

> displays the words ENTER A on the Casio *CFX*-9850G screen and waits for you to input a value that will be assigned to the variable A

"ENTER B"? → B	SHIFT A-LOCK F2 *["]* E N T E R SPACE B F2 *["]* SHIFT PRGM F4 *[?]* → ALPHA B EXE
"ENTER C"? → C	SHIFT A-LOCK F2 *["]* E N T E R SPACE C F2 *["]* SHIFT PRGM F4 *[?]* → ALPHA C EXE
$B^2 - 4AC$ → D	ALPHA B x^2 – 4 ALPHA A ALPHA C → ALPHA D EXE

> calculates the discriminant and stores its value as D

$(-B + \sqrt{D}) \div (2A)$ → M	((–) ALPHA B + SHIFT $\sqrt{\ }$ ALPHA D) ÷ (2 ALPHA A) → ALPHA M EXE

> calculates one root and stores it as M

$(-B - \sqrt{D}) \div (2A)$ → N	((–) ALPHA B – SHIFT $\sqrt{\ }$ ALPHA D) ÷ (2 ALPHA A) → ALPHA N EXE
D < 0 ⇒ Goto 1	ALPHA D F6 *[▷]* F3 *[REL]* F4 *[<]* 0 EXIT F6 *[▷]* F3 *[JUMP]* F3 *[⇒]* F2 *[Goto]* 1 EXE

> tests to see if the discriminant is negative;
>
> if the discriminant is negative, jumps to the line Lbl 1 below; if the discriminant is not negative, continues on to the next line

Casio *CFX*-9850G Color Power Graphic Calculator

D = 0 ⇒ Goto 2	ALPHA D EXIT F6 [▷] F3 [REL] F1 [=] 0 EXIT F6 [▷] F3 [JUMP] F3 [⇒] F2 [Goto] 2 EXE

tests to see if the discriminant is zero;

if the discriminant is zero, jumps to the line Lbl 2 below; if the discriminant is not zero, continues on to the next line

"TWO REAL ROOTS"	SHIFT A-LOCK F2 ["] T W O SPACE R E A L SPACE R O O T S F2 ["] ALPHA EXE
M◢	ALPHA M EXIT F5 [◢]

displays one root and pauses

N	ALPHA N EXE
Goto 3	F3 [JUMP] F2 [Goto] 3 EXE

jumps to the end of the program

Lbl 1	F1 [Lbl] 1 EXE

jumping point for the Goto command above

"COMPLEX ROOTS"	SHIFT A-LOCK F2 ["] C O M P L E X SPACE R O O T S F2 ["] ALPHA EXE

displays a message in case the roots are complex numbers

M◢	ALPHA M EXIT F5 [◢]
N	ALPHA N EXE
Goto 3	F3 [JUMP] F2 [Goto] 3 EXE
Lbl 2	F1 [Lbl] 2 EXE
"DOUBLE ROOT"	SHIFT A-LOCK F2 ["] D O U B L E SPACE R O O T F2 ["] ALPHA EXE

displays a message in case there is a double root

M	ALPHA M EXE
Lbl 3	F1 [Lbl] 3

When you have finished, press EXIT three times to leave the program editor and reenter the program list screen.

If you want to clear a program, enter the program list screen. Move to the program you want to delete, and when the cursor is blinking next to its name, press F4 [DEL] F1 [YES] to remove it from the calculator's memory.

8.10.2 Executing a Program: To execute the program you have entered, go to the program list screen, highlight the program name and then press F1 *[EXE]* or EXE.

The program has been written to prompt you for values of the coefficients a, b, and c in a quadratic equation $ax^2 + bx + c = 0$. Input a value, then press EXE to continue the program. After the program has run, you can press EXE to run the program again.

Note that when you execute a program, the calculator puts you into the calculation mode, so if you need to return to the program list you must press MENU B.

8.11   Differentiation

8.11.1 Limits: Suppose you need to find this limit: $\lim\limits_{x \to 0} \dfrac{\sin 4x}{x}$. Plot the graph of $f(x) = \dfrac{\sin 4x}{x}$ in a convenient viewing rectangle that contains the point where the function appears to intersect the line $x = 0$ (because you want the limit as $x \to 0$). Your graph should support the conclusion that $\lim\limits_{x \to 0} \dfrac{\sin 4x}{x} = 4$ (Figure 8.104).

To test whether the conclusion that $\lim\limits_{x \to \infty} \dfrac{2x-1}{x+1} = 2$ is reasonable, evaluate $f(x) = \dfrac{2x-1}{x+1}$ for several large positive values of x (since you want the limit as $x \to \infty$). For example, evaluate $f(100)$, $f(1000)$, and $f(10,000)$. Another way to test the conclusion is to examine the graph of $f(x) = \dfrac{2x-1}{x+1}$ in a viewing rectangle that extends over large values of x. See, as in Figure 8.105 (where the viewing rectangle extends horizontally from 0 to 100), whether the graph is asymptotic to the horizontal line $y = 2$. Enter $\dfrac{2x-1}{x+1}$ for Y1 and 2 for Y2.

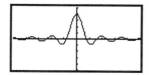

Figure 8.104: Checking $\lim\limits_{x \to 0} \dfrac{\sin 4x}{x} = 4$

Figure 8.105: Checking $\lim\limits_{x \to \infty} \dfrac{2x-1}{x+1} = 2$

8.11.2 Numerical Derivatives: The derivative of a function f at x can be defined as the limit of the slopes of secant lines, so $f'(x) = \lim\limits_{\Delta x \to 0} \dfrac{f(x + \Delta x) - f(x - \Delta x)}{2\Delta x}$. And for small values of Δx, the expression $\dfrac{f(x + \Delta x) - f(x - \Delta x)}{2\Delta x}$ gives a good approximation to the limit.

The Casio *CFX*-9850G has a function in the function analysis menu that will calculate the *symmetric differ-ence*, $\dfrac{f(x+\Delta x)-f(x-\Delta x)}{2\Delta x}$. To activate the function analysis menu in the calculation mode (MENU 1) press OPTN F4 *[CALC]*. The function you want is activated by pressing F2 *[d/dx]*. So to find a numerical approximation to $f'(2.5)$ when $f(x)=x^3$ and with $\Delta x=0.001$, press F2 *[d/dx]* x,θ,T ∧ 3, 2.5 , .001) EXE as shown in Figure 8.106. The format of this command is d/dx(*expression, variable, value,* Δx). The same derivative is also approximated in Figure 8.105 using $\Delta x=0.0001$. For most purposes, $\Delta x=0.001$ gives a very good approximation to the derivative.

Figure 8.106: Using *d/dx*

Technology Tip: It is sometimes helpful to plot both a function and its derivative together. In Figure 8.108, the function $f(x)=\dfrac{5x-2}{x^2+1}$ and its numerical derivative (actually, an approximation to the derivative given by the symmetric difference) are graphed on viewing window that extends from −6 to 6 vertically and horizon-tally. In the graph mode (MENU 5), enter $\dfrac{5x-2}{x^2+1}$ for Y1. Move the highlight to Y2 and press OPTN F2 *[CALC]* F1 *[d/dx]* VARS F4 *[GRPH]* F1 *[Y]* 1, x,θ,T , .001) EXE (Figure 8.107).

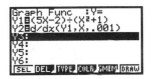

Figure 8.107: Entering $f(x)$ and $f'(x)$

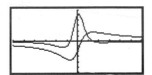

Figure 8.108: Graphs of $f(x)$ and $f'(x)$

The Casio *CFX*-9850G can compute the derivative of a point on a graph drawn in the graph mode. Enter the graph mode (MENU 5), press SHIFT SET UP, scroll down to Derivative, and press F1 *[On]*. Return to the graph function screen. For example, graph the function $f(x)=\dfrac{5x-2}{x^2+1}$ in the viewing window that extends from −6.3 to 6.3 horizontally and from −6.2 to 6.2 vertically. Then press F1 *[Trace]*. The coordinates for the left-most x-value of the range will appear, along with the derivative at that point above the y-coordinate. You can use the left and right arrow keys to move to a specific point, say $x=-2.3$. Figure 8.109 shows the deriva-tive at that point to be about −0.774692.

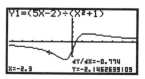

Figure 8.109: Derivative of $f(x) = \dfrac{5x-2}{x^2+1}$ at $x = -2.3$

If more than one function is graphed you can use ▲ and ▼ to scroll between the functions.

8.11.3 Newton's Method: With the Casio *CFX*-9850G, you may iterate using Newton's method to find the zeros of a function. Recall that Newton's Method determines each successive approximation by the formula

$$x_{n+1} = x_n - \frac{f(x_n)}{f'(x_n)}.$$

As an example of the technique, consider $f(x) = 2x^3 + x^2 - x + 1$. In the calculation mode (MENU 1) enter this function in the function memory as f1 (refer back to Section 8.2.1). Set the range to the standard viewing window and clear any graphs from the window (SHIFT SKETCH F1 *[CLS]* EXE). Graph the function in the calculation mode by pressing SHIFT Sketch F5 *[GRPH]* F1 *[Y =]* OPTN F6 *[▷]* F6 *[▷]* F3 *[FMEM]* F3 *[fn]* F1 *[f1]* EXE. A look at the graph suggests that it has a zero near $x = -1$, so return to the calculation screen by pressing F6 *[G↔T]* and start the iteration by storing -1 as x. Then, press these keystrokes (assuming the list of functions is still at the bottom of the screen): x,θ,T − F1 *[f1]* ÷ OPTN F4 *[CALC]* F2 *[d/dx]* OPTN F6 *[▷]* F6 *[▷]* F3 *[FMEM]* F3 *[fn]* F1 *[f1]* , x,θ,T , .001) → x,θ,T EXE EXE (Figure 8.110) to calculate the first two iterations of Newton's method. Press EXE repeatedly until two successive approximations differ by less than some predetermined value, say 0.0001. Note that each time you press EXE, the Casio will use the *current* value of x, and that value is changing as you continue the iteration.

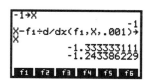

Figure 8.110: Newton's method

Technology Tip: Newton's Method is sensitive to your initial value for x, so look carefully at the function's graph to make a good first estimate. Also, remember that the method sometimes fails to converge!

You may want to write a short program for Newton's Method. See your calculator's manual for further information.

8.12   Integration

8.12.1 Approximating Definite Integrals: The Casio *CFX*-9850G has a function which is accessed from the function analysis menu that will approximate a definite integral. For example, to find a numerical approximation to $\int_0^1 \cos x^2 dx$ first enter the calculation mode (**MENU** 1) and then press **OPTN** F4 *[CALC]* ⊢4 *[∫dx]* cos X,θ,T x^2 , 0 , 1) **EXE** (first two lines in Figure 8.111). The Casio *CFX*-9850G uses a method known as Simpson's Rule to perform the calculation. The format of this command is ∫*(expression, lower limit, upper limit, n)*, where the last optional entry *n* is an integer from 1 to 9 which gives the number of intervals 2^n used in computing the integral. If *n* is not specified the calculator will automatically assign a value. The last two lines in Figure 8.111 shows the calculation when *n* is 7.

Figure 8.111: Using ∫*dx*

8.12.2 Areas: You may approximate the area under the graph of a function $y = f(x)$ between $x = A$ and $x = B$ with your Casio *CFX*-9850G. To do this you must be in the calculation mode (**MENU** 1) and access the command by pressing **SHIFT** Sketch F5 *[GRPH]* F5 *[G-∫dx]*. For example, here are the keystrokes for finding the area under the graph of the function $y = \cos x^2$ between $x = 0$ and $x = 1$. The area is represented by the definite integral $\int_0^1 \cos x^2 dx$. First clear any existing graphs and select an appropriate viewing window. Then press **SHIFT** Sketch F5 *[GRPH]* F5 *[G-∫dx]* cos X,θ,T x^2 , 0 , 1 (Figure 8.112) followed by **EXE** to draw the graph. Notice that these keystrokes are nearly identical to the keystrokes used above in computing the integral; the difference is from using the command G-∫dx instead of ∫dx. You can also add a third value *n* to your command to specify the number of intervals used in computing the integral. The range in Figure 8.113 extends from –2 to 2 horizontally and from –2 to 2 vertically. If you need to change your range press F3 *[V-Window]*, enter your new values, press **EXIT** to enter the calculation screen, and then press **EXE** to redraw.

Figure 8.112: Using G-∫*dx*

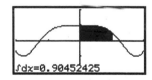

Figure 8.113: Graph and area

Technology Tip: Suppose that you want to find the area between two functions, $y = f(x)$ and $y = g(x)$ from $x = A$ and $x = B$. If $f(x) \geq g(x)$ for $A \leq x \leq B$, then graph the expression $f(x) - g(x)$ in the manner described above to find the required area.

Chapter 9

Sharp EL-9200/9300

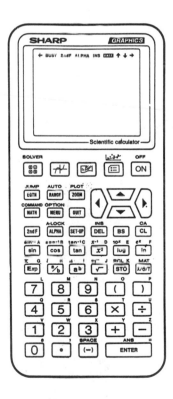

9.1 Getting started with the Sharp EL-9200/9300

9.1.1 Basics: Press the ON key to begin using your Sharp EL-9200/9300 calculator. If you need to adjust the display contrast, first press 2ndF and then press OPTION. Next press + (the *plus* key) to increase the contrast or − (the *minus* key) to decrease the contrast. Leave this menu by pressing QUIT. When you have finished with the calculator, turn it off to conserve battery power by pressing 2ndF and then OFF.

Figure 9.1: Operation mode keys

The four keys left of ON are used to set the Sharp EL-9200/9300's operation mode: for calculations, graphs, programming, statistics, statistical graphs, and solving equations (EL-9300 only). Press the first mode key, the one with arithmetic operators, for *calculation* mode. You need to press the calculation mode key before performing computations or evaluations.

Check the Sharp EL-9200/9300's settings by pressing SETUP. If necessary, use ▲ (the *up* arrow key) or ▼ (the *down* arrow key) to highlight a setting you want to change; you may also jump to a setting by pressing its letter. Next press ENTER or ▶ (the *right* arrow key) to move to a sub-menu of options; use an arrow key to move to your choice and press ENTER to put it into effect. Once again, you may just jump to an option by pressing its number. To start with, select these options as illustrated in Figure 9.2 by pressing the indicated keys: radian measure, press B 2; floating decimal point, press C 1; rectangular coordinates, E 1; one-line editing, F 2; decimal answers, G 1. Details on alternative options will be given later in this guide. For now, leave the SETUP menu by pressing QUIT. You may return to this menu at any time.

Figure 9.2: SETUP menu

Now press MENU 1 for real-number calculation mode.

9.1.2 Editing: One advantage of one-line editing on the Sharp EL-9200/9300 is that up to 8 lines are visible at a time , so you can *see* a long calculation. For example, type this sum (see Figure 9.3):

$$1 + 2 + 3 + 4 + 5 + 6 + 7 + 8 + 9 + 10 + 11 + 12 + 13 + 14 + 15 + 16 + 17 + 18 + 19 + 20$$

Then press ENTER to see the answer, too.

Often we do not notice a mistake until we see how unreasonable an answer is. The Sharp EL-9200/9300 permits you to re-display an entire calculation, edit it easily, then execute the *corrected* calculation.

Suppose you had typed 12 + 34 + 56 as in Figure 9.3 but had *not* yet pressed ENTER, when you realize that 34 should have been 74. Simply press ◀ (the *left* arrow key) as many times as necessary to move the blinking cursor left to 3, then type 7 to write over it. On the other hand, if 34 should have been 384, move the cursor back to 4, press 2ndF INS (the cursor changes to a blinking arrow) and then type 8 (inserts at the cursor position and other characters are pushed to the right). Press 2ndF INS again to cancel insert mode. If the 34 should have been 3 only, either move the cursor *onto* 4 and press DEL to delete it, or move the cursor just *after* the 4 and press BS to back space over it.

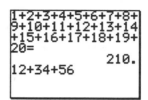

Figure 9.3: Editing expressions

Technology Tip: To move quickly to the *beginning* of an expression you are currently editing, press 2ndF ◀ ; to jump to the *end* of that expression, press 2ndF ▶ .

Even if you had pressed ENTER, you may still edit the previous expression. Press CL and then any arrow key to *recall* the last expression that was entered. Pressing the *up* or *left* arrow key restores the previous expression with the cursor at the *end* of the line; pressing the *down* or *right* arrow key restores the last expression with the cursor at the *beginning* of the line. Now you can change it. In fact, the Sharp EL-9200/9300 retains many prior entries in a "last entry" storage area. Press 2ndF and an arrow key repeatedly to cycle through previous command lines that the calculator has remembered.

Technology Tip: When you need to evaluate a formula for different values of a variable, use the editing feature to simplify the process. For example, suppose you want to find the balance in an investment account if there is now $5000 in the account and interest is compounded annually at the rate of 8.5%. The formula for the balance is $P\left(1+\frac{r}{n}\right)^{nt}$, where P = principal, r = rate of interest (expressed as a decimal), n = number of times interest is compounded each year, and t = number of years. In our example, this becomes $5000(1+.085)^t$. Here are the keystrokes for finding the balance after $t = 3$, 5, and 10 years.

Years	Keystrokes	Balance
3	CL 5000 (1 + .085) a^b 3 ENTER	$6386.45
5	◀ ◀ 5 ENTER	$7518.28
10	◀ ◀ 10 ENTER	$11,304.92

Then to find the balance from the same initial investment but after 5 years when the annual interest rate is 7.5%, press these keys to change the last calculation above: ◀ ◀ DEL ◀ 5 ◀ ◀ ◀ ◀ ◀ 7 ENTER.

9.1.3 Key Functions: Most keys on the Sharp EL-9200/9300 offer access to more than one function, just as the keys on a computer keyboard can produce more than one letter ("g" and "G") or even quite different characters ("5" and "%"). The primary function of a key is indicated on the key itself, and you access that function by a simple press on the key.

To access the *second* function indicated in *yellow* above a key, first press 2ndF (an indicator appears at the top of the screen) and *then* press the key. For example, to calculate 5^{-1}, press 5 2ndF x^{-1} ENTER.

When you want to use a letter or other character printed in *blue* above a key, first press ALPHA (another indicator appears at the top of the screen) and then the key. For example, to use the letter K in a formula, press ALPHA K. If you need several letters in a row, press 2ndF A-LOCK, which is like CAPS LOCK on a computer keyboard, and then press all the letters you want. Remember to press ALPHA when you are finished and want to restore the keys to their primary functions.

9.1.4 Order of Operations: The Sharp EL-9200/9300 performs calculations according to the standard algebraic rules. Working outwards from inner parentheses, calculations are performed from left to right. Powers and roots are evaluated first, followed by multiplications and divisions, and then additions and subtractions.

Note that the Sharp EL-9200/9300 distinguishes between *subtraction* and the *negative sign*. If you wish to enter a negative number, it is necessary to use the (-) key. For example, you would evaluate $-5-(4\cdot-3)$ by pressing (-) 5 − (4 × (-) 3) ENTER to get 7.

Enter these expressions to practice using your Sharp EL-9200/9300.

Expression	Keystrokes	Display
$7-5\cdot3$	7 − 5 × 3 ENTER	-8
$(7-5)\cdot3$	(7 − 5) × 3 ENTER	6
$20-10^2$	120 − 10 x^2 ENTER	20
$(120-10)^2$	(120 − 10) x^2 ENTER	12100
$\dfrac{24}{2^3}$	24 ÷ 2 a^b 3 ENTER	3

$\left(\dfrac{24}{2}\right)^3$	(24 ÷ 2) aᵇ 3 ENTER	1728
$(7--5)\cdot-3$	(7 − (-) 5) × (-) 3 ENTER	-36

9.1.5 Algebraic Expressions and Memory: Your calculator can evaluate expressions such as $\dfrac{N(N+1)}{2}$ *after* you have entered a value for *N*. Suppose you want *N* = 200. Press 200 STO N to store the value 200 in memory location *N*. (The STO key prepares the Sharp EL-9200/9300 for an alphabetical entry, so it is *not* necessary to press ALPHA also. And there is no need to press ENTER at the end.) Whenever you use *N* in an expression, the calculator will substitute the value 200 until you make a change by storing *another* number in *N*. Next enter the expression $\dfrac{N(N+1)}{2}$ by typing ALPHA N (ALPHA N + 1) ÷ 2 ENTER. For *N* = 200, you will find that $\dfrac{N(N+1)}{2} = 20100$.

The contents of any memory location may be revealed by typing just its letter name and then ENTER. Another way to recall the value of *N* is to press 2ndF RCL N. And the Sharp EL-9200/9300 retains memorized values even when it is turned off, so long as its batteries are good.

9.1.6 Repeated Operations with ANS: The result of your *last* calculation is always stored in memory location ANS and replaces any previous result. This makes it easy to use the answer from one computation in another computation. For example, press 30 + 15 ENTER so that 45 is the last result displayed. Then press 2ndF ANS ÷ 9 ENTER and get 5 because $\frac{45}{9} = 5$.

With a function like division, you press the ÷ key *after* you enter an argument. For such functions, whenever you would start a new calculation with the previous answer followed by pressing the function key, you may press just the function key. So instead of 2ndF ANS ÷ 9 in the previous example, you could have pressed simply ÷ 9 to achieve the same result. This technique also works for these functions: + - × x² aᵇ %
2ndF x⁻¹.

Here is a situation where this is especially useful. Suppose a person makes $5.85 per hour and you are asked to calculate earnings for a day, a week, and a year. Execute the given keystrokes to find the person's incomes during these periods (results are shown in Figure 9.4):

Pay period	Keystrokes	Earnings
8-hour day	5.85 × 8 ENTER	$46.80
5-day week	× 5 ENTER	$234
52-week year	× 52 ENTER	$12,168

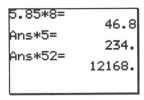

```
5.85*8=
                    46.8
Ans*5=
                    234.
Ans*52=
                 12168.
```

Figure 9.4: ANS variable

9.1.7 The MATH Menu: Operators and functions associated with a scientific calculator are available either immediately from the keys of the Sharp EL-9200/9300 or by 2ndF keys. You have direct key access to common arithmetic operations (x^2, $\sqrt{\ }$, 2ndF x^{-1}, a^b, 2ndF $\sqrt[a]{\ }$), trigonometric functions (sin, cos, tan) and their inverses (2ndF $\sin^{-1}$, 2ndF $\cos^{-1}$, 2ndF $\tan^{-1}$), exponential and logarithmic functions (log, 2ndF 10^x, ln, 2ndF e^x), and a famous constant (2ndF π).

A significant difference between the Sharp EL-9200/9300 and many scientific calculators is that the Sharp EL-9200/9300 requires the argument of a function *after* the function, as you would see a formula written in your textbook. For example, on the Sharp EL-9200/9300 you calculate $\sqrt{16}$ by pressing the keys $\sqrt{\ }$ 16 in that order.

Here are keystrokes for basic mathematical operations. Try them for practice on your Sharp EL-9200/9300.

Expression	Keystrokes	Display
$\sqrt{3^2+4^2}$	$\sqrt{\ }$ (3 x^2 + 4 x^2) ENTER	5
$2\frac{1}{3}$	2 + 3 2ndF x^{-1} ENTER	2.333333333
$\log 200$	LOG 200 ENTER	2.301029996
$2.34 \cdot 10^5$	2.34 × 2ndF 10^x 5 ENTER	234000
$\sqrt[3]{125}$	3 2ndF $\sqrt[a]{\ }$ 125 ENTER	5

Back in Section 9.1.1, you used the SETUP menu to select *one-line* editing. The Sharp EL-9200/9300 calculator also supports *equation* editing, so that mathematical expressions appear on the screen as they do in your textbook (see Figure 9.5 and Figure 9.6 for a comparison). Press SETUP F 1 ENTER to enable equation editing and once again execute the keystrokes listed above.

The equation editor displays "built-up" expressions and reduces the need for parentheses, but the calculator's response to keystrokes can be slower.

In equation editing, you terminate the scope of some functions by pressing ▶, instead of adding extra parentheses. For example, evaluate $\sqrt{3^2+4^2}+5^2$ with the following keystrokes in one-line editing: $\sqrt{\ }$ (3 x^2 + 4 x^2) + 5 x^2 ENTER. The answer is 50. Try these keystrokes again in equation editing and the answer is different because you are evaluating $\sqrt{(3^2+4^2)+5^2}$ and not $\sqrt{3^2+4^2}+5^2$ here. Now press these keys in the equation editor: $\sqrt{\ }$ 3 x² + 4 x² ▶ + 5 x² ENTER to get 50 again. In practice, simply watch the screen and the cursor to see when ▶ is needed during equation editing.

Additional mathematical operations and functions are available from the MATH menu (Figure 9.7). Press MATH to see the various options. You will learn in your mathematics textbook how to apply many of them. As an example, calculate |−5| by pressing MATH A 1 (-) 5 ENTER. To leave the MATH menu and take no other action, press QUIT.

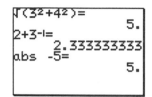

Figure 9.5: One-line editing

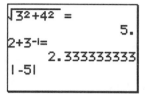

Figure 9.6: Equation editing

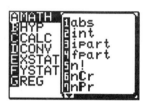

Figure 9.7: MATH menu

The *factorial* of a non-negative integer is the *product* of *all* the integers from 1 up to the given integer. The symbol for factorial is the exclamation point. So 4! (pronounced *four factorial*) is $1 \cdot 2 \cdot 3 \cdot 4 = 24$. You will learn more about applications of factorials in your textbook, but for now use the Sharp EL-9200/9300 to calculate 4! Press these keystrokes: 4 MATH A 5 ENTER.

The Sharp EL-9200/9300 can do arithmetic with complex numbers. In calculation mode, press MENU A 4; then press SETUP H and either 1 for rectangular complex coordinates or 2 for polar complex coordinates. Choose rectangular coordinates to divide $2+3i$ by $4-2i$; press (2 + 3 2ndF i) ÷ (4 − 2 2ndF i) ENTER to get $.1+.8i$.

9.2   Functions and Graphs

9.2.1 Evaluating Functions: Suppose you receive a monthly salary of $1975 plus a commission of 10% of sales. Let x = your sales in dollars; then your wages W in dollars are given by the equation $W = 1975 + .10x$.

If your January sales were \$2230 and your February sales were \$1865, what was your income during those months?

Here's how to use your Sharp EL-9200/9300 to perform this task. Let $x = 2230$ by pressing CL 2230 STO X/θ/T. (The X/θ/T key lets you enter the variable x easily without having to use the ALPHA key.) Then evaluate the expression $1975 + .10x$ for January's wages by pressing these keys: 1975 + .10 X/θ/T ENTER. Now set $x = 1865$ by pressing 1865 STO X/θ/T. These steps are shown in Figure 9.8. Recall the expression $1975 + .10x$ by pressing 2ndF ▲ *twice*, then press ENTER (see Figure 9.9) to find the February wages.

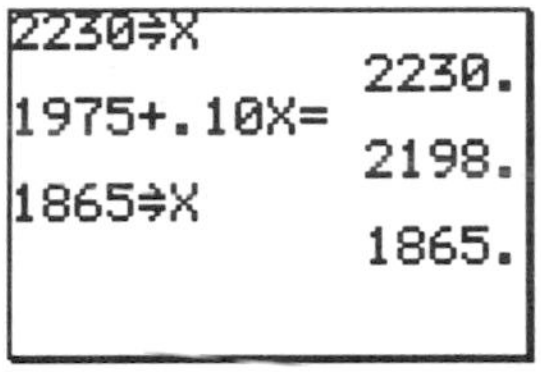

Figure 9.8: Evaluating a function

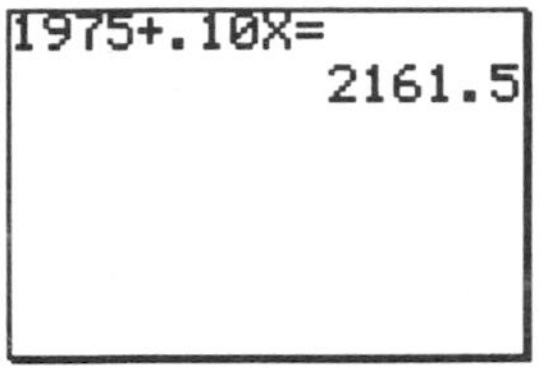

Figure 9.9: February's wages

Each time the Sharp EL-9200/9300 evaluates the function $1975 + .10x$, it uses the *current* value of x.

Technology Tip: The Sharp EL-9200/9300 does not require multiplication to be expressed between variables, so xxx means x^3. It is often easier to press two or three x's together than to search for the square key or the power key. Of course, expressed multiplication is also not required between a constant and a variable. Hence to enter $2x^3 + 3x^2 - 4x + 5$ in the Sharp EL-9200/9300, you might save keystrokes and press just these keys: 2 X/θ/T X/θ/T X/θ/T + 3 X/θ/T X/θ/T - 4 X/θ/T + 5.

9.2.2 Functions in a Graph Window: Enter the graph mode of the Sharp EL-9200/9300 by pressing the second key in the top row. The ability to draw a graph contributes substantially to our ability to solve problems.

For example, here is how to graph $y = -x^3 + 4x$. First press the graph mode key and delete anything that may be there by moving with the up or down arrow key (in equation editing, press 2ndF and then the arrow key) to Y1 or to any of the other functions and pressing CL wherever necessary. Then, with the cursor on the top line Y1, press (-) X/θ/T a^b 3 + 4 X/θ/T to enter the function (as in Figure 9.10 for one-line editing and Figure 9.11 for equation editing).

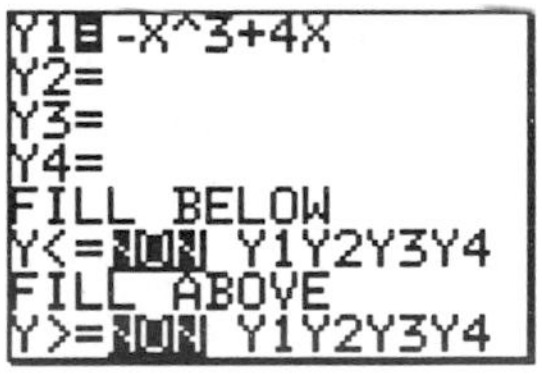

Figure 9.10: Graph entry - one-line editing

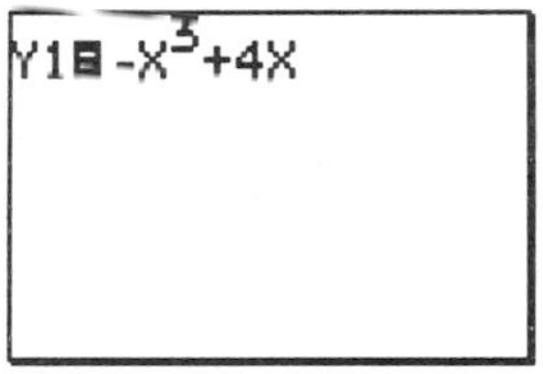

Figure 9.11: Graph entry - equation editing

Now press the graph mode key again and the Sharp EL-9200/9300 changes to a window with the graph of $y = -x^3 + 4x$. You may return to edit the function by pressing either EQTN or MENU A and a number.

Your graph window may look like the one in Figure 9.12 or it may be different. Since the graph of $y = -x^3 + 4x$ extends infinitely far left and right and also infinitely far up and down, the Sharp EL-9200/9300 can display only a piece of the actual graph. This displayed rectangular part is called a *viewing rectangle*. You can easily change the viewing rectangle to enhance your investigation of a graph.

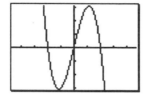

Figure 9.12: Graph of $y = -x^3 + 4x$

The viewing rectangle in Figure 9.12 shows the part of the graph that extends horizontally from -4.7 to 4.7 and vertically from -3.1 to 3.1. Press RANGE to see information about your viewing rectangle; press arrow keys to move between the X RANGE and the Y RANGE screens. Figures 9.13 and 9.14 show the RANGE screens that correspond to the viewing rectangle in Figure 9.12. This is the *default* viewing rectangle for the Sharp EL-9200/9300.

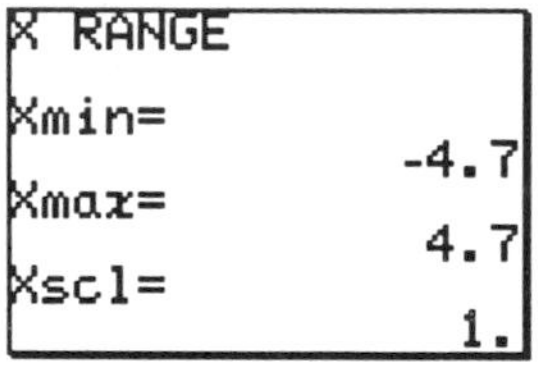

```
X RANGE
Xmin=
              -4.7
Xmax=
               4.7
Xscl=
                1.
```

Figure 9.13: Default X RANGE

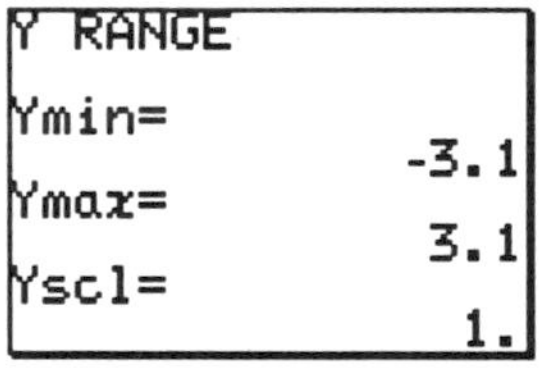

```
Y RANGE
Ymin=
              -3.1
Ymax=
               3.1
Yscl=
                1.
```

Figure 9.14: Default Y RANGE

The variables Xmin and Xmax are the minimum and maximum *x*-values of the viewing rectangle; Ymin and Ymax are its minimum and maximum *y*-values. Xscl and Yscl set the spacing between tick marks on the axes.

Use the arrow keys ▲ and ▼ to move up and down from one line to another in these lists; pressing the ENTER key will move down the list. Input a new value. The Sharp EL-9200/9300 will *not* permit a maximum that is *less* than the corresponding minimum. Also, remember to use the (-) key, not - (which is subtraction), when you want to enter a negative value. The following figures show different viewing rectangles with their ranges.

Sharp EL-9200/9300 Graphing Scientific Calculator

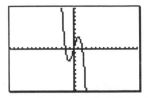

Figure 9.15: Window [-15, 15] by [-10, 10]

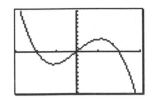

Figure 9.16: Window [-3, 3] by [-10, 10]

To set the range quickly to default values (see Figures 9.13 and 9.14), press RANGE MENU A ENTER. You may also set range values suitable for graphing particular functions, such as power and root functions (RANGE MENU B), exponential and logarithmic functions (RANGE MENU C), and trigonometric functions (RANGE MENU D).

Technology Tip: After you input a function, press 2ndF AUTO to draw the graph in a window with the current X RANGE values but automatically scaled in the vertical direction. This is an advantage when you are not sure how tall a viewing rectangle to set.

9.2.3 Piecewise-Defined Functions: The *greatest integer function*, written [[x]], gives the greatest *integer* less than or equal to a number x. On the Sharp EL-9200/9300, the greatest integer function is called int and is located by pressing MATH A 2 (see Figure 7.7). So calculate [[6.78]] = 6 by pressing the calculation mode key and then MATH A 2 6.78 ENTER.

To graph y = [[x]], press the graph mode key, move beside Y1, and press CLEAR MATH A 2 X/θ/T. Then press the graph mode key again to draw the graph. Figure 9.17 shows this graph in a viewing rectangle from -5 to 5 in both directions.

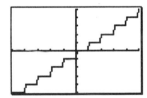

Figure 9.17: Connected graph of y = [[x]]

Figure 9.18: Dot graph of y = [[x]]

The true graph of the greatest integer function is a step graph, like the one in Figure 9.18. For the graph of y = [[x]], a segment should *not* be drawn between every pair of successive points. You can change from Connected line to Dot graph on the Sharp EL-9200/9300 by pressing MENU C 2. Restore the calculator to Connected by pressing MENU C 1.

9.2.4 Graphing a Circle: Here is a useful technique for graphs that are not functions, but that can be "split" into a top part and a bottom part, or into multiple parts. Suppose you wish to graph the circle whose equation is $x^2 + y^2 = 36$. First solve for y and get an equation for the top semicircle, $y = \sqrt{36 - x^2}$, and for the bottom semicircle, $y = -\sqrt{36 - x^2}$. Then graph the two semicircles simultaneously.

The keystrokes to draw this circle's graph follow. Enter $\sqrt{36 - x^2}$ as Y1 and $-\sqrt{36 - x^2}$ as Y2 (see Figure 9.19) by pressing the graph mode key and then $\sqrt{}$ (36 - X/θ/T x²) ENTER (-) $\sqrt{}$ (36 - X/θ/T x²). Press the graph mode key again to draw them both.

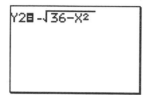

Figure 9.19: Bottom semicircle

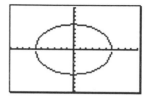

Figure 9.20: Circle's graph

If your range were set so that the viewing rectangle extends from -10 to 10 in both directions, your graph would look like Figure 9.20. Now this does *not* look like a circle, because the units along the axes are not the same. Press RANGE and change the viewing rectangle to extend from -12 to 12 in the horizontal direction and from -8 to 8 in the vertical direction and see a graph (Figure 9.21) that appears more circular.

Technology Tip: The way to get a circle's graph to look circular is to change the range variables so that the value of Ymax - Ymin is $\frac{2}{3}$ times Xmax - Xmin. The method works because the dimensions of the Sharp EL-9200/9300's display are such that the ratio of vertical to horizontal is approximately $\frac{2}{3}$.

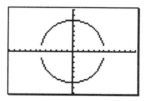

Figure 9.21: A circle

The two semicircles in Figure 9.21 do not meet because of an idiosyncrasy in the way the Sharp EL-9200/9300 plots a graph.

Technology Tip: A square viewing rectangle, in which units along both axes are the same, is also important when you want to judge whether two lines are perpendicular. The intersection of perpendicular lines will always *look* like a right angle in a square viewing rectangle.

9.2.5 *TRACE:* Graph $y = -x^3 + 4x$ in the default viewing rectangle. Press either of the arrow keys ◀ or ▶ and see the cursor move along the graph. The coordinates of the cursor's location are displayed at the bottom of the screen, as in Figure 9.22, in floating decimal format. The cursor is constrained to the function. The coordinates that are displayed belong to points on the function's graph, so the y-coordinate is the calculated value of the function at the corresponding x-coordinate.

Figure 9.22: Trace on $y = -x^3 + 4x$

Press 2ndF ◀ or 2ndF ▶ to jump to the endpoints of the part of the graph that is displayed in a window. Remove the trace cursor and its coordinates from the graph window by pressing CL.

Now plot a second function, $y = -.25x$, along with $y = -x^3 + 4x$. Press MENU A 2 and enter $-.25x$ for Y2, then press the graphing mode key.

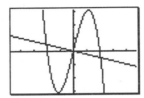

Figure 9.23: $y = -x^3 + 4x$ and $y = -.25x$

Note that the equal signs next to Y1 and Y2 are *both* highlighted. This means *both* functions will be graphed. Press MENU A 1, move the cursor directly on top of the equal sign next to Y1 and press ENTER. This equal sign should no longer be highlighted. Now press the graph mode key and see that only Y2 is plotted.

So up to 4 different functions may be stored in the Y= list and any combination of them may be graphed simultaneously. You can make a function active or inactive for graphing by pressing ENTER on its equal sign to highlight (activate) or remove the highlight (deactivate). Go back and do what is needed in order to graph Y1 but not Y2.

Now activate Y2 again so that both graphs are plotted. Press ◀ or ▶ and the cursor appears first on the graph of $y = -x^3 + 4x$ because it is Y1. Press ▲ to move the cursor vertically to the graph of $y = -.25x$; move the cursor back to $y = -x^3 + 4x$ by pressing ▼. Next press the right and left arrow keys to trace along the graph of $y = -.25x$. When more than one function is plotted, you can move the trace cursor vertically from one graph to another in this way.

Technology Tip: By the way, trace along the graph of $y = -.25x$ and press and hold either ◀ or ▶. Eventually you will reach the left or right edge of the window. Keep pressing the arrow key and the Sharp EL-

9200/9300 will allow you to continue the trace by panning the viewing rectangle. Check the RANGE screen to see that Xmin and Xmax are automatically updated.

The Sharp EL-9200/9300's display has 95 horizontal columns of pixels and 63 vertical rows. So when you trace a curve across a graph window, you are actually moving from Xmin to Xmax in 94 equal jumps, each called Δx. You would calculate the size of each jump to be $\Delta x = \dfrac{\text{Xmax} - \text{Xmin}}{94}$. Sometimes you may want the jumps to be friendly numbers like .1 or .25 so that, when you trace along the curve, the x-coordinates will be incremented by such a convenient amount. Just set your viewing rectangle for a particular increment Δx by making Xmax = Xmin + 94·Δx. For example, if you want Xmin = -5 and Δx = .3, set Xmax = -5 + 94·.3 = 23.2. Likewise, set Ymax = Ymin + 62·Δy if you want the vertical increment to be some special Δy.

To center your window around a particular point, say (h, k), and also have a certain Δx, set Xmin = h - 47·Δx and Xmax = h + 47·Δx. Likewise, make Ymin = k - 31·Δy and Ymax = k + 31·Δy. For example, to center a window around the origin, (0, 0), with both horizontal and vertical increments of .25, set the range so that Xmin = 0 - 47·.25 = -11.75, Xmax = 0 + 47·.25 = 11.75, Ymin = 0 - 31·.25 = -7.75, and Ymax = 0 + 31·.25 = 7.75.

The Sharp EL-9200/9300's standard window is already a friendly viewing rectangle, centered at the origin (0, 0) with $\Delta x = \Delta y = 0.1$.

See the benefit by first plotting $y = x^2 + 2x + 1$ in a graphing window extending from -5 to 5 in both directions. Trace near its y-intercept, which is (0, 1), and move towards its x-intercept, which is (-1, 0). Then initialize the range to the default window and trace again near the intercepts.

9.2.6 ZOOM: Plot again the two graphs, for $y = -x^3 + 4x$ and for $y = -.25x$. There appears to be an intersection near $x = 2$. The Sharp EL-9200/9300 provides several ways to enlarge the view around this point. You can change the viewing rectangle directly by pressing RANGE and editing the values of Xmin, Xmax, Ymin, and Ymax. Figure 9.22 shows a new window extending from 1 to 3 horizontally and from -2 to 1 vertically. Trace has been turned on and the coordinates of a point on $y = -x^3 + 4x$ that is close to the intersection are displayed.

A more efficient method for enlarging the view is to draw a new viewing rectangle with the cursor. Start again with a graph of the two functions $y = -x^3 + 4x$ and $y = -.25x$ in a default viewing rectangle (press RANGE MENU ENTER for the default window, from -4.7 to 4.7 along the x-axis and from -3.1 to 3.1 along the y-axis).

Figure 9.24: Closer view

Now imagine a small rectangular box around the intersection point, near $x = 2$. Press ZOOM 1 (Figure 9.25) to draw a box to define this new viewing rectangle. Use the arrow keys to move the cursor, whose coordinates are displayed at the bottom of the window, to one corner of the new viewing rectangle you imagine.

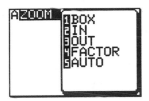

Figure 9.25: ZOOM menu

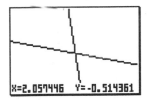

Figure 9.26: New viewing rectangle

Press ENTER to fix the corner where you have moved the cursor. Use the arrow keys again to move the cursor to the diagonally opposite corner of the new rectangle. If this box looks all right to you, press ENTER. The rectangular area you have enclosed will now enlarge to fill the graph window (Figure 9.26).

You may cancel the zoom any time *before* this last ENTER by pressing CL.

You can also gain a quick magnification of the graph around the cursor's location. Return once more to the default range for the graph of the two functions $y = -x^3 + 4x$ and $y = -.25x$. Trace as close as you can to the point of intersection near $x = 2$ (see Figure 9.22). Then press ZOOM 2 and the calculator draws a magnified graph, centered at the cursor's position (Figure 9.27). The range variables are changed to reflect this new viewing rectangle. Look in the RANGE menu to check.

As you see in the ZOOM menu (Figure 9.25), the Sharp EL-9200/9300 can zoom in (press ZOOM 2) or zoom out (press ZOOM 3). Zoom out to see a larger view of the graph, centered at the cursor position. You can change the horizontal and vertical scale of the magnification by pressing ZOOM 4 and editing X-FACTOR and Y-FACTOR, the horizontal and vertical magnification factors.

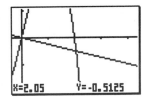

Figure 9.27: After a zoom in

It is not necessary for **X-FACTOR** and **Y-FACTOR** to be equal. Sometimes, you may prefer to zoom in one direction only, so the other factor should be set to 1. As usual, press **QUIT** to leave the **ZOOM** menu.

Technology Tip: If you should zoom in too much and lose the curve, press **ZOOM 5** for auto scaling and start again.

An advantage of zooming in from the default viewing rectangle or from a friendly viewing rectangle is that subsequent windows will also be friendly.

9.2.7 Relative Minimums and Maximums: Graph $y = -x^3 + 4x$ once again in the default viewing rectangle (Figure 9.12). This function appears to have a relative minimum near $x = -1$ and a relative maximum near $x = 1$. You may zoom and trace to approximate these extreme values.

First trace along the curve near the local minimum. Notice by how much the x-values and y-values change as you move from point to point. Trace along the curve until the y-coordinate is as *small* as you can get it, so that you are as close as possible to the local minimum, and zoom in (press **ZOOM 2** or use a zoom box). Now trace again along the curve and, as you move from point to point, see that the coordinates change by smaller amounts than before. Keep zooming and tracing until you find the coordinates of the local minimum point as accurately as you need them, approximately (-1.15, -3.08).

Follow a similar procedure to find the local maximum. Trace along the curve until the y-coordinate is as *great* as you can get it, so that you are as close as possible to the local maximum, and zoom in. The local maximum point on the graph of $y = -x^3 + 4x$ is approximately (1.15, 3.08).

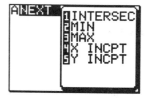

Figure 9.28: JUMP menu

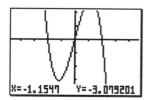

Figure 9.29: Relative minimum on $y = -x^3 + 4x$

The Sharp EL-9200/9300 automates the search for relative minimum and relative maximum points. Trace along the curve until the cursor is to the *left* of a local extreme point. Then press **2ndF JUMP** (Figure 9.28) and choose **2** for a minimum value of the function or **3** for a maximum value. The calculator searches from left to right for the next relative minimum or maximum and displays the *approximate* coordinates of the relative minimum/maximum point (see Figure 9.29).

9.3   Solving Equations and Inequalities

9.3.1 Intercepts and Intersections: Tracing and zooming are also used to locate an x-intercept of a graph, where a curve crosses the x-axis. For example, the graph of $y = x^3 - 8x$ crosses the x-axis three times (see

Figure 9.30). After tracing over to the x-intercept point that is furthest to the left, zoom in (Figure 9.31). Continue this process until you have located all three intercepts with as much accuracy as you need. The three x-intercepts of $y = x^3 - 8x$ are approximately -2.828, 0, and 2.828.

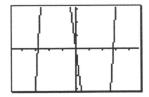

Figure 9.30: Graph of $y = x^3 - 8x$

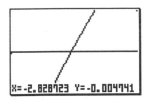

Figure 9.31: An x-intercept of $y = x^3 - 8x$

Technology Tip: As you zoom in, you may also wish to change the spacing between tick marks on the x-axis so that the viewing rectangle shows scale marks near the intercept point. Then the accuracy of your approximation will be such that the error is less than the distance between two tick marks. Change the x-scale on the Sharp EL-9200/9300 from the **RANGE** menu. Move the cursor down to Xscl and enter an appropriate value.

The Sharp EL-9200/9300 automates the search for x-intercepts. First trace along the graph until the cursor is just left of an x-intercept. Press **2ndF JUMP** (Figure 9.28) and choose **4** to find the next x-intercept of this function. Repeat until you have located all x-intercepts of this graph.

An x-intercept of a function's graph is a *root* of the equation $f(x) = 0$. So these techniques for locating x-intercepts also serve to find the roots of an equation.

TRACE and **ZOOM** are especially important for locating the intersection points of two graphs, say the graphs of $y = -x^3 + 4x$ and $y = -.25x$. Trace along one of the graphs until you arrive close to an intersection point. Then press ▲ or ▼ to jump to the other graph. Notice that the x-coordinate does not change, but the y-coordinate is likely to be different (see Figures 9.32 and 9.33).

When the two y-coordinates are as close as they can get, you have come as close as you now can to the point of intersection. So zoom in around the intersection point, then trace again until the two y-coordinates are as close as possible. Continue this process until you have located the point of intersection with as much accuracy as necessary.

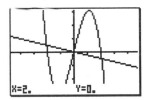

Figure 9.32: Trace on $y = -x^3 + 4x$

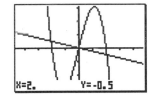

Figure 9.33: Trace on $y = -.25x$

Automate the search for points of intersection by tracing along one curve until you are left of an intersection. Then press 2ndF JUMP 1 to locate the next intersection point. The calculator displays *approximate* coordinates; zoom in to improve the approximation.

9.3.2 Solving Equations by Graphing: Suppose you need to solve the equation $24x^3 - 36x + 17 = 0$. First graph $y = 24x^3 - 36x + 17$ in a window large enough to exhibit *all* its x-intercepts, corresponding to all the equation's roots. Then use trace and zoom, or the Sharp EL-9200/9300's JUMP menu, to locate each one. In fact, this equation has just one solution, approximately $x = -1.414$.

Remember that when an equation has more than one root, it may be necessary to change the viewing rectangle a few times to locate all of them.

Technology Tip: To solve an equation like $24x^3 + 17 = 36x$, you may first transform it into standard form, $24x^3 - 36x + 17 = 0$, and proceed as above. However, you may also graph the *two* functions $y = 24x^3 + 17$ and $y = 36x$, then zoom and trace to locate their point of intersection.

9.3.3 Solving Systems by Graphing: The solutions to a system of equations correspond to the points of intersection of their graphs (Figure 9.34). For example, to solve the system $y = x^2 - 3x - 4$ and $y = x^3 + 3x^2 - 2x - 1$, first graph them together. Then zoom and trace, or 2ndF JUMP, to locate their point of intersection, approximately (-2.17, 7.25).

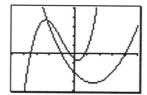

Figure 9.34: Solving a system of equations

You must judge whether the two current y-coordinates are sufficiently close for $x = -2.17$ or whether you should continue to zoom and trace to improve the approximation.

The solutions of the system of two equations $y = x^3 + 3x^2 - 2x - 1$ and $y = x^2 - 3x - 4$ correspond to the solutions of the single equation $x^3 + 3x^2 - 2x - 1 = x^2 - 3x - 4$, which simplifies to $x^3 + 2x^2 + x + 3 = 0$. So you may also graph $y = x^3 + 2x^2 + x + 3$ and find its x-intercepts to solve the system.

9.3.4 Solving Inequalities by Graphing: Consider the inequality $-\frac{3x}{2} \geq x - 4$. To solve it with your Sharp EL-9200/9300, graph the two functions $y = 1 - \frac{3x}{2}$ and $y = x - 4$ (Figure 9.35). First locate their point of in-

Sharp EL-9200/9300 Graphing Scientific Calculator

tersection, at $x = 2$. The inequality is true when the graph of $y = 1 - \dfrac{3x}{2}$ lies *above* the graph of $y = x - 4$, and that occurs for $x < 2$. So the solution is the half-line $x \le 2$, or $(-\infty, 2]$.

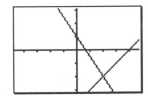

Figure 9.35: Solving $\quad -\dfrac{3x}{2} \ge x - 4$

The Sharp EL-9200/9300 is capable of shading the region above or below a graph or between two graphs. For example, to graph $y \ge x^2 - 1$, first input the function $y = x^2 - 1$ as Y1. Then press MENU 5 (see Figure 9.36). Move the cursor to Y1 in the FILL ABOVE part and press ENTER to highlight it. These keystrokes instruct the calculator to shade the region *above* $y = x^2 - 1$. The result is shown in Figure 9.37.

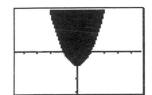

Figure 9.36: FILL menu   Figure 9.37: Graph of $y \ge x^2 - 1$

To clear the shading, press MENU 5 and highlight NON for both FILL BELOW and FILL ABOVE.

Now use shading to solve the previous inequality, $\quad -\dfrac{3x}{2} \ge x - 4$. Input $y = 1 - \dfrac{3x}{2}$ as Y1 and $y = x - 4$ as Y2. Then press MENU 5 and highlight Y1 for FILL BELOW and Y2 for FILL ABOVE. The shading extends left from $x = 2$, hence the solution to $\quad -\dfrac{3x}{2} \ge x - 4$ is the half-line $x \le 2$, or $(-\infty, 2]$.

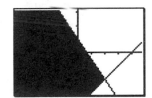

Figure 9.38: FILL menu   Figure 9.39: Graph of $\quad -\dfrac{3x}{2} \ge x - 4$

9.4.1 Degrees and Radians: The trigonometric functions can be applied to angles measured either in radians or degrees, but you should take care that the Sharp EL-9200/9300 is configured for whichever measure you need. Back in calculation mode (press the calculation mode key first), press SETUP to see the current settings. Next press B and then 1 for degrees or 2 for radians. To leave the SETUP menu, press QUIT.

It's a good idea to check the angle measure setting before executing a calculation that depends on a particular measure. You may change a mode setting at any time and not interfere with pending calculations. Try the following keystrokes to see this in action.

Expression	Keystrokes	Display
$\sin 45^\circ$	SETUP B 1 ENTER sin 45 ENTER	0.707106781
$\sin \pi^\circ$	sin 2ndF π ENTER	0.054803665
$\sin \pi$	sin 2ndF π SETUP B 2 ENTER ENTER	0.
$\sin 45$	sin 45 ENTER	0.850903524
$\sin \frac{\pi}{6}$	sin 2ndF π % 6 ENTER	0.5

The first line of keystrokes sets the Sharp EL-9200/9300 in degree mode and calculates the sine of 45 *degrees*. While the calculator is still in degree mode, the second line of keystrokes calculates the sine of π *degrees*, 3.1415°. The third line changes to radian mode just before calculating the sine of π *radians*. The fourth line calculates the sine of 45 *radians* (the calculator is already in radian mode).

Technology Tip: Here's how to mix degrees and radians in a calculation. Execute these keystrokes to calculate $\tan 45^\circ + \sin \frac{\pi}{6}$ as shown in Figure 9.40: tan 45 SETUP B 1 ENTER ENTER + sin 2ndF π % 6 SETUP B 2 ENTER ENTER. Do you get 1.5 whether your calculator began *either* in degree mode *or* in radian mode?

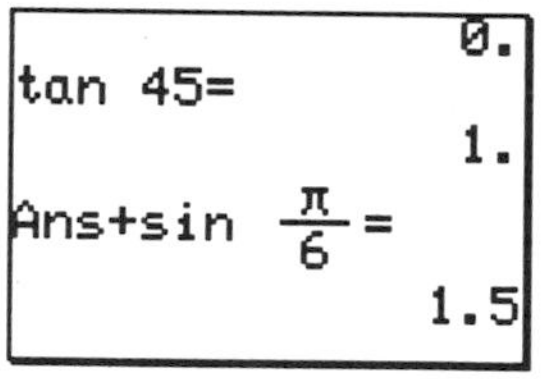

Figure 9.40: Angle measure

9.4.2 Graphs of Trigonometric Functions: When you graph a trigonometric function, you need to pay careful attention to the choice of graph window. For example, graph $y = \dfrac{\sin 30x}{30}$ in the default viewing rectangle. Trace along the curve to see where it is. Zoom in to a better window, or use the period and amplitude to establish better RANGE values.

Technology Tip: Create a good viewing rectangle for a trigonometric graph by pressing RANGE MENU D and selecting from the list of trigonometric functions.

9.5   Scatter Plots

9.5.1 Entering Data: This table shows total prize money (in millions of dollars) awarded at the Indianapolis 500 race from 1981 to 1989. (*Source:* Indianapolis Motor Speedway Hall of Fame.)

Year	1981	1982	1983	1984	1985	1986	1987	1988	1989
Prize ($ million)	$1.61	$2.07	$2.41	$2.80	$3.27	$4.00	$4.49	$5.03	$5.72

We'll now use the Sharp EL-9200/9300 to construct a scatter plot that represents these points and to find a linear model that approximates the given data.

To enter statistics mode, press the fourth operation mode key, the one with the image of a data card. Before entering the data, press MENU D 2 ENTER to clear away any previous data. Then press 3 to select two-variable data format (Figure 7.41).

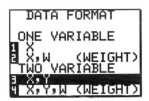

Figure 9.41: DATA FORMAT menu

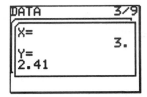

Figure 9.42: DATA card

A DATA card (Figure 9.42) is now on the screen. Instead of entering the full year 198x, enter only x. Here are the keystrokes for the first three years: 1 ENTER 1.61 ENTER 2 ENTER 2.07 ENTER 3 ENTER 2.41 ENTER and so on. Continue to enter all the data.

Use the left and right arrow keys to browse through the data cards. MENU B enables you to jump to the first data card (2ndF ◀ is a shortcut), the last data card (2ndF ▶ is a shortcut), or any data card that you specify by its number. You may edit statistical data in the same way you edit expressions in the home screen. Move the cursor to the *x* or *y* value for any data point you wish to change, then type the correction.

9.5.2 Plotting Data: Once all the data points have been entered, press 2ndF and the statistics graphing key (the fourth key in the operation mode row) then E ENTER to draw a scatter plot. Your viewing rectangle is important, so you may wish to change the RANGE to improve the view of the data. Figure 9.43 shows the scatter plot in a window extending from 0 to 10 horizontally and from 0 to 6 vertically.

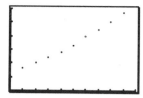

Figure 9.43: Scatter plot

9.5.3 Regression Line: The Sharp EL-9200/9300 calculates the slope and *y*-intercept for the line that best fits all the data. After the data points have been entered, press the statistics mode key and MENU 2 to calculate a linear regression model. As you see in Figure 9.44, the Sharp EL-9200/9300 names the *y*-intercept a and calls the slope b. The number r (between -1 and 1) is called the *correlation coefficient* and measures the goodness of fit of the linear regression equation with the data. The closer |r| is to 1, the better the fit; the closer |r| is to 0, the worse the fit.

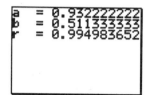

Figure 9.44: Linear regression model

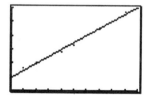

Figure 9.45: Linear regression line

Graph the line $y = a + bx$ by pressing 2nF and the statistics graphing key, then F 1. See how well this line fits with your data points (see Figure 9.45).

Technology Tip: In calculation mode, you gain access to the quantities a, b, and r from the MATH menu, option G.

9.6   Matrices

9.6.1 Making a Matrix: The Sharp EL-9200/9300 can display and use 26 different matrices, each identified by a letter of the alphabet. Here's how to create this 3×4 matrix $\begin{bmatrix} 1 & -4 & 3 & 5 \\ -1 & 3 & -1 & -3 \\ 2 & 0 & -4 & 6 \end{bmatrix}$ in your calculator.

In calculation mode, press MENU A 3 for matrix mode, then MENU B 0 1 to edit matrix A. If some other matrix A were already in the calculator's memory, first press MENU D 0 1 ENTER to clear it away, then MENU B 0 1 to create a new one. You will be prompted for matrix A's dimensions, so press 3 ENTER 4 ENTER. If you need to change the dimensions of matrix A, press MENU C 0 1 and input its new dimensions.

Use the arrow keys to move the cursor directly to a matrix element you want to change. If you press ENTER, you will move down a column and then right to the next row. Continue to enter all the elements of matrix A.

Leave matrix editing by pressing QUIT and return to the home screen.

9.6.2 Matrix Math:
With your Sharp EL-9200/9300 set in matrix mode, you can perform many calculations with matrices. First, let's see matrix A itself (the one you created in the preceding section) by pressing MAT A ENTER.

Technology Tip: Now, because the calculator is in matrix mode, press the variable key X/θ/T for the mat prefix. And this MAT key prepares your Sharp EL-9200/9300 for an alphabetical entry, so it is *not* necessary to press ALPHA before typing the letter name for a matrix.

Calculate the scalar multiplication 2A by pressing 2 MAT A ENTER. To replace matrix B by 2A, press 2 MAT A STO MAT B ENTER. Press MENU to verify that the dimensions of matrix B have been changed automatically to reflect this new value.

Add two matrices A and B by pressing MAT A + MAT B ENTER. Subtraction is similar.

Now set the dimensions of matrix C to 2×3 and enter this as C: $\begin{bmatrix} 2 & 0 & 3 \\ 1 & -5 & -1 \end{bmatrix}$. For matrix multiplication of C by A, press MAT C × MAT A ENTER. You should get $\begin{bmatrix} 8 & -8 & -6 & 28 \\ 4 & -19 & 12 & 14 \end{bmatrix}$.

Technology Tip: If you tried to multiply A by C, your Sharp EL-9200/9300 would signal an error because the dimensions of the two matrices do not permit multiplication this way.

You may use x^2 to abbreviate multiplying a matrix M by itself, but take care that M is a *square* matrix or such multiplication is not possible. For example, to calculate $M \cdot M$, press MAT M x^2 ENTER.

The *transpose* of a matrix A is another matrix with the rows and columns interchanged. The symbol for the transpose of A is A^T. To calculate A^T, press MATH E 5 MAT A ENTER.

9.6.3 Row Operations:
Here are the keystrokes necessary to perform elementary row operations on a matrix. Your textbook provides more careful explanation of the elementary row operations and their uses.

To interchange the second and third rows of a matrix A, press MATH F 1 ALPHA A ALPHA , 2 ALPHA , 3) ENTER. The format of this command is row swap(*matrix, row1, row2*).

To add row 2 and row 3 and store the results in row 3, press MATH F 2 ALPHA A ALPHA , 2 ALPHA , 3) ENTER. The format of this command is row plus(*matrix, row1, row2*).

And to multiply row 2 by -4 and *store* the results in row 2, thereby replacing row 2 with new values, press MATH F 3 (-) 4 ALPHA , ALPHA A ALPHA , 2) ENTER. The format of this command is row mult(*scalar, matrix, row*).

To multiply row 2 by -4 and *add* the results to row 3, thereby replacing row 3 with new values, press MATH F 4 (-) 4 ALPHA , ALPHA A ALPHA , 2 ALPHA , 3) ENTER. The format of this command is row m.p.(*scalar, matrix, row1, row2*).

Technology Tip: It is important to remember that your Sharp EL-9200/9300 does *not* store a matrix obtained as the result of any row operations. The calculator places a result only in the temporary ANS matrix. So when you need to perform several row operations in succession, it is a good idea to store the result of each one in a temporary place. You may wish to use matrix Z to hold such intermediate results.

For example, use elementary row operations to solve this system of linear equations: $\begin{cases} x - 2y + 3z = 9 \\ -x + 3y = -4 \\ 2x - 5y + 5z = 17 \end{cases}$.

First enter this *augmented matrix* as A in your Sharp EL-9200/9300: $\begin{bmatrix} 1 & -2 & 3 & 9 \\ -1 & 3 & 0 & -4 \\ 2 & -5 & 5 & 17 \end{bmatrix}$. Next store this matrix in C (press MAT A STO MAT C ENTER) so you may keep the original in case you need to recall it.

Here are the row operations and their associated keystrokes. At each step, the result is stored in C and replaces the previous matrix C.

Row Operation	*Keystrokes*
row plus(C, 1, 2)	MATH F 2 ALPHA C ALPHA , 1 ALPHA , 2) STO MAT C
row m.p.(-2, C, 1, 3)	MATH F 4 (-) 2 ALPHA , ALPHA C ALPHA , 1 ALPHA , 3) STO MAT C
row plus(C, 2, 3)	MATH F 2 ALPHA C ALPHA , 2 ALPHA , 3) STO MAT C
row mult(½, C, 3)	MATH F 3 1 ÷ 2 ALPHA , ALPHA C ALPHA , 3) STO MAT C

You should see that $z = 2$, so $y = -1$ and $x = 1$.

9.6.4 Determinants and Inverses: Enter this 3×3 square matrix as A: $\begin{bmatrix} 1 & -2 & 3 \\ -1 & 3 & 0 \\ 2 & -5 & 5 \end{bmatrix}$. To calculate its deter-

minant, $\begin{vmatrix} 1 & -2 & 3 \\ -1 & 3 & 0 \\ 2 & -5 & 5 \end{vmatrix}$, press MATH E 6 MAT A ENTER. You should find that $|A| = 2$.

Since the determinant of this matrix A is not zero, it has an inverse, A^{-1}. Press MAT A 2ndF x⁻¹ ENTER to calculate the inverse of matrix A.

Now let's solve a system of linear equations by matrix inversion. Once more, consider $\begin{cases} x - 2y + 3z = 9 \\ -x + 3y = -4 \\ 2x - 5y + 5z = 17 \end{cases}$.

The coefficient matrix for this system is the matrix $\begin{bmatrix} 1 & -2 & 3 \\ -1 & 3 & 0 \\ 2 & -5 & 5 \end{bmatrix}$ that was entered in the previous example. If

necessary, enter it again as A in your Sharp EL-9200/9300 and enter the matrix of constants $\begin{bmatrix} 9 \\ -4 \\ 17 \end{bmatrix}$ as B. Then

press MAT A x⁻¹ × MAT B ENTER to calculate the solution matrix, $\begin{bmatrix} 1 \\ -1 \\ 2 \end{bmatrix}$. The solutions are still $x = 1$, $y = $

-1, and $z = 2$.

9.7   Sequences

9.7.1 Iteration with the ANS Key: The 2ndF ANS feature enables you to perform *iteration*, the process of evaluating a function repeatedly. As an example, calculate $\dfrac{n-1}{3}$ for $n = 27$. Then calculate $\dfrac{n-1}{3}$ for $n = $ the answer to the previous calculation. Continue to use each answer as n in the *next* calculation. Here are key-strokes to accomplish this iteration on the Sharp EL-9200/9300 calculator (see the results in Figure 9.46). Notice that when you use Ans in place of n in a formula, it is sufficient to press ENTER to continue an iteration.

Assure that you are in the correct mode by pressing the calculation mode key, then MENU 1.

Iteration	Keystrokes	Display
1	27 ENTER	27.
2	(2ndF ANS - 1) ÷ 3 ENTER	8.666666667
3	ENTER	2.555555556
4	ENTER	0.518518518
5	ENTER	-0.160493827

Press ENTER several more times and see what happens with this iteration. You may wish to try it again with a different starting value.

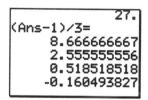

Figure 9.46: Iteration

9.7.2 Arithmetic and Geometric Sequences: Use iteration with the Ans variable to determine the *n*-th term of a sequence. For example, find the 18th term of an *arithmetic* sequence whose first term is 7 and whose common difference is 4. Enter the first term 7, then start the progression with the recursion formula, 2ndF ANS + 4 ENTER. This yields the 2nd term, so press ENTER sixteen more times to find the 18th term. For a *geometric* sequence whose common ratio is 4, start the progression with 2ndF ANS × 4 ENTER.

Of course, you could also use the *explicit* formula for the *n*-th term of an arithmetic sequence, $t_n = a + (n-1)d$. First enter values for the variables *a*, *d*, and *n*, then evaluate the formula by pressing ALPHA A + (ALPHA N - 1) ALPHA D ENTER. For a geometric sequence whose *n*-th term is given by $t_n = a \cdot r^{n-1}$, enter values for the variables *a*, *r*, and *n*, then evaluate the formula by pressing ALPHA A ALPHA R a^b (ALPHA N - 1) ENTER.

9.8   Parametric and Polar Graphs

9.8.1 Graphing Parametric Equations: The Sharp EL-9200/9300 plots parametric equations as easily as it plots functions. Just use the SETUP menu (Figure 9.2), go to COORD, and select XYT (press SETUP E 3). Be sure, if the independent parameter is an angle measure, that DRG is set to whichever you need, Rad or Deg.

For example, here are the keystrokes needed to graph the parametric equations $x = \cos^3 t$ and $y = \sin^3 t$. First check that angles are currently being measured in radians. Change to parametric mode and press the graphing mode key. Enter the two parametric equations by pressing (cos X/θ/T) a^b 3 ENTER for X1T and (sin X/θ/T) a^b 3 ENTER for Y1T . Now, when you press the variable key X/θ/T, you get a T because the calculator is in parametric mode.

Also look at the new RANGE menu. In the default window, the values of T go from 0 to π in steps of $\frac{\pi}{94}$ =.0334 , since the Sharp EL-9200/9300 plots in 94 increments across the screen, and the default view extends from -4.7 to 4.7 in the horizontal direction and from -3.1 to 3.1 in the vertical direction. But here T has been changed to continue to 2π (the step size Tstp is automatically changed to $\frac{2\pi}{94}$ =.0668). And a good viewing rectangle extends from -2 to 2 in both directions. Press the graphing mode key again to see the parametric graph (Figure 9.47).

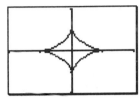

Figure 9.47: Parametric graph of $x = \cos^3 t$ and $y = \sin^3 t$

You may ZOOM and TRACE along parametric graphs just as you did with function graphs. As you trace along this graph, notice that the cursor moves in the *counterclockwise* direction as T increases.

9.8.2 Rectangular-Polar Conversion: The MATH menu (Figure 9.7) provides functions for converting between rectangular and polar coordinate systems. These functions use the current settings, so it is a good idea to check the default angle measure before any conversion. For the following examples, the Sharp EL-9200/9300 is set to radian measure; check that it is in calculation mode, too.

Given rectangular coordinates $(x, y) = (4, -3)$, convert *from* these rectangular coordinates *to* polar coordinates (r, θ) by pressing 4 ALPHA , (-) 3 MATH D 3. The value of r is displayed; press ALPHA θ ENTER to display the value of θ.

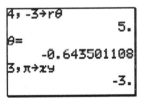

Figure 9.48: Coordinate conversions

Suppose $(r, \theta) = (3, \pi)$. To convert *from* these polar coordinates *to* rectangular coordinates (x, y), press 3 ALPHA , 2ndF π MATH D 4. The x-coordinate is displayed; press ALPHA Y ENTER to display the y-coordinate.

9.8.3 Graphing Polar Equations: The Sharp EL-9200/9300 graphs a polar function after you press SETUP E 2.

For example, to graph $r = 4 \sin \theta$, press the graphing mode key and input 4 sin X/θ/T; note that the X/θ/T key produces θ in polar mode. Choose a good viewing rectangle and an appropriate interval and increment for θ (in particular, use 2π for θmax).

9.9   Probability

9.9.1 Random Numbers: The command random generates a number between 0 and 1. In real number calculation mode, press MATH A 8 ENTER to generate a random number. Press ENTER to generate another number; keep pressing ENTER to generate more of them.

If you need a random number between, say, 0 and 10, then press 10 MATH A 8 ENTER. To get a random number between 5 and 15, press 5 + 10 MATH A 8 ENTER.

9.9.2 Permutations and Combinations: To calculate the number of *permutations* of 12 objects taken 7 at a time, denoted by $_{12}P_7$, press 12 MATH A 7 7 ENTER. Thus $_{12}P_7 = 3{,}991{,}680$.

For the number of *combinations* of 12 objects taken 7 at a time, denoted by $_{12}C_7$, press 12 MATH A 6 7 ENTER. So $_{12}C_7 = 792$.

9.9.3 Probability of Winning: A state lottery is configured so that each player chooses six different numbers from 1 to 40. If these six numbers match the six numbers drawn by the State Lottery Commission, the player wins the top prize. There are $_{40}C_6$ ways for the six numbers to be drawn. If you purchase a single lottery ticket, your probability of winning is 1 in $_{40}C_6$. Press 1 ÷ 40 MATH A 6 6 ENTER to calculate your chances, but don't be disappointed.

9.10   Programming

9.10.1 Entering a Program: The Sharp EL-9200/9300 is a programmable calculator that can store sequences of commands for later replay. Here's an example to show you how to enter a useful program that solves quadratic equations by the quadratic formula.

Press the programming mode key (in the middle of the top row) to access the programming menu, where you will find a list of any programs that were input previously. The Sharp EL-9200 has space for up to 55 programs, and the Sharp EL-9300 has space for up to 99 programs.

To create a new program, press C ENTER, then 4 so this program will run in complex mode. The ALPHA indicator is on, so press letter keys to name this program quadratic. Then press ▼ to continue.

Notice that the name quadratic is in *lowercase* letters. In programming mode, pressing ALPHA or 2ndF A·LOCK allows you to enter a lowercase letter. For uppercase letters, press ALPHA 2ndF or 2ndF A·LOCK 2ndF.

A single *uppercase* letter, when used for a variable name, refers to a memory location. It is called a *global* variable. For any other program, or even outside programming mode, this memory location retains the value you store there until you store something else. However a *lowercase* letter names a *local* variable that exists only during the current program. Values stored in local variables cannot be passed to another program and are not available outside the program in which they are created.

Lowercase variable names can be longer than one letter. For example, length is a valid local variable name. So multiplication between local variables must be expressed: you must enter length × width for a product.

Any command you could enter directly in the Sharp EL-9200/9300's home screen can be entered as a line in a program. There are also special programming commands.

You *must* press ENTER or ▼ after each line to complete the entry. Press CL to clear a single line; press 2ndF CA to delete an entire program.

Enter the program quadratic by pressing the keystrokes given in the listing below. You may interrupt program input at any time by pressing a mode key. To return later for more editing, press the programming mode key, then B, use the arrow keys to locate the program's name in the listing, and press ENTER. Take care, however, that the keystrokes given here assume that you are entering these lines of code sequentially. Frequently, for example, when you press 2ndF A·LOCK to put the calculator in alphabetic entry mode for one line of code, subsequent lines will assume that the calculator is already in that state.

Program Line	Keystrokes
Print "Enter a	2ndF COMMAND A 1 2ndF COMMAND 2 2ndF A·LOCK 2ndF E N T E R SPACE A ▼

displays the words *Enter a* on the Sharp EL-9200/9300 screen

Input a	2ndF COMMAND 3 A ▼

waits for you to input a value that will be assigned to the variable a

Print "Enter b	2ndF COMMAND 1 2ndF COMMAND 2 2ndF E N T E R SPACE B ▼
Input b	2ndF COMMAND 3 B ▼
Print "Enter c	2ndF COMMAND 1 2ndF COMMAND 2 2ndF E N T E R SPACE C ▼
Input c	2ndF COMMAND 3 C ▼

d = b²-4a*c   D = B ALPHA x² − 4 ALPHA A × ALPHA C ⏷

calculates the discriminant and stores its value as d

m = (-b+√d)/(2a)   ALPHA M ALPHA = ((-) ALPHA B + √ ALPHA D) ÷
(2 ALPHA A) ⏷

calculates one root and stores it as m

n = (-b-√d)/(2a)   ALPHA N ALPHA = ((-) ALPHA B − √ ALPHA D) ÷
(2 ALPHA A) ⏷

If d<0 Goto 1   2ndF COMMAND B 3 ALPHA D 2ndF COMMAND C 2 0
ALPHA SPACE 2ndF COMMAND B 2 1 ⏷

tests to see if the discriminant is negative;

in case the discriminant is negative, jumps to the line Label 1 below;
if the discriminant is not negative, continues on to the next line

If d=0 Goto 2   2ndF COMMAND 3 ALPHA D 2ndF COMMAND C 1 0
ALPHA SPACE 2ndF COMMAND B 2 2 ⏷

tests to see if the discriminant is zero;

in case the discriminant is zero, jumps to the line Label 2 below;
if the discriminant is not zero, continues on to the next line

Print "Two real roots   2ndF COMMAND A 1 2ndF COMMAND 2 2ndF A·LOCK
2ndF T W O SPACE R E A L SPACE R O O T S ⏷

Print m   2ndF COMMAND 1 M ⏷

displays one root

Print n   2ndF COMMAND 1 N ALPHA ⏷

End   2ndF COMMAND 6 ⏷

stops program execution

Label 1   2ndF COMMAND B 1 1 ⏷

jumping point for the Goto command above

Print "Complex roots   2ndF COMMAND A 1 2ndF COMMAND 2
2ndF A·LOCK 2ndF C O M P L E X SPACE R O O T S ⏷

displays a message in case the roots are complex numbers

Print m   2ndF COMMAND 1 M ⏷

displays one root

Print n	2ndF COMMAND 1 N ALPHA ▾
End	2ndF COMMAND 6 ▾
Label 2	2ndF COMMAND B 1 2 ▾
Print "Double root	2ndF COMMAND A 1 2ndF COMMAND 2
	2ndF A·LOCK 2ndF D O U B L E SPACE R O O T ▾

displays a message in case there is a double root

Print m	2ndF COMMAND 1 M ▾
End	2ndF COMMAND 6 ▾

When you have finished, press any mode key to leave the program editor.

9.10.2 Running a Program: To run the program just entered, press the programming mode key and ENTER. Go to its name in the program listing, then press ENTER to select this program and to execute it.

The program has been written to prompt you for values of the coefficients a, b, and c in a quadratic equation $ax^2 + bx + c = 0$. Input a value, then press ENTER to continue the program.

If you need to interrupt a program during execution, press QUIT.

The instruction manual for your Sharp EL-9200/9300 gives detailed information about programming. Refer to it to learn more about programming and how to use other features of your calculator.

9.11 Differentiation

9.11.1 Limits: Suppose you need to find this limit: $\lim\limits_{x\to 0}\dfrac{\sin 4x}{x}$. Plot the graph of $f(x)=\dfrac{\sin 4x}{x}$ in a convenient viewing rectangle that contains the point where the function appears to intersect the line $x = 0$ (because you want the limit as $x \to 0$). Your graph should lend support to the conclusion that $\lim\limits_{x\to 0}\dfrac{\sin 4x}{x}=4$.

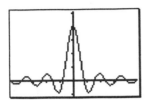

Figure 9.48: Checking $\lim\limits_{x\to 0}\dfrac{\sin 4x}{x}=4$

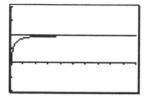

Figure 9.49: Checking $\lim\limits_{x\to\infty}\dfrac{2x-1}{x+1}=2$

To test the reasonableness of the conclusion that $\lim\limits_{x\to\infty}\dfrac{2x-1}{x+1}=2$, evaluate the function $f(x)=\dfrac{2x-1}{x+1}$ for several large positive values of x (since you want the limit as $x \to \infty$). For example, evaluate $f(100)$, $f(1000)$, and $f(10,000)$. Another way to test the reasonableness of this result is to examine the graph of $f(x)=\dfrac{2x-1}{x+1}$ in a viewing rectangle that extends over large values of x. See, as in Figure 9.49 (where the viewing rectangle extends horizontally from 0 to 100), whether the graph is asymptotic to the horizontal line $y = 2$ (enter $\dfrac{2x-1}{x+1}$ for Y1 and 2 for Y2).

9.11.2 Numerical Derivatives: The derivative of a function f at x can be defined as the limit of the slopes of secant lines through the point $(x, f(x))$, so $f'(x)=\lim\limits_{\Delta x\to 0}\dfrac{f(x+\Delta x)-f(x)}{\Delta x}$. And for small values of Δx, the expression $\dfrac{f(x+\Delta x)-f(x)}{\Delta x}$ gives a good approximation to the limit.

The Sharp EL-9200/9300 has a function d/dx(in the MATH CALC menu (Figure 9.50) to calculate the *difference quotient*, $\dfrac{f(x+\Delta x)-f(x)}{\Delta x}$. So to find a numerical approximation to $f'(2.5)$ when $f(x)=x^3$ and with $\Delta x = 0.001$, press (for one-line editing) MATH C 8 X/θ/T ∧ 3 ALPHA , 2.5 ALPHA , .001) ENTER as shown in the top line of Figure 9.51. The format of this command is d/dx(*expression, value, Δx*). The same derivative is also approximated in Figure 9.50 using $\Delta x = 0.00001$. For most purposes, $\Delta x = 0.00001$ gives a

Sharp EL-9200/9300 Graphing Scientific Calculator

very good approximation to the derivative and is the Sharp EL-9200/9300 calculator's default. So if you do use this value, just enter d/dx(*expression, value*) as in the final calculation of Figure 9.51.

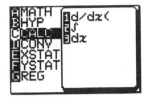

Figure 9.50: MATH CALC menu

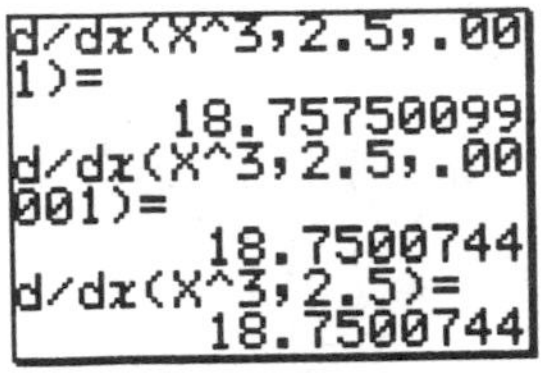

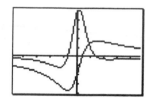

Figure 9.51: Using d/dx(

Technology Tip: It is sometimes helpful to plot both a function and its derivative together. In Figure 9.53, the function $f(x) = \dfrac{5x-2}{x^2+1}$ and its numerical derivative (actually, an approximation to the derivative given by the difference quotient) are graphed. You can duplicate this graph by first entering $\dfrac{5x-2}{x^2+1}$ for Y1. Then enter its numerical derivative for Y2 by pressing MATH C 1, the keystrokes for $\dfrac{5x-2}{x^2+1}$ again, and finally ALPHA , X/θ/T) as you see in Figure 9.52.

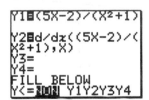

Figure 9.52: Entering $f(x)$ and $f'(x)$

Figure 9.53: Graphs of $f(x)$ and $f'(x)$

9.11.3 Newton's Method: With your Sharp EL-9200/9300, you may iterate using Newton's method to find the zeros of a function. Recall that Newton's Method determines each successive approximation by the formula $x_{n+1} = x_n - \dfrac{f(x_n)}{f'(x_n)}$.

As an example of the technique, consider $f(x) = 2x^3 + x^2 - x + 1$. A look at its graph suggests that it has a zero near $x = -1$, so start the iteration by storing -1 as x (see figure 9.54). Then execute these keystrokes: X/θ/T − (followed by the keystrokes for $2x^3 + x^2 - x + 1$; next press) ÷ MATH C 1 again followed by the keystrokes for $2x^3 + x^2 - x + 1$; and finally press ALPHA , X/θ/T) STO X/θ/T. Press ENTER repeatedly until two successive approximations differ by less than some predetermined value, say 0.0001. Note that each time you press ENTER, the Sharp EL-9200/9300 will use its *current* value of x, and that value is changing as you continue the iteration.

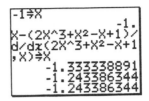

Figure 9.54: Newton's method

Technology Tip: Newton's Method is sensitive to your seed value for x, so look carefully at the function's graph to make a good first estimate. Also, remember that the method sometimes fails to converge!

You may want to save the Newton's Method formula as a short program. See your calculator's manual for further information on programming the Sharp EL-9200/9300.

9.12   Integration

9.12.1 *Approximating Definite Integrals:*
The Sharp EL-9200/9300 has the function ∫ in the MATH CALC menu to approximate an integral by Simpson's Rule,

$$\int_a^b f(x)dx \approx \frac{b-a}{3n}[f(x_0) + 4f(x_1) + 2f(x_2) + 4f(x_3) + \cdots + 4f(x_{n-1}) + f(x_n)].$$

So to find a numerical approximation to $\int_0^1 \cos x^2\, dx$ with 20 subdivisions (a number that controls the accuracy of the approximation), press these keystrokes for one-line editing: MATH C 2 COS X/θ/T x^2 ALPHA , 0 ALPHA , 1 ALPHA , 10 MATH 3 ENTER as shown in Figure 9.55. For equation editing, press MATH C 2 0 ◼ 1 ▶ COS X/θ/T x^2 ALPHA , 10 MATH 3 ENTER as illustrated in Figure 9.56. The format of this command is ∫ *expression, lower limit, upper limit, 1/2 the number of subdivisions* dx.

Note that this calculator requires that integration begin with an integral sign and conclude with a dx. Also, since the usual statement of Simpson's Rule requires an *even* number of subdivisions, you enter any integer and the Sharp EL-9200/9300 uses *double* that number of subdivisions. The same integral is also approximated in Figure 9.55 using 100 subdivisions, this calculator's default that is used when no other number is specified.

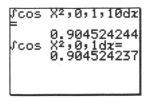

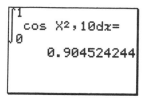

Figure 9.55: One-line editing: ∫ dx   Figure 9.56: Equation editing: ∫ dx

9.12.2 Areas: You may approximate the area under the graph of a function $y = f(x)$ between $x = A$ and $x = B$ with your Sharp EL-9200/9300. For example, the area under the graph of the function $y = \cos x^2$ between $x = 0$ and $x = 1$ is represented by the definite integral $\int_0^1 \cos x^2 \, dx$, which is approximated as shown above.

Technology Tip: Suppose that you want to find the area between two functions, $y = f(x)$ and $y = g(x)$, from $x = A$ to $x = B$. If $f(x) \geq g(x)$ for $A \leq x \leq B$, then approximate $\int_A^B (f(x) - g(x)) dx$ to find the required area.

Chapter 10

Hewlett Packard HP 48G

10.1 Getting started with the HP-48G

10.1.1 Basics: Press the ON key to begin using your HP-48G calculator. If you need to adjust the display contrast, first press and hold ON, then press + to darken the display or − to lighten the display. When you have finished with the calculator, turn it off to conserve battery power by pressing the green ➦ key and then OFF.

Check the HP-48G's settings by pressing ➦ and then MODES. At the bottom of the screen is a menu line of options, each corresponding to the white function key below it (see Figure 10.1).

Throughout this chapter of the Guide, we write *[COMMAND]* to represent the white function key directly below the menu item COMMAND. For example, *[OK]* will represent the far-right function key, below OK in the menu line of Figure 10.1.

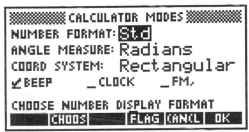

Figure 10.1: CALCULATOR MODES

Use the arrow keys, ▲ ▼ ◀ and ▶, to move the highlight to a mode you want to change. A prompt is displayed at the bottom of the window, just above the menu options. To change the number format, for example, move the highlight to the top of the window and press the function key *[CHOOS]*. Then move the highlight to Standard and press the function key *[OK]* to choose the standard display mode. Change other modes as necessary to configure your calculator for number format, angle measure, and coordinate system as illustrated in Figure 10.1. Details on alternative options will be given later in this guide.

Another way to change a mode setting is to highlight what you wish to change and press the +/− key to cycle among alternatives until the setting you want is displayed.

Accept the displayed settings by pressing *[OK]*. Cancel any changes and return to the home screen by pressing *[CANCL]*.

The four numbered lines in the home screen (Figure 10.2) correspond to the first four *cards* in the HP-48G's *stack*. Mathematical objects - numbers, expressions, equations, matrices - are stored in this stack, which may contain a very large number of cards. New cards are entered at the front of the stack, and existing cards are renumbered. The stack level of a card increases by one as each new card is created. The reverse occurs when you use values from the stack.

Press these keys to observe stack entry: 5.1 ENTER 6.2 ENTER 7.3 ENTER 8.4 ENTER 9.5 ENTER. Now 5.1 has been pushed up to stack level 5; to see it again, press the purple ↰ key and then DROP to drop every card one level (see Figure 10.2). The number 9.5 that was at stack level 1 is now discarded. (When there is no command line displayed, it is sufficient to press just the DROP key without the ← key before it.)

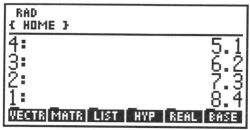

Figure 10.2: The card stack

Clear the entire stack by pressing ← CLEAR or just CLEAR. (Since there is no command line displayed, it is sufficient to press CLEAR without the preliminary ←.) It's a good idea to clear the stack before each new sequence of calculations.

The HP-48G offers two ways of performing arithmetic calculations. The *stack method* is useful for quick calculations. Arguments, or numbers, are entered into the stack first, followed by an operation. For example, to compute $\sqrt{16}$, press 16. The stack moves up to make room for a *command line*. Next press √x and the answer 4 is put in stack level 1. To calculate 12 + 34, press 12 ENTER 34 ENTER +. Actually, the second ENTER, after 34, is not really necessary, because mathematical keys like the + key automatically cause an ENTER before performing the indicated mathematical operation. For example, press 126 ENTER 6 ÷ and get 21.

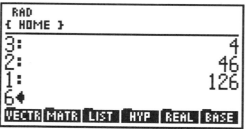

Figure 10.3: Arithmetic by the stack method

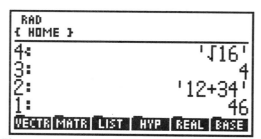

Figure 10.4: Arithmetic by the algebraic method

In the *algebraic method*, you enter an expression first and then you evaluate it. Begin by pressing ' because algebraic expressions must be surrounded by "tick marks." For example, calculate $\sqrt{16}$ again by pressing CLEAR ' √x 16 ENTER. Press ENTER once more to duplicate the expression into stack level 2. Finally, press EVAL to evaluate the expression in stack level 1 and get $\sqrt{16}$ = 4 (see Figure 10.4). Next, to calculate 12 + 34, press ' 12 + 34 ENTER ENTER EVAL.

The HP-48G provides an EquationWriter application to simplify the creation of algebraic expressions. To access the EquationWriter, press ← EQUATION. Use the following keystrokes to input $\sqrt{3^2 + 4^2}$: √x 3 yˣ 2 ▶ + 4 yˣ 2 ENTER (see Figure 10.5). The right arrow ▶ key is used to advance the cursor and conclude exponents, roots, fractions, and others. The HP-48G User's Guide provides detailed examples to illustrate its EquationWriter.

The expression is now properly formatted with tick marks and entered in stack level 1. So press EVAL to evaluate it.

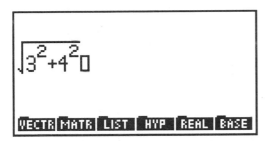

Figure 10.5: EquationWriter

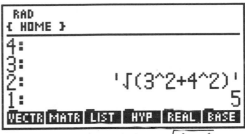

Figure 10.6: Calculating $\sqrt{3^2 + 4^2}$

Technology Tip: Erase a command line or cancel any command by pressing ON, which serves as the HP-48G's general-purpose CANCEL key.

10.1.2 Editing: Often we do not notice a mistake until we see how unreasonable an answer is. The HP-48G permits you to re-display an entire calculation, edit it easily, then execute the *corrected* calculation.

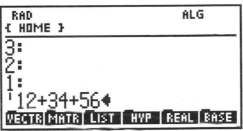

Figure 10.7: Editing the command line

Suppose you had typed 12 + 34 + 56 as in Figure 10.7 but had *not* yet pressed ENTER, when you realize that 34 should have been 74. Simply press ◀ (the *left* arrow key) as many times as necessary to move the blinking cursor left to 3, then press DEL 7 to delete the 3 and insert 7. You might also move the cursor to just right of 3, press ← to backspace over it, then press 7. On the other hand, if 34 should have been 384, move the cursor back to 4 and type 8 (inserts at the cursor position and other characters are pushed to the right). If the 34 should have been 3 only, move the cursor to 4 and press DEL to delete it.

You may edit an expression so long as it remains in the stack. Press ON to clear the command line and ← EDIT to edit the expression currently in stack level 1. Use the left and right arrow keys to move through the expression; press DEL or ← to delete unwanted characters; directly enter new characters as required. While editing, you may press ↱ ◀ to jump to the *left* end of an expression; press ↱ ▶ to jump to the *right* end.

When your editing is completed, press ENTER to put the edited expression into stack level 1; otherwise, press CANCEL to quit without saving any changes.

To restore the stack to its state before the last command was executed, press ↱ UNDO.

If the line you wish to edit is higher up in the stack, press DROP (if there is no command line, otherwise press ← DROP) as many times as necessary to move the line down the stack until it is at level 1; then press ← EDIT as before. Another technique is to press ON (to clear the command line) and ▲ as many times as needed to move the cursor up to the stack level you want; then press ← EDIT. When you have completed editing, press ENTER to put the edited expression in the stack level from which you began.

You may prefer to move an expression from the stack into EquationWriter for editing. Consult the User's Guide for instructions on how to edit in EquationWriter.

Technology Tip: When you need to evaluate a formula for different values of a variable, use the editing feature to simplify the process. For example, suppose you want to find the balance in an investment account if there is now \$5000 in the account and interest is compounded annually at the rate of 8.5%. The formula for the balance is $P\left(1+\frac{r}{n}\right)^{nt}$, where P = principal, r = rate of interest (expressed as a decimal), n = number of times interest is compounded each year, and t = number of years. In our example, this becomes $5000(1+.085)^{t}$. Here are the keystrokes for finding the balance after $t = 3$, 5, and 10 years.

Years	Keystrokes	Balance
	' 5000 × ← ( ) 1 + .085 ▶ y^x 3	
	ENTER ENTER ENTER ENTER	
	▲ ▲ ← EDIT ↱ ▶ ◀ ← 5 ENTER	
	▲ ← EDIT ↱ ▶ ◀ ← 10 ENTER	
3	ENTER EVAL	\$6386.45
5	DROP EVAL	\$7518.28
10	DROP EVAL	\$11,304.92

Then to find the balance from the same initial investment but after 5 years when the annual interest rate is 7.5%, press these keys after the last calculation above: DROP ← EDIT ↱ ▶ ◀ ← 5 ◀ ◀ ◀ ◀ ← 7 ENTER EVAL. The balance you get should be \$7178.15.

```
RAD
{ HOME }
4:
3:      '5000*(1+.085)^3'
2:      '5000*(1+.085)^10'
1:             7518.2834509
VECTR MATR LIST  HYP  REAL BASE
```

Figure 10.8: Editing expressions

10.1.3 Key Functions: Most keys on the HP-48G offer access to more than one function, just as the keys on a computer keyboard can produce more than one letter ("g" and "G") or even quite different characters ("5" and "%"). The primary function of a key is indicated on the key itself, and you access that function by a simple press on the key. For example, to calculate $\sqrt{25}$ by the stack method, press 25 √x .

To access the *purple* function indicated to the *left* above a key, first press ← (the ← annunciator appears at the top of the window) and *then* press the key. For example, to calculate 0^3, press 3 ← 10ˣ.

For the *green* function indicated to the *right* above a key, first press ↱ (now the ↱ annunciator appears at the top of the window) and *then* press the key. For example, to calculate log 1000, press 1000 ↱ LOG.

When you want to use an uppercase letter printed *below* a key, first press α (the α annunciator appears at the top of the window) and then the key. For example, to use the letter K in a formula, press α K. For a lowercase letter, press α ← before you press the letter key. So press α ← K for the letter k. If you need several uppercase letters in a row, press α α, which is like CAPS LOCK on a computer keyboard, and then press all the letters you want. You may also *press and hold* the α key and then press as many letter keys as you wish. To lock in lowercase alpha mode, press α α ← α. Remember to press α when you are finished and want to restore the keys to their primary functions; ENTER and CANCEL also terminate alpha lock.

10.1.4 Order of Operations: When you put mathematical objects directly into the stack, the operations follow the order of entry. For example, press 3 ENTER 4 ENTER 5 + × and get 27, because you are calculating $3 \cdot (4 + 5)$. Next press 3 ENTER 4 ENTER 5 × + and get 23, because this time you are computing 3 + (4·5).

When you use algebraic editing, the HP-48G performs calculations according to the standard algebraic rules. Working outwards from inner parentheses, calculations are performed from left to right. Powers and roots are evaluated first, followed by multiplications and divisions, and then additions and subtractions.

Note that the HP-48G distinguishes between *subtraction* and the *negative sign*. If you wish to enter a negative number, it is necessary to use the +/- key. Press +/- to change the sign *after* the number is entered. For example, you would evaluate $-5 - (4 \cdot -3)$ by pressing ' 5 +/- – ← ( ) 4 × 3 +/- EVAL to get 7.

Enter these expressions to practice using your HP-48G.

Expression	Keystrokes	Display
$7 - 5 \cdot 3$	' 7 − 5 × 3 EVAL	-8
$(7 - 5) \cdot 3$	' ← ( ) 7 − 5 ▶ × 3 EVAL	6
$20 - 10^2$	' 120 − 10 y^x 2 EVAL	20
$(120 - 10)^2$	' ← ( ) 120 − 10 ▶ y^x 2 EVAL	12100
$\dfrac{24}{2^3}$	' 24 ÷ 2 y^x 3 EVAL	3
$\left(\dfrac{24}{2}\right)^3$	' ← ( ) 24 ÷ 2 ▶ y^x 3 EVAL	1728
$(7 - -5) \cdot -3$	' ← ( ) 7 − 5 +/- ▶ × 3 +/- EVAL	-36

10.1.5 Algebraic Expressions and Memory: Your calculator can evaluate expressions such as $\dfrac{N(N+1)}{2}$ *after* you have entered a value for N. Suppose you want $N = 200$. Press ' α N ← = 200 ← DEF to store the value 200 in memory location N. Whenever you use N in an expression, the calculator will substitute the value 200 until you make a change by storing *another* number in N. Next enter the expression $\dfrac{N(N+1)}{2}$ by typing ' α N × ← ( ) α N + 1 ▶ ÷ 2 EVAL. For $N = 200$, you will find that $\dfrac{N(N+1)}{2} = 20100$.

The contents of any memory location may be revealed by typing just its letter name and then ENTER. And the HP-48G retains memorized values even when it is turned off, so long as its batteries are good.

10.1.6 Repeated Operations: The result of your *last* calculation is always stored in stack level 1. This makes it easy to use the answer from one computation in another computation. For example, press 30 ENTER 15 + so that 45 is the last result displayed. Then press 9 ÷ and get 5 because $\frac{45}{9} = 5$.

Here is a situation where this is especially useful. Suppose a person makes $5.85 per hour and you are asked to calculate earnings for a day, a week, and a year. Execute the given keystrokes to find the person's incomes during these periods.

Pay period	Keystrokes	Earnings
8-hour day	5.85 ENTER 8 ×	$46.80
5-day week	5 ×	$234
52-week year	52 ×	$12,168

10.1.7 The MATH Menu: Operators and functions associated with a scientific calculator are available either immediately from the keys of the HP-48G or by the purple ← or green → shift keys. You have direct key access to common arithmetic operations ($\sqrt{x}$, ← x^2, y^x, $\frac{1}{x}$), trigonometric functions (SIN, COS, TAN) and their inverses (ASIN, ACOS, ATAN), exponential and logarithmic functions (→ LOG, ← 10^x, → LN, ← e^x), and a famous constant (← π).

A significant difference between the HP-48G and many scientific calculators is that the HP-48G's algebraic calculation method requires the argument of a function *after* the function, as you would see a formula written in your textbook. For example, on the HP-48G you calculate $\sqrt{16}$ by pressing the keys ' $\sqrt{x}$ 16 in that order.

Here are keystrokes for basic mathematical operations. Try them for practice on your HP-48G.

Expression	Keystrokes	Display
$\sqrt{3^2 + 4^2}$	' $\sqrt{x}$ ← ( ) 3 y^x 2 + 4 y^x 2 EVAL	5
$2\frac{1}{3}$	' 2 + 1 ÷ 3 EVAL	2.3333
$\log 200$	' → LOG 200 EVAL	2.3010
$2.34 \cdot 10^5$	' 2.34 × 10 y^x 5 EVAL	234000

Additional mathematical operations and functions are available from the math menu (Figure 10.8). Press MTH to see the various options. You will learn in your mathematics textbook how to apply many of them. As an example, calculate $|-5|$ by pressing 5 +/- MTH *[REAL]* NXT *[ABS]* to see 5.

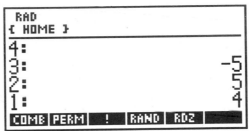

Figure 10.9: Factorial in the MTH menu

The *factorial* of a non-negative integer is the *product* of *all* the integers from 1 up to the given integer. The symbol for factorial is the exclamation point. So 4! (pronounced *four factorial*) is $1 \cdot 2 \cdot 3 \cdot 4 = 24$. You will learn more about applications of factorials in your textbook, but for now use the HP-48G to calculate 4! Press these keystrokes: 4 MTH NXT *[PROB]* *[!]*.

The complex number $a + bi$ is represented by the HP-48G as an ordered pair (a,b). Perform arithmetic with complex numbers by using this ordered pair notation. So to divide $2 + 3i$ by $4 - 2i$, first put $(2,3)$ in the stack, next enter $(4, -2)$, then divide them to get $(.1, .8)$ for $.1 + .8i$.

Use the calculator's polynomial root-finder to determine all the zeros, both real and complex, for a polynomial. So start to find the zeros of $f(x) = x^4 - 3x^3 + 6x^2 + 2x - 60$ by pressing ↱ SOLVE ▾ ▾ [OK]. Move the highlight to COEFFICIENTS and enter the polynomial's coefficients between square brackets by pressing ← [] 1 SPC 3 +/- SPC 6 SPC 2 SPC 60 +/- [OK]. Then press [SOLVE] and see that its zeros are 3, -2, $+3i$, and $-3i$; they are also entered in stack level 1.

Technology Tip: You must enter *all* the coefficients of a polynomial, especially a coefficient 0 for any "missing" term. So to find the zeros of $x^2 + 1$, you would enter its coefficients as $[1\,0\,1]$.

10.2 ´ Functions and Graphs

10.2.1 Evaluating Functions: Suppose you receive a monthly salary of $1975 plus a commission of 10% of sales. Let x = your sales in dollars; then your wages W in dollars are given by the equation $W = 1975 + .10x$. If your January sales were $2230 and your February sales were $1865, what was your income during those months?

Here's how to use your HP-48G to perform this task. Input the equation $W = 1975 + .10x$ by pressing these keys: ' α W ← = 1975 + .10 × α ← X ← DEF (see Figure 10.10).

Figure 10.10: Entering a formula

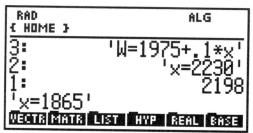

Figure 10.11: Evaluating the formula

Assign the value 2230 to the variable x by these keystrokes: ' α ← X ← = 2230 ← DEF. Next press α W EVAL and find January's wages (see Figure 10.11). Repeat these steps to find the February wages. Each time the HP-48G evaluates W, it uses the *current* value of x.

Another way to accomplish this is to define the function $W(x) = 1975 + .10x$ by pressing ' α W ← () α ← X ▸ ← = 1975 + .10 × α ← X ← DEF (see Figure 10.12). Then press ' α W ← () 2230 EVAL to evaluate $W(2230)$. Repeat for $W(1865)$.

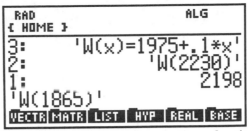

Figure 10.12: Defining and evaluating a function

Technology Tip: The HP-48G allows variable names to be longer than a single letter. You may define such variables, for example, as *length* and *width*. For this reason, the HP-48G requires multiplication to be expressed, so *3x* must be input as 3·*x*. The calculator would interpret *xxx* as a three-letter variable name, not as x^3.

10.2.2 Functions in a Graph Window: On the HP-48G, once you have entered a function, you can easily generate its graph. The ability to draw a graph contributes substantially to our ability to solve problems.

For example, here is how to graph $y = -x^3 + 4x$. First press ➜ PLOT. If the plot type is not Function, then move the highlight to the TYPE: field, press *[CHOOS]*, move to Function, and press *[OK]*. The independent variable is set for *X* by default; you may change it if you wish. You can get the independent variable quickly by pressing a function key. Move the highlight, if necessary, to EQ: and enter the function (as in Figure 10.13) by pressing ' - *[X]* y^x 3 + 4 × *[X]* *[OK]*. Now press NXT *[ERASE] [DRAW]* and the HP-48G switches to a window with the graph of $y = -x^3 + 4x$.

While the HP-48G is calculating coordinates for a plot, it displays a busy indicator at the top of the graph window.

Your graph window may look like the one in Figure 10.14 or it may be different. Since the graph of $y = -x^3 + 4x$ extends infinitely far left and right and also infinitely far up and down, the HP-48G can display only a piece of the actual graph. This displayed rectangular part is called a *viewing rectangle*. You can easily change the viewing rectangle to enhance your investigation of a graph.

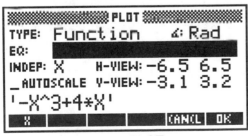

Figure 10.13: PLOT screen

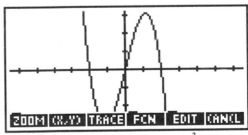

Figure 10.14: Graph of $y = -x^3 + 4x$

The viewing rectangle in Figure 10.14 shows the part of the graph that extends horizontally from x_{min} = -6.5 to x_{max} = 6.5 and vertically from y_{min} = -3.1 to y_{max} = 3.2. This is the *default* viewing rectangle for the HP-48G. The PLOT screen has information about your viewing rectangle.

Use the arrow keys to move around the PLOT menu and enter new values for the horizontal dimensions x_{min} and x_{max} and for the vertical dimensions y_{min} and y_{max}. Remember to use the +/- key, not − (which is subtraction), when you want to enter a negative value. The following figures show different viewing rectangles for the same function, $y = -x^3 + 4x$.

In Figures 10.15 and 10.16, the menu was turned off by pressing *[EDIT]* NXT *[MENU]*. Restore the menu by pressing any function key.

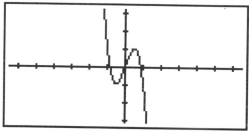

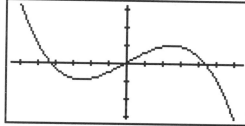

Figure 10.15: Window [-15, 15] by [-10, 10]   Figure 10.16: Window [-3, 3] by [-10, 10]

To set the plot parameters quickly back to their default values (see Figure 10.13), press *[ZOOM]* *[ZDFLT]* while in a plot window.

Leave the plot window and return to the home screen by pressing CANCEL.

10.2.3 Piecewise-Defined Functions:
The greatest integer function, written [[x]], gives the greatest *integer* less than or equal to a number x. On the HP-48G, the greatest integer function is called FLOOR and is located under the REAL sub-menu of the MTH menu (see Figure 10.12). So calculate [[6.78]] = 6 by pressing 6.78 MTH *[REAL]* NXT NXT *[FLOOR]*.

To graph y = [[x]], go in the PLOT menu, move beside EQ:, and press ' MTH *[REAL]* NXT NXT *[FLOOR]* α X ENTER. Figure 10.16 shows this graph in a viewing rectangle from -5 to 5 in both directions.

The true graph of the greatest integer function is a step graph, like the one in Figure 10.18. For the graph of y = [[x]], a segment should *not* be drawn between every pair of successive points. You can change from a connected line graph to a dot graph on the HP-48G from the PLOT menu. Press *[OPTS]* and move to CONNECT; toggle *[✓CHK]* to remove the check mark next to CONNECT; then press *[OK]*. Erase the previous graph and draw a new one.

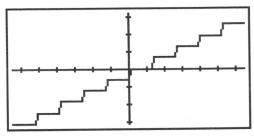

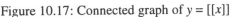

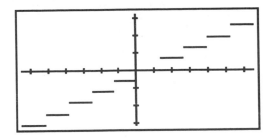

Figure 10.17: Connected graph of $y = [[x]]$ Figure 10.18: Dot graph of $y = [[x]]$

You should also change to Dot graph when plotting a piecewise-defined function. For example, to plot the graph of $f(x) = \begin{cases} x^2 + 1, & x < 0 \\ x - 1, & x \ge 0 \end{cases}$, enter the expression $(x^2 + 1)(x < 0) + (x - 1)(x \ge 0)$ for **EQ:** in your **PLOT** menu by pressing ' ← () [X] y^x 2 + 1 ▶ × ← () [X] PRG [TEST] [<] 0 ▶ + ← () α X − 1 ▶ × ← () α X [≥] 0 ENTER. Then *uncheck* the CONNECT option, erase any prior graph, and draw the new one.

Usually you want a connected graph, so go back to the plot options menu and toggle CONNECT on again.

10.2.4 Graphing a Circle: Here is a useful technique for graphs that are not functions, but that can be "split" into a top part and a bottom part, or into multiple parts. Suppose you wish to graph the circle whose equation is $x^2 + y^2 = 36$. First solve for y and get an equation for the top semicircle, $y = \sqrt{36 - x^2}$, and for the bottom semicircle, $y = -\sqrt{36 - x^2}$. Then graph the two semicircles simultaneously.

The keystrokes to draw this circle's graph follow. In the PLOT menu, execute these keystrokes to fill the **EQ:** field with a *list* of the two semicircles: ← {} ' √x ← () 36 - [X] y^x 2 ▶ ▶ SPC ' - √x ← () 36 - [X] y^x 2 [OK]. Then press [ERASE] [DRAW] to draw them both.

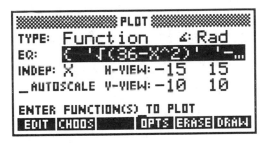

Figure 10.19: Two semicircles

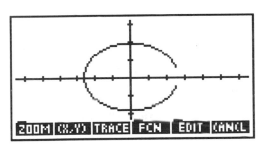

Figure 10.20: Circle: [-15, 15] by [-10, 10]

In the viewing rectangle of Figure 10.20, the graph does *not* look like a circle, because the units along the axes are not the same. Change the window so that the value of y_{max} - y_{min} is $\frac{1}{2}$ times x_{max} - x_{min}. For example, see Figure 10.21 and the corresponding graph in Figure 10.22. The method works because the dimensions of the HP-48G's display are such that the ratio of vertical to horizontal is approximately $\frac{1}{2}$.

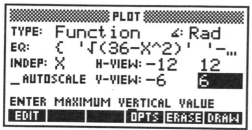

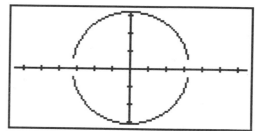

Figure 10.21: $\frac{\text{vertical}}{\text{horizontal}} = \frac{12}{24} = \frac{1}{2}$

Figure 10.22: A "square" circle

The two semicircles in Figure 10.22 do not meet because of an idiosyncrasy in the way the HP-48G plots a graph.

10.2.5 *TRACE:* Graph $y = -x^3 + 4x$ in the default viewing rectangle. Press any of the arrow keys ▲ ▼ ◄ ► and see the cursor move from the center of the viewing rectangle. Press *[(X, Y)]* so that the coordinates of the cursor's location are displayed at the bottom of the screen, as in Figure 10.23, in floating decimal format. This cursor is called a *free-moving cursor* because it can move from dot to dot *anywhere* in the graph window.

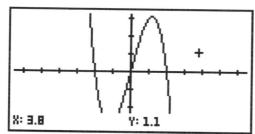

Figure 10.23: Free-moving cursor

Remove the cursor's coordinates and restore the menu by pressing NXT.

Toggle *[TRACE]* on (a white square replaces the "E") to enable the left ◄ and right ► arrow keys to move the cursor along the function. Also press *[(X, Y)]* to display the cursor's coordinates. The cursor is no longer free-moving, but is now constrained to the function. The coordinates that are displayed belong to points on the function's graph, so the *y*-coordinate is the calculated value of the function at the corresponding *x*-coordinate.

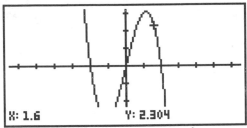

Figure 10.24: Trace on $y = -x^3 + 4x$

Now plot a second function, $y = -.25x$, along with $y = -x^3 + 4x$. In the **PLOT** menu's **EQ:** field, press the following keys: ← { } ' - *[X]* yˣ 3 + 4 × *[X]* ▶ **SPC** ' .25 +/- × *[X]* (see Figure 10.25). Their graphs are plotted together in Figure 10.26.

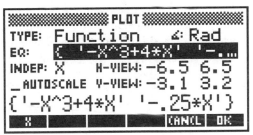

Figure 10.25: Two functions

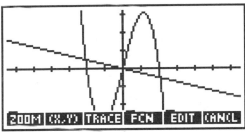

Figure 10.26: $y = -x^3 + 4x$ and $y = -.25x$

Toggle **TRACE** on, with the cursor's coordinates displayed. The cursor appears first on the graph of $y = -x^3 + 4x$ because it is defined first in the list. Press the up ▲ or down ▼ arrow key to move the cursor vertically to the graph of $y = -.25x$. Next press the right and left arrow keys to trace along the graph of $y = -.25x$. When more than one function is plotted, you can move the trace cursor vertically from one graph to another in this way.

Technology Tip: By the way, to remind you of the function being graphed, press ← and hold **VIEW**. This is especially helpful when you are tracing along two or more graphs.

The HP-48G's display has 131 horizontal columns of pixels and 64 vertical rows. So when you trace a curve across a graph window, you are actually moving from Xmin to Xmax in 130 equal jumps, each called Δx. You would calculate the size of each jump to be $\Delta x = \dfrac{\text{Xmax} - \text{Xmin}}{130}$. Sometimes you may want the jumps to be friendly numbers like .1 or .25 so that, when you trace along the curve, the x-coordinates will be incremented by such a convenient amount. Just set your viewing rectangle for a particular increment Δx by making Xmax = Xmin + 130·Δx. For example, if you want Xmin = -5 and Δx = .3, set Xmax = -5 + 130·.3 = 34. Likewise, set Ymax = Ymin + 63·Δy if you want the vertical increment to be some special Δy.

To center your window around a particular point, say (h, k), and also have a certain Δx, set Xmin = h - 65·Δx and Xmax = h + 65·Δx. Likewise, make Ymin = k - 31·Δy and Ymax = k + 32·Δy. For example, to center a window around the origin, (0, 0), with both horizontal and vertical increments of .25, set the range so that Xmin = 0 - 65·.25 = -16.25, Xmax = 0 + 65·.25 = 16.25, Ymin = 0 - 31·.25 = -7.75, and Ymax = 0 + 32·.25 = 8.

The HP-48G's default window is already a friendly viewing rectangle, centered at the origin (0, 0) with $\Delta x = \Delta y = 0.1$.

See the benefit by first plotting $y = x^2 + 2x + 1$ in a graphing window extending from -10 to 10 in both directions. Trace near its y-intercept, which is (0, 1), and move towards its x-intercept, which is (-1, 0). Then change to the default viewing rectangle, and trace again near the intercepts.

10.2.6 ZOOM: Plot again the two graphs, for $y = -x^3 + 4x$ and for $y = -.25x$. There appears to be an intersection near $x = 2$. The HP-48G provides several ways to enlarge the view around this point. You can change the viewing rectangle directly by changing the horizontal and vertical view parameters in the PLOT menu. Figure 10.28 shows a new viewing rectangle for the parameters displayed in Figure 10.27. Trace has been turned on and the coordinates of a point on $y = -x^3 + 4x$ that is close to the intersection are displayed.

A more efficient method for enlarging the view is to draw a new viewing rectangle with the cursor. Start again with a graph of the two functions $y = -x^3 + 4x$ and $y = -.25x$ in a default viewing rectangle (press [ZOOM] [ZDFLT] for the default window).

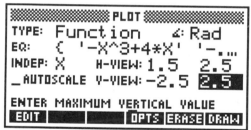

Figure 10.27: New view parameters

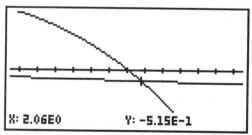

Figure 10.28: Closer view

Now imagine a small rectangular box around the intersection point, near $x = 2$. First move the cursor to one corner of the new viewing rectangle you imagine. Then press [ZOOM] [BOXZ] (Figure 10.29) to draw a box to define this new viewing rectangle. Use the arrow keys to move the cursor to the diagonally opposite corner of the new rectangle (Figure 10.30). If this box looks all right to you, press [ZOOM]. The rectangular area you have enclosed will now enlarge to fill the graph window (Figure 10.31).

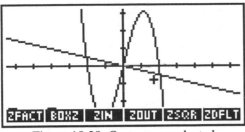

Figure 10.29: One corner selected

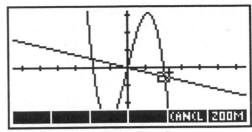

Figure 10.30: Box drawn

You may cancel the zoom at any time by pressing *[CANCL]*.

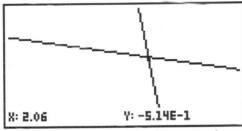

Figure 10.31: New viewing rectangle

You can also gain a quick magnification of the graph around the cursor's location. Return once more to the standard range for the graph of the two functions $y = -x^3 + 4x$ and $y = -.25x$. Use arrow keys to move the cursor as close as you can to the point of intersection near $x = 2$ (see Figure 10.32). Then press *[ZOOM]* *[ZIN]* and the calculator draws a magnified graph, centered at the cursor's position (Figure 10.33). The view variables are changed to reflect this new viewing rectangle. Look in the PLOT menu to check.

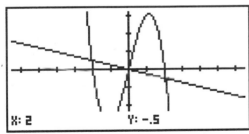

Figure 10.32: Before a zoom in

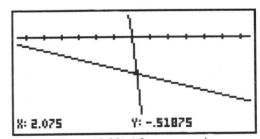

Figure 10.33: After a zoom in

As you see in the ZOOM menu (Figure 10.29), the HP-48G can zoom in (press *[ZOOM]* *[ZIN]*) or zoom out (press *[ZOOM]* *[ZOUT]*). Zoom out to see a larger view of the graph, centered at the cursor position. You can change the horizontal and vertical scale of the magnification by pressing *[ZOOM]* *[ZFACT]* (see Figure 10.34) and editing X-FACTOR and Y-FACTOR, the horizontal and vertical magnification factors.

The default zoom factor is 4 in both directions. It is not necessary for X-FACTOR and Y-FACTOR to be equal. Sometimes, you may prefer to zoom in one direction only, so the other factor should be set to 1. As usual, press [OK] or [CANCL] to leave the ZOOM FACTORS menu.

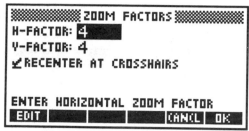

Figure 10.34: Set zoom factors

Technology Tip: If you should zoom in too much and lose the curve, zoom back to the default viewing rectangle and start over. Also, you may wish to use *[ZOOM] [ZSQR]* to make a viewing rectangle in which the vertical scale matches the horizontal scale. This would be helpful if you want to graph the two halves of a circle, as in Section 10.2.4.

10.2.7 Relative Minimums and Maximums: Graph $y = -x^3 + 4x$ once again in the standard viewing rectangle (Figure 10.9). This function appears to have a relative minimum near $x = -1$ and a relative maximum near $x = 1$. You may zoom and trace to approximate these extreme values.

First trace along the curve near the local minimum. Notice by how much the x-values and y-values change as you move from point to point. Trace along the curve until the y-coordinate is as *small* as you can get it, so that you are as close as possible to the local minimum, and zoom in (press *[ZOOM] [ZIN]* or use a zoom box). Now trace again along the curve and, as you move from point to point, see that the coordinates change by smaller amounts than before. Keep zooming and tracing until you find the coordinates of the local minimum point as accurately as you need them, approximately (-1.15, -3.08).

Follow a similar procedure to find the local maximum. Trace along the curve until the y-coordinate is as *great* as you can get it, so that you are as close as possible to the local maximum, and zoom in. The local maximum point on the graph of $y = -x^3 + 4x$ is approximately (1.15, 3.08).

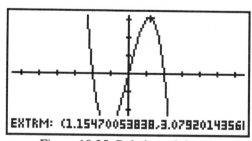

Figure 10.35: Relative minimum

The HP-48G automates the search for relative minimum and relative maximum points. Trace along the curve until the cursor is near a local extreme point. Then press *[FCN] [EXTR]*. The calculator searches for the nearest relative minimum or maximum (actually, it searches for the nearest "critical point" of the graph) and displays the coordinates of the local extreme point it finds (see Figure 10.35).

10.3   Solving Equations and Inequalities

10.3.1 Intercepts and Intersections: Tracing and zooming are also used to locate an x-intercept of a graph, where a curve crosses the x-axis. For example, the graph of $y = x^3 - 8x$ crosses the x-axis three times (see Figure 10.36). After tracing over to the x-intercept point that is furthest to the left, zoom in (Figure 10.37). Continue this process until you have located all three intercepts with as much accuracy as you need. The three x-intercepts of $y = x^3 - 8x$ are approximately -2.828, 0, and 2.828.

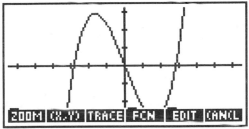

Figure 10.36: Graph of $y = x^3 - 8x$

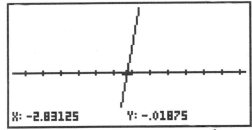

Figure 10.37: An x-intercept of $y = x^3 - 8x$

Technology Tip: As you zoom in, you may also wish to change the spacing between tick marks on the x-axis so that the viewing rectangle shows scale marks near the intercept point. Then the accuracy of your approximation will be such that the error is less than the distance between two tick marks. Go to the PLOT OPTIONS sub-menu (press *[OPTS]* in the PLOT menu). Move the cursor down and enter an appropriate spacing for the ticks.

An x-intercept of a function's graph is a *root* of the equation $f(x) = 0$. So these techniques for locating x-intercepts also serve to find the roots of an equation.

Once more, the HP-48G automates the search for x-intercepts. First trace along the graph until the cursor is close to an x-intercept. Press *[FCN] [ROOT]* to find the nearest x-intercept of this function. Repeat until you have located all x-intercepts of this graph.

TRACE and ZOOM are especially important for locating the intersection points of two graphs, say the graphs of $y = -x^3 + 4x$ and $y = -.25x$. Trace along one of the graphs until you arrive close to an intersection point. Then press ▲ or ▼ to jump to the other graph. Notice that the x-coordinate does not change, but the y-coordinate is likely to be different (see Figures 10.38 and 10.39).

When the two y-coordinates are as close as they can get, you have come as close as you now can to the point of intersection. So zoom in around the intersection point, then trace again until the two y-coordinates are as close as possible. Continue this process until you have located the point of intersection with as much accuracy as necessary.

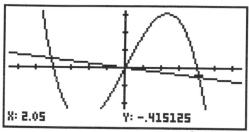

Figure 10.38: Trace on $y = -x^3 + 4x$

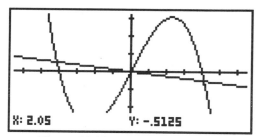

Figure 10.39: Trace on $y = -.25x$

Automate the search for points of intersection by tracing close to an intersection. Then press *[FCN] [ISECT]* to locate the nearest intersection point.

10.3.2 Solving Equations by Graphing:

10.3.2 Solving Equations by Graphing: Suppose you need to solve the equation $24x^3 - 36x + 17 = 0$. First graph $y = 24x^3 - 36x + 17$ in a window large enough to exhibit *all* its x-intercepts, corresponding to all its roots. Then use trace and zoom to locate each one. In fact, this equation has just one solution, approximately $x = -1.414$.

Remember that when an equation has more than one x-intercept, it may be necessary to change the viewing rectangle a few times to locate all of them.

Technology Tip: To solve an equation like $24x^3 + 17 = 36x$, you may first transform it into standard form, $24x^3 - 36x + 17 = 0$, and proceed as above. However, you may also graph the *two* functions $y = 24x^3 + 17$ and $y = 36x$, then zoom and trace to locate their point of intersection. On the HP-48G, when you enter an equation like $24x^3 + 17 = 36x$ in the EQ: field of the PLOT menu, the calculator graphs both sides of the equation.

10.3.3 Solving Systems by Graphing: The solutions to a system of equations correspond to the points of intersection of their graphs. For example, to solve the system $y = x^2 - 3x - 4$ and $y = x^3 + 3x^2 - 2x - 1$, first graph them together. Then zoom and trace to locate their point of intersection, approximately (-2.17, 7.25).

You must judge whether the two current y-coordinates are sufficiently close for $x = -2.17$ or whether you should continue to zoom and trace to improve the approximation.

The solutions of the system of two equations $y = x^3 + 3x^2 - 2x - 1$ and $y = x^2 - 3x - 4$ correspond to the solutions of the single equation $x^3 + 3x^2 - 2x - 1 = x^2 - 3x - 4$, which simplifies to $x^3 + 2x^2 + x + 3 = 0$. So you may also graph $y = x^3 + 2x^2 + x + 3$ and find its x-intercepts to solve the system.

10.3.4 Solving Inequalities by Graphing: Consider the inequality $-\dfrac{3x}{2} \geq x - 4$. To solve it with your HP-48G, graph the two functions $y = 1 - \dfrac{3x}{2}$ and $y = x - 4$ (Figure 10.40). First locate their point of intersection, at $x = 2$. The inequality is true when the graph of $y = 1 - \dfrac{3x}{2}$ lies *above* the graph of $y = x - 4$, and that occurs for $x < 2$. So the solution is the half-line $x \leq 2$, or $(-\infty, 2]$.

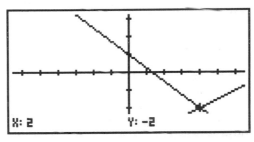

Figure 10.40: Solving $-\dfrac{3x}{2} \geq x - 4$

10.4   Trigonometry

10.4.1 Degrees and Radians: The trigonometric functions can be applied to angles measured either in radians or degrees, but you should take care that the HP-48G is configured for whichever measure you need. If your calculator is currently set for radian measure, the RAD annunciator is displayed at the top left of the home screen. If the calculator is set for degree measure, there is no annunciator. You may change the default in the MODES menu (see Section 10.1.1). Or toggle quickly between radians and degrees by pressing ↰ RAD.

It's a good idea to check the angle measure setting before executing a calculation that depends on a particular measure. You may change a mode setting at any time and not interfere with pending calculations. Try the following keystrokes to see this in action.

Expression	Keystrokes	Display
$\sin 45°$	↱ MODES ▼	
	press +/- until Degrees is displayed, then [OK]	
	45 SIN	0.7071

$\sin \pi^\circ$	↰ π SIN ↰ ▸NUM	0.0548
$\sin \pi$	↰ RAD ↰ π SIN	0.0000
$\sin 45$	45 SIN	0.8509
$\sin \frac{\pi}{6}$	↰ π 6 ÷ SIN ↰ ▸NUM	0.5000

The first line of keystrokes sets the HP-48G in degree mode and calculates the sine of 45 *degrees*. While the calculator is still in degree mode, the second line of keystrokes calculates the sine of π *degrees*, 3.1415°. The third line toggles to radian mode just before calculating the sine of π *radians*. The fourth line calculates the sine of 45 *radians* (the calculator is already in radian mode).

Technology Tip: Here's how to mix degrees and radians in a calculation. For example, calculate $\tan 45^\circ + \sin \frac{\pi}{6}$. First evaluate $\tan 45^\circ$ in degree mode, then press ENTER to place this value in the stack. Next toggle to radian mode, evaluate $\sin \frac{\pi}{6}$, and enter its value in the stack. Finally, press + to add the two latest stack entries and get 1.5.

10.4.2 Graphs of Trigonometric Functions:
When you graph a trigonometric function, you need to pay careful attention to the choice of graph window. For example, graph $y = \dfrac{\sin 30x}{30}$ in the default viewing rectangle. Trace along the curve to see where it is. Zoom in to a better window, or use the period and amplitude to establish a better view.

Technology Tip: The HP-48G has a quick way to make a good window for graphing trigonometric functions. Press *[ZOOM]* NXT NXT *[ZTRIG]* for a window in which $\Delta x = \frac{\pi}{20}$ and $\Delta y = 0.1$.

Next graph $y = \tan x$ in the default window. The HP-48G plots consecutive points and then connects them with a segment, so the graph is not exactly what you should expect. You may wish to change from connected line to dot graph (see Section 10.2.3) when you plot the tangent function.

10.5   Scatter Plots

10.5.1 Entering Data:
This table shows total prize money (in millions of dollars) awarded at the Indianapolis 500 race from 1981 to 1989. (*Source:* Indianapolis Motor Speedway Hall of Fame.)

Year	1981	1982	1983	1984	1985	1986	1987	1988	1989
Prize ($ million)	$1.61	$2.07	$2.41	$2.80	$3.27	$4.00	$4.49	$5.03	$5.72

We'll now use the HP-48G to construct a scatter plot that represents these points and to find a linear model that approximates the given data. Go to the PLOT menu and change the type to Scatter. Check Autoscale so the calculator will display all the data in the largest possible window (see Figure 10.41).

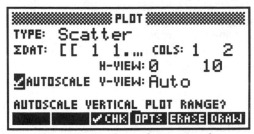
Figure 10.41: Setup for scatter plot

Now move the highlight to ΣDAT: to input the data table; press *[CHOOS] [NEW]* and name this data table INDY. Then move to the Object field and press ↱ MATRIX. Instead of entering the full year 198x, enter only x. Here are the keystrokes for the first three years: 1 ENTER 1.61 ENTER ▼ 2 ENTER 2.07 ENTER 3 ENTER 2.41 ENTER and so on (see Figure 10.42). Pressing the down arrow key after entering the first row is a signal that the last *column* of data has been reached; hereafter, press ENTER and move *across* rows and down columns. Continue to enter all the given data. Press ENTER when you have finished.

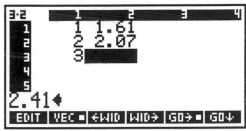
Figure 10.42: Data table

10.5.2 Plotting Data: Once all the data points have been entered, press *[ERASE] [DRAW]* to draw a scatter plot. As with other plots, you may zoom and also display the coordinates of the cursor's position.

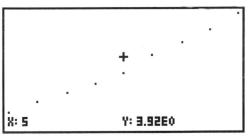
Figure 10.43: Scatter plot

10.5.3 Regression Line: The HP-48G calculates the slope and *y*-intercept for the line that best fits all the data. After the data points have been entered, go to the home screen and press ↱ STAT to calculate a linear regression model. Select Fit data... as you see in Figure 10.44, then verify that the MODEL: choice is Linear Fit. Press *[OK]* and the linear regression model is put in stack level 3. Below it in stack level 2 is the number r (between -1 and 1), called the *correlation coefficient*. It measures the goodness of fit of the linear regression equation with the data. The closer |r| is to 1, the better the fit; the closer |r| is to 0, the worse the fit. Press DROP twice for a better look at the linear regression model.

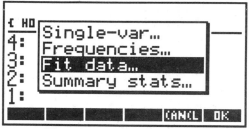

Figure 10.44: Statistics menu

Figure 10.45: Linear regression model

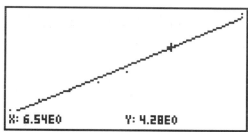

Figure 10.46: Linear regression line

Graph the linear regression model over the data by returning to the scatter plot and pressing *[STATL]*. See how well this line fits with your data points (see Figure 10.46).

10.5.4 Exponential Growth Model: After data points have been entered, press ➮ STAT and select Fit data... again, but now choose Exponential Fit for MODEL: to calculate an exponential growth model $y = a \cdot e^{mx}$ for the data. Convert this to the familiar form $y = a \cdot b^x$ by setting $b = e^m$.

10.6 Matrices

10.6.1 Making a Matrix: The HP-48G can display and use many different matrices. Here's how to create

this 3×4 matrix $\begin{bmatrix} 1 & -4 & 3 & 5 \\ -1 & 3 & -1 & -3 \\ 2 & 0 & -4 & 6 \end{bmatrix}$ as matrix A in your calculator. Since we're using only integers in these examples, change to the standard number display format by pressing ↱ MODES and then +/- until Std appears; next press *[OK]*.

Press ↱ MEMORY *[NEW]* to create a new variable (Figure 10.47 shows the new variable menu); then with the OBJECT: field highlighted, press ↱ MATRIX to switch to the MatrixWriter application (Figure 10.48).

Figure 10.47: Matrix A named

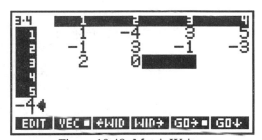

Figure 10.48: MatrixWriter

Starting in the 1st row and 1st column, press 1 ENTER. Note the box to the right of GO♦ in the menu in Figure 10.48. This signifies that pressing ENTER results in a move to the *right* along a *row*. If you would prefer ENTER to cause the cursor to move *down* the current *column*, press the last function key. For now, set your calculator so that entry moves right. Here are the keystrokes you need to continue to input the first row: 4 +/- ENTER 3 ENTER 5 ENTER. When you reach the end of the first row, press ▼ to mark the final column of this matrix. Hereafter, input the remaining elements of the matrix and press only ENTER after each; the cursor will move to the right along each row, then jump back to the *first* row in the *next* column. You must press 0 ENTER to input 0 as an element of the matrix. When the whole matrix is entered, press ENTER once more.

With the cursor in the NAME field (see Figure 10.47), press α A ENTER to call this matrix A. Finally, press *[OK]* NXT *[OK]* and return to the home screen.

10.6.2 Matrix Math: From the home screen you can perform many calculations with matrices. First, let's see matrix A itself by pressing α A ENTER (Figure 10.49).

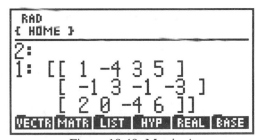

Figure 10.49: Matrix A

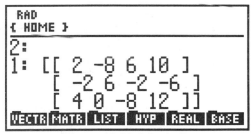

Figure 10.50: Matrix B

Calculate the scalar multiplication 2A by pressing 2 ENTER α A ×. To set matrix B equal to 2A, press ➡
MEMORY *[NEW]* NXT *[CALC]* 2 ENTER α A × *[OK]*. Name this matrix B and return to the home screen;
press α B ENTER to see it (Figure 10.50).

Add the two matrices A and B by pressing α A ENTER α B +. Subtraction is similar.

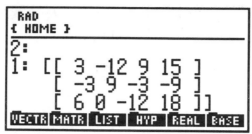

Figure 10.51: Matrix A + matrix B

Now create this 2×3 matrix as C: $\begin{bmatrix} 2 & 0 & 3 \\ 1 & -5 & -1 \end{bmatrix}$. For matrix multiplication of C by A, press α C ENTER α
A ×. If you tried to multiply A by C, your HP-48G would signal an error because the dimensions of the two
matrices do not permit multiplication this way.

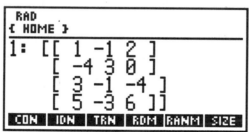

Figure 10.52: Aᵀ

The *transpose* of a matrix A is another matrix with the rows and columns interchanged. The symbol for the
transpose of A is Aᵀ. To calculate Aᵀ, press α A ENTER MTH *[MATR]* *[MAKE]* *[TRN]*.

10.6.3 Row Operations: Here are the keystrokes necessary to perform elementary row operations on a matrix. Your textbook provides more careful explanation of the elementary row operations and their uses.

To interchange the second and third rows of the matrix A that was defined above, press α A ENTER 2
ENTER 3 MTH *[MATR]* *[ROW]* NXT *[RSWP]* (see Figure 10.53). The format of this command is *matrix*
ENTER *row1* ENTER *row2* MTH *[MATR]* *[ROW]* NXT *[RSWP]*.

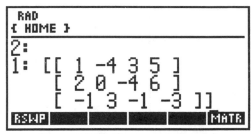

Figure 10.53: Swap rows 2 and 3

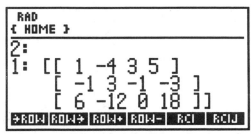

Figure 10.54: Add -4 times row 2 to row 3

To multiply row 2 by -4 and *store* the results in row 2, thereby replacing row 2 with new values, press α A ENTER 4 +/- ENTER 2 MTH *[MATR] [ROW] [RCI]*. The format of this command is *matrix* ENTER *factor* ENTER *row* MTH *[MATR] [ROW] [RCI]*.

To multiply row 2 by -4 and *add* the results to row 3, thereby replacing row 3 with new values, press α A ENTER 4 +/- ENTER 2 ENTER 3 MTH *[MATR] [ROW] [RCIJ]* (see Figure 10.54). The format of this command is *matrix* ENTER *factor* ENTER *row1* ENTER *row2* MTH *[MATR] [ROW] [RCIJ]*.

Technology Tip: It is important to remember that your HP-48G does *not* store a matrix obtained as the result of any row operations. So when you need to perform several row operations in succession, you may wish to store the result of each operation in a temporary place.

For example, use elementary row operations to solve this system of linear equations:
$$\begin{cases} x - 2y + 3z = 9 \\ -x + 3y = -4 \\ 2x - 5y + 5z = 17 \end{cases}.$$

First enter this *augmented matrix* as A in your HP-48G:
$$\begin{bmatrix} 1 & -2 & 3 & 9 \\ -1 & 3 & 0 & -4 \\ 2 & -5 & 5 & 17 \end{bmatrix}.$$

Here are the row operations and their associated keystrokes. Note that steps 2, 3, and 4 assume the previously calculated matrix is already in stack level 1. The solution is shown in Figure 10.55.

Row Operation	Keystrokes
add row 1 to row 2	α A ENTER 1 ENTER 1 ENTER 2 MTH *[MATR] [ROW] [RCIJ]*
add -2 times row 1 to row 3	2 +/- ENTER 1 ENTER 3 *[RCIJ]*
add row 2 to row 3	1 ENTER 2 ENTER 3 *[RCIJ]*
multiply row 3 by ½	1 ENTER 2 ÷ 3 *[RCI]*

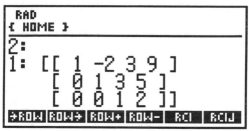

Figure 10.55: Final matrix after row operations

Thus $z = 2$, so $y = -1$ and $x = 1$.

10.6.4 Determinants and Inverses: Enter this 3×3 square matrix as A: $\begin{bmatrix} 1 & -2 & 3 \\ -1 & 3 & 0 \\ 2 & -5 & 5 \end{bmatrix}$. To calculate its de-

terminant, $\begin{vmatrix} 1 & -2 & 3 \\ -1 & 3 & 0 \\ 2 & -5 & 5 \end{vmatrix}$, press α A MTH *[MATR] [NORM]* NXT *[DET]*. You should find that $|A| = 2$.

Since the determinant of matrix A is not zero, it has an inverse, A^{-1}. Press α A $\frac{1}{x}$ to calculate the inverse of matrix A, as seen in Figure 10.56.

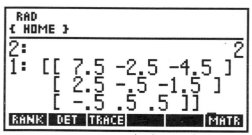

Figure 10.56: $|A|$ and A^{-1}

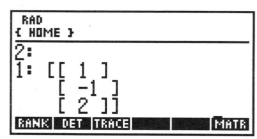

Figure 10.57: Solution matrix

Now let's solve a system of linear equations by matrix inversion. Once more, consider $\begin{cases} x - 2y + 3z = 9 \\ -x + 3y = -4 \\ 2x - 5y + 5z = 17 \end{cases}$.

HP-48G Series Calculator

The coefficient matrix for this system is the matrix $\begin{bmatrix} 1 & -2 & 3 \\ -1 & 3 & 0 \\ 2 & -5 & 5 \end{bmatrix}$ that was entered in the previous example. If

necessary, enter it again as A in your HP-48G. Enter the matrix $\begin{bmatrix} 9 \\ -4 \\ 17 \end{bmatrix}$ as B. Then press α A ÷ α B × to calcu-

late the solution matrix (Figure 10.57). The solutions are still $x = 1$, $y = -1$, and $z = 2$.

10.7   Sequences

10.7.1 Iteration: Iteration is the process of evaluating a function repeatedly. As an example, calculate $\dfrac{n-1}{3}$
for $n = 27$. Then calculate $\dfrac{n-1}{3}$ for n = the answer to the previous calculation. Continue to use each answer
as n in the *next* calculation. Here are keystrokes to accomplish this iteration on the HP-48G calculator (see the
results in Figure 10.58).

Iteration	Keystrokes	Display
	' α Y ← = ← ( ) α N - 1 ▶ ÷ 3 ← DEF	
1	27 ENTER ENTER ' α N STO	27
2	α Y EVAL ENTER ' α N STO	8.66666666667
3	α Y EVAL ENTER ' α N STO	2.55555555556
4	α Y EVAL ENTER ' α N STO	.51851851852
5	α Y EVAL ENTER ' α N STO	-.16049382716

Continue several more times and see what happens with this iteration. You may wish to try it again with a
different starting value.

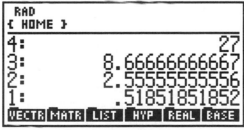

Figure 10.58: Iteration

10.7.2 Arithmetic and Geometric Sequences: It is easy to use direct stack input to determine the n-th term of a sequence. For example, find the 18th term of an *arithmetic* sequence whose first term is 7 and whose common difference is 4. Enter the first term 7 in the stack, then start the progression by pressing 4 +. This yields the 2nd term, so repeat 4 + sixteen more times to find the 18th term. For a *geometric* sequence whose common ratio is 4, start the progression with 4 ×.

Of course, you could also use the *explicit* formula for the n-th term of an arithmetic sequence, $t_n = a + (n-1)d$. First store values for the variables a, d, and n, then evaluate the formula by pressing ' α ← A + ← ( ) α ← N - 1 ▶ × α ← D EVAL. For a geometric sequence whose n-th term is given by $t_n = a \cdot r^{n-1}$, enter values for the variables a, r, and n, then evaluate the formula by pressing ' α ← A × α ← R y^x ← ( ) α ← N - 1 EVAL.

10.7.3 Sums of Sequences: Calculate the sum $\sum_{n=1}^{12} 4(0.3)^n$ in the HP-48G's EquationWriter application by pressing ↱ Σ α N ▶ 1 ▶ 12 ▶ 4 × .3 y^x α N ENTER. Then press EVAL; you should get 1.71428480324. The format of this command is 'Σ(*variable=begin, end, expression*)'.

10.8   Parametric and Polar Graphs

10.8.1 Graphing Parametric Equations: The HP-48G plots parametric equations as easily as it plots functions. Just go to the PLOT menu (Figure 10.59) and change the type setting to Parametric. Be sure, if the independent parameter is an angle measure, that your calculator is set to whichever you need, radians or degrees.

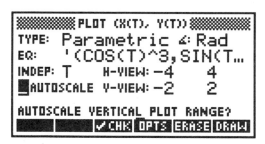

Figure 10.59: Parametric PLOT menu

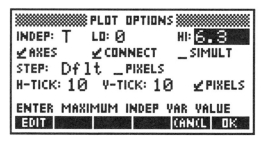

Figure 10.60: Parametric plot options

For example, here are the keystrokes needed to graph the parametric equations $x = \cos^3 t$ and $y = \sin^3 t$. First check that angles are currently being measured in radians. After setting plot type to parametric, change the independent variable to T. Also change the view to extend horizontally from -4 to 4 and vertically from -2 to 2. Press *[OPTS]* so that you may change LO: and HI: values for T so that T ranges from 0 to 6.3 (approximately 2π). Press *[OK]* to accept these changes and return to the main PLOT menu.

Move the cursor to EQ: (Figure 10.59) and enter the two parametric equations in the form $(x(t), y(t))$ by pressing ' ⬅ ( ) COS α T ▶ y^x 3 ⬅ , ⬅ ( ) SIN α T ▶ y^x 3 ENTER. Press *[ERASE] [DRAW]* to see Figure 10.61.

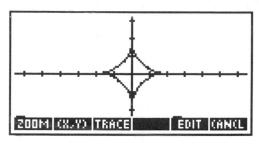

Figure 10.61: Parametric graph of $x = \cos^3 t$ and $y = \sin^3 t$

You may ZOOM and TRACE along parametric graphs just as you did with function graphs. As you trace along this graph, notice that the cursor moves in the *counterclockwise* direction as T increases.

10.8.2 Rectangular-Polar Coordinate Conversion: The HP-48G converts easily between rectangular and polar coordinate systems. Since the conversion uses the current MODES settings, it is a good idea to check the default angle measure beforehand. Of course, you may press ⬅ RAD to change the current angle measure setting at any time, as explained in Section 10.4.1. For the following examples, the HP-48G is first set to *degree* measure and *rectangular* coordinates.

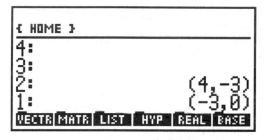

Figure 10.62: Polar coordinates   Figure 10.63: Rectangular coordinates

Given rectangular coordinates $(x, y) = (4, -3)$, convert *from* these rectangular coordinates *to* polar coordinates (r, θ) by pressing ⬅ ( ) 4 ⬅ , 3 +/- ENTER to input (4, -3). Then press ➡ POLAR to toggle the calculator into polar display mode. Notice the polar display annunciator at the top left of the screen (Figure 10.62).

Suppose $(r, \theta) = (3, 180°)$. To convert *from* these polar coordinates *to* rectangular coordinates (x, y), press ⬅ ( ) 3 ⬅ , ➡ ∠ 180 ENTER to input (3, 180°). Once again, press ➡ POLAR to toggle the HP-48G *off* polar display mode. There should no longer be a polar display annunciator at the top left of the screen (see Figure 10.63).

10.8.3 Graphing Polar Equations: The HP-48G graphs a polar function in a polar plot. To graph $r = 4 \sin \theta$, go to the PLOT menu and change plot type to Polar. Make θ the independent variable by pressing α ↱ F in the INDEP: field. Choose a good viewing rectangle and appropriate options LO: and HI: for the parameter θ. In Figure 10.64, the graphing window extends horizontally from -6.5 to 6.5 and vertically from -1.1 to 5.2.

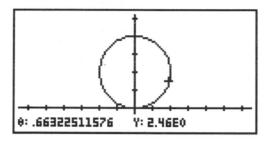

Figure 10.64: Polar graph of $r = 4 \sin \theta$

10.9   Probability

10.9.1 Random Numbers: The command RAND generates a number between 0 and 1. You will find this command in the probability sub-menu of the MTH menu. Press MTH NXT [PROB] [RAND] to generate a random number. Press [RAND] to generate another number; keep pressing [RAND] to generate more of them.

If you need a random number between, say, 0 and 10, then press 10 ENTER [RAND] ×. To get a random number between 5 and 15, press 5 ENTER 10 ENTER [RAND] × +.

10.9.2 Permutations and Combinations: To calculate the number of *permutations* of 12 objects taken 7 at a time, $_{12}P_7$, press 12 ENTER 7 MTH NXT [PROB] [PERM]. Then $_{12}P_7 = 3{,}991{,}680$, as shown in Figure 10.65.

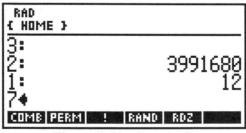

Figure 10.65: $_{12}P_7$ and $_{12}C_7$

For the number of *combinations* of 12 objects taken 7 at a time, $_{12}C_7$, press 12 ENTER 7 MTH NXT *[PROB]* *[COMB]*. So $_{12}C_7 = 792$.

10.9.3 Probability of Winning: A state lottery is configured so that each player chooses six different numbers from 1 to 40. If these six numbers match the six numbers drawn by the State Lottery Commission, the player wins the top prize. There are $_{40}C_6$ ways for the six numbers to be drawn. If you purchase a single lottery ticket, your probability of winning is 1 in $_{40}C_6$. Press 40 ENTER 6 MTH NXT *[PROB]* *[COMB]* ⅟ₓ to calculate your chances, but don't be disappointed.

10.10   Programming

10.10.1 Entering a Program: The HP-48G is a programmable calculator that can store sequences of commands for later replay. Its programming language is structured, like other modern computer programming languages you may know.

An HP-48G program is an object in the stack and can be stored in a variable. Programs are sequences of commands and numbers, the same as you would enter directly in the calculator. The elements of a program are contained between double angle brackets, which you get by pressing ← « », and they are separated by spaces, using the SPC key.

The HP-48G has a collection of sample programs that you can see and use and even modify. To get them from the calculator's memory, enter the word TEACH in the command line, then press ENTER. This loads the EXAMPLES directory. There is a listing of the sample programs in your User's Guide. To view any program, enter its name into stack level 1 and press ← EDIT.

10.10.2 Executing a Program: To execute a program, enter its name in stack level 1 and press ENTER.

The instruction manual for your HP-48G gives detailed information about programming. Refer to it to learn more about programming and how to use other features of your calculator.

10.11 Differentiation

10.11.1 Limits: Suppose you need to find this limit: $\lim\limits_{x\to0}\dfrac{\sin 4x}{x}$. Plot the graph of $f(x)=\dfrac{\sin 4x}{x}$ in a convenient viewing rectangle that contains the point where the function appears to intersect the line $x = 0$ (because you want the limit as $x \to 0$). Your graph should lend support to the conclusion that $\lim\limits_{x\to0}\dfrac{\sin 4x}{x}=4$.

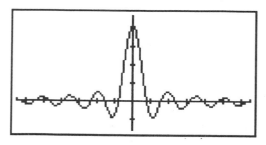

Figure 10.66: Checking $\lim\limits_{x\to0}\dfrac{\sin 4x}{x}=4$

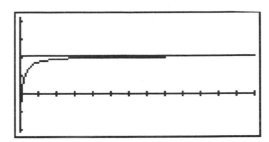

Figure 10.67: Checking $\lim\limits_{x\to\infty}\dfrac{2x-1}{x+1}=2$

To test the reasonableness of the conclusion that $\lim\limits_{x\to\infty}\dfrac{2x-1}{x+1}=2$, evaluate the function $f(x)=\dfrac{2x-1}{x+1}$ for several large positive values of x (since you want the limit as $x \to \infty$). For example, evaluate $f(100)$, $f(1000)$, and $f(10{,}000)$. Another way to test the reasonableness of this result is to examine the graph of $f(x)=\dfrac{2x-1}{x+1}$ in a viewing rectangle that extends over large values of x. See, as in Figure 10.67 (where the viewing rectangle extends horizontally from 0 to 100), whether the graph is asymptotic to the horizontal line $y = 2$ (plot both $y=\dfrac{2x-1}{x+1}$ and $y = 2$).

10.11.2 Numerical Derivatives: The HP-48G calculates derivatives numerically as well as symbolically. For example, to find a numerical approximation to $f'(2.5)$ when $f(x)=x^3$, first press ⟶ SYMBOLIC and select Differentiate. Enter x^3 as the expression and specify x as the variable with respect to which you want to differentiate. Press the +/– key to cycle among alternatives until the Numeric result type is displayed and enter 2.5 as the value of x for which you want to find the derivative (see Figure 10.68). Finally, press *[OK]* for the answer (Figure 10.69).

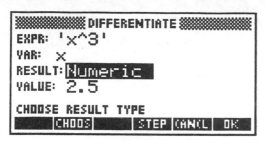

Figure 10.68: DIFFERENTIATE Numeric

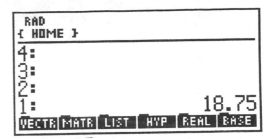

Figure 10.69: Numerical derivative

If you need to differentiate symbolically, first purge your variable from memory by recalling the variable to the stack and pressing ⬅ PURG. Next follow the preceding steps except choose the Symbolic result type (Figure 10.70). The press *[OK]* for the symbolic derivative (Figure 10.71). If you would like to see the intermediate steps taken by the HP-48G as it performs a differentiation, press *[step]* to place in the stack the result of the first step. Then press EVAL for subsequent steps in the process of symbolically differentiating the expression.

Figure 10.70: DIFFERENTIATE Symbolic

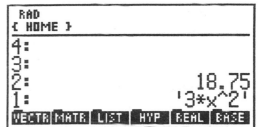

Figure 10.71: Symbolic Derivative

To find a derivative in EquationWriter, press ➡ ∂ and enter x (be sure you have already purged the variable x from memory) as the variable of differentiation (see Figure 10.72). Press ▶ and enter the expression x^3, then press ENTER to place this in the stack. Now press EVAL a couple of times for the result (Figure 10.73).

Figure 10.72: Differentiation with EquationWriter

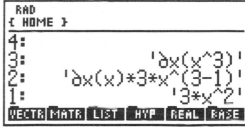

Figure 10.73: Symbolic derivative

Technology Tip: It is sometimes helpful to plot a function and its derivative together. In Figure 10.74, the function $f(x) = \dfrac{5x - 2}{x^2 + 1}$ and its numerical derivative are both graphed.

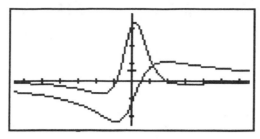

Figure 10.74: Graphs of $f(x)$ and $f'(x)$

You can duplicate this graph by creating the equation $f(x) = \dfrac{5x-2}{x^2+1}$ (Figure 10.75) with these keystrokes: ' α

↰F ↰ ( ) α ↰ X ▶ ↰ = ↰ ( ) 5 × α ↰ X − 2 ▶ ÷ ↰ ( ) α ↰ X y^x 2 + 1 ENTER. Then press ↰DEF to make this equation the definition of a new function.

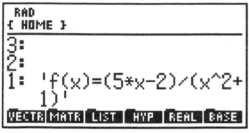

Figure 10.75: Defining $f(x) = \dfrac{5x-2}{x^2+1}$

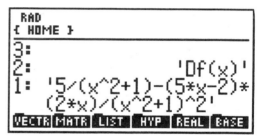

Figure 10.76: Defining $Df(x) = f'(x)$

Next create the equation $Df(x) = f'(x)$ by entering ' α D α ↰F ↰ ( ) α ↰ X directly into the stack.

Be sure to purge the variable x from memory. Use EquationWriter to create the expression $\dfrac{\partial}{\partial x} f(x)$ and put it in the stack. Press EVAL several times until the derivative is fully simplified (Figure 10.76). Then press ↰ = ↰DEF to make an equation and define a new function.

The final step is to plot the function and its derivative simultaneously. Go into the PLOT menu, move beside EQ:, and press ↰ { } ' α ↰F ↰ ( ) α ↰ X ▶ ▶ SPC ' α D α ↰F ↰ ( ) α ↰ X [OK]. Also enter α ↰ X beside INDEP: to establish x as the independent variable. Figure 10.74 shows the result after you press [ERASE] [DRAW].

You may also approximate a derivative while you are examining the graph of a function. When you are in a graph window, trace along the curve to a point where you want the derivative (see Figure 10.77 for the graph of $f(x) = \dfrac{5x-2}{x^2+1}$ at $x = -2.3$) and press *[FCN] [SLOPE]*. The value of the derivative is displayed at the bottom of the graph window (as in figure 10.78) and also entered in the stack.

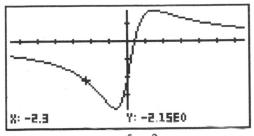

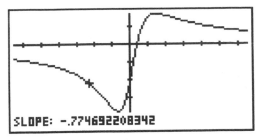

Figure 10.77: $f(x) = \dfrac{5x-2}{x^2+1}$ at $x = -2.3$

Figure 10.78: $f'(-2.3)$

Technology Tip: With the HP-48G, you can graph a function along with the tangent line at one of its points. So plot the function $f(x) = \dfrac{5x-2}{x^2+1}$ once again. Then trace with the cursor to a point of interest and press *[FCN] NXT [TANL]*. The equation of the tangent line is displayed in the graph window and also entered into the stack. You may continue to draw more tangent lines.

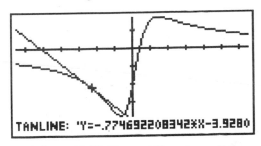

Figure 10.79: Tangent line at $x = -2.3$

10.11.3 Newton's Method: With your HP-48G, you may iterate using Newton's Method to find the zeros of a function. Recall that Newton's Method determines each successive approximation by the formula

$$x_{n+1} = x_n - \frac{f(x_n)}{f'(x_n)}.$$

As an example of the technique, consider $f(x) = 2x^3 + x^2 - x + 1$. First define this function and also define its derivative as Df (see Figure 10.80). A look at the graph of the function suggests that it has a zero near $x = -1$, so start the iteration by storing -1 as the variable x. Enter the Newton's Method formula

$$y = x - \frac{f(x)}{f'(x)} = x - \frac{f(x)}{Df(x)}$$ as shown in stack level 4 of Figure 10.81. Then press these keys: α Y EVAL
ENTER ' α X STO. Execute this sequence of keystrokes repeatedly until two successive approximations differ by less than some predetermined value, say 0.0001. Note that each time you execute these keystrokes, the HP-48G will use its *current* value of x, and that value is changing as you continue the iteration.

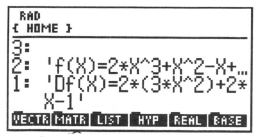

Figure 10.80: Defining $f(x)$ and $Df(x)$

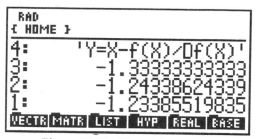

Figure 10.81: Newton's method

Technology Tip: Newton's Method is sensitive to your seed value for x, so look carefully at the function's graph to make a good first estimate. Also, remember that the method sometimes fails to converge!

You may want to save the Newton's Method formula as a short program. See your calculator's manual for further information on programming the HP-48G.

10.12   Integration

10.12.1 Approximating Definite Integrals: The HP-48G approximates integrals numerically and also evaluates them symbolically. For example, to find $\int_0^1 \cos x \, dx$, first press ⌐➤ SYMBOLIC and select Integrate. Enter $\cos x$ as the expression and specify x as the variable for this integration.

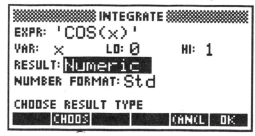

Figure 10.82: INTEGRATE Numeric

Set 0 as the lower limit and 1 as the upper limit. Press the +/– key to cycle among alternatives until the **Numeric** result type is displayed (see Figure 10.82). Finally, press *[OK]* for the answer (stack level 3 in Figure 10.84).

If you need to integrate symbolically, first purge your variable from memory by recalling the variable to the stack and pressing ↰ PURG. Next follow the preceding steps except choose the Symbolic result type. Then press *[OK]* for the symbolic integral (stack level 1 in Figure 10.83) and EVAL to evaluate it (stack level 1 in Figure 10.84).

Figure 10.83: INTEGRATE Symbolic

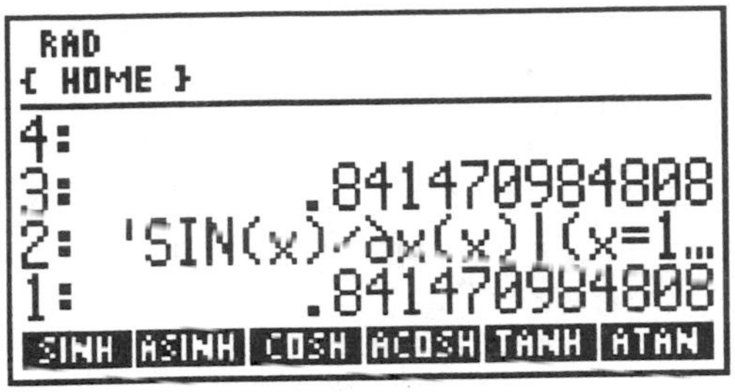

Figure 10.84: $\int_0^1 \cos x \, dx$

Technology Tip: Suppose you want to find an indefinite integral, say $\int \cos x \, dx$. Purge the variable x from memory and integrate symbolically. Enter 0 for the lower limit and the variable x for the upper limit. Press *[OK]* to compute a closed-form expression and place it on stack level 1 (Figure 10.85). Now press PRG *[TYPE] [OBJ→]* 3 ↰ STACK NXT *[DRPN]* to discard the lower limits and press EVAL to evaluate the result for the upper limit and obtain the indefinite integral (Figure 10.86).

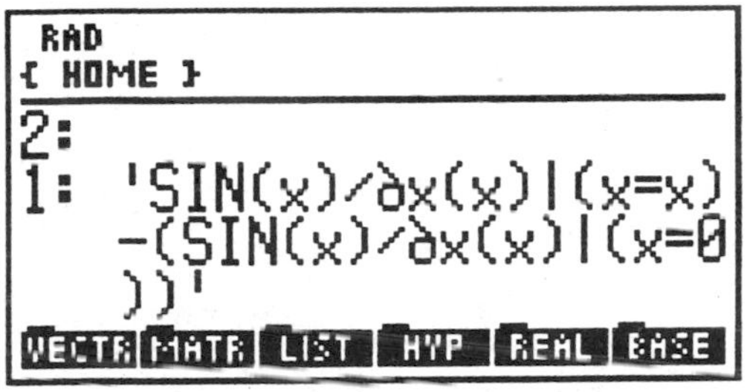

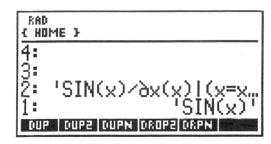

Figure 10.85: Indefinite integral

Figure 10.86: $\int \cos x \, dx$

To find a definite integral in EquationWriter, press ↱ ∫ and then enter the lower limit followed by ▶, the upper limit followed by ▶, and the integrand followed by ▶. Next enter the variable of integration (see Figure 10.87 for $\int_0^1 \cos x \, dx$) and press ENTER to place this integral in the stack. Now press EVAL a couple of times for the result (Figure 10.88).

Figure 10.87: Integration with EquationWriter

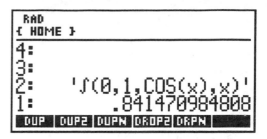

Figure 10.88: Symbolic integration

10.12.2 Areas: You may approximate the area under the graph of a function $y = f(x)$ between $x = A$ and $x = B$ with your HP-48G. For example, here are keystrokes for finding the area under the graph of the function $y = \cos x^2$ between $x = 0$ and $x = 1$. This area is represented by the definite integral $\int_0^1 \cos x^2 dx$. So graph $f(x) = \cos x^2$ and trace along the curve to where $x = 0$, the lower limit. Press the multiplication key $\times$ to mark this place (see Figure 10.89). Now trace again to the upper limit, where $x = 1$, and press NEXT *[FCN] [AREA]*. The area of region under the graph between the lower limit and the upper limit is calculated and displayed.

Technology Tip: When approximating the area under $f(x) = \cos x^2$ between $x = 0$ and $x = 1$, you must trace along the curve to *exactly* where $x = 0$ and $x = 1$. Now to trace along the curve to $x = a$, the viewing rectangle must be chosen so that the function is evaluated at $x = a$. The window shown in Figure 10.89 was made first by zooming to a decimal window, and then changing its vertical dimensions to appropriate values. By contrast, find the area under $f(x) = \cos x^2$ between $x = 0$ and $x = 1$ in a window that extends horizontally from -4 to 4.

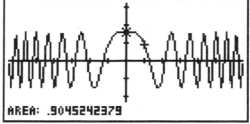

Figure 10.89: Area under a curve

Technology Tip: Suppose that you want to find the area between two functions, $y = f(x)$ and $y = g(x)$, from $x = A$ to $x = B$. If $f(x) \geq g(x)$ for $A \leq x \leq B$, then use $f(x) - g(x)$ for the integrand and proceed as before to find the required area.

Chapter 11

Hewlett Packard HP 38G

11.1 Getting started with the HP 38G

11.1.1 Basics: Press the ON key to begin using your HP 38G. If you need to adjust the display contrast, first press and hold ON, then press + to increase the contrast or press - to decrease the contrast.

Midway between the ON key and the ENTER key is turquoise key that functions like the SHIFT key on a computer keyboard. We use the symbol ■ for this turquoise shift key. After you press ■, the symbol ↰ appears at the top left corner of the screen.

Across the top of the HP 38G's keypad are six black function keys that assume different functions in different contexts. When pressing a black key has the effect COMMAND within the current context, we use the symbol [COMMAND] to refer to that key.

When you have finished with the calculator, turn it off to conserve battery power by pressing the turquoise shift key ■ and then OFF. Power is automatically switched off after several minutes of inactivity.

Check the HP 38's settings by pressing ■ MODES. If necessary, use the arrow keys to move the blinking cursor to a setting you want to change, Press [CHOOSE] to select a new setting. To start with, select the options illustrated in Figure 11.1: degrees, standard number format, and dot (period) for the decimal point. Now return the home screen by pressing HOME

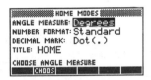

Figure 11.1: HOME MODES

11.1.2 Editing: The HP 38G's home screen has labels for the six black function keys across the bottom (not every function key may be active, in which case its label is blank). Above the function key labels is the edit line, where input first appears; and above the edit line is a history area, where previous input and results are stored. For example, enter the home screen and type this sum:

$$1+2+3+4+5+6+7+8+9+10+11+12+13+14+15+16+17+18+19+20$$

Then press ENTER to move this up into the history area and see the answer (Figure 11.2).

Figure 11.2: Home screen

Often we do not notice a mistake until we see how unreasonable an answer is. The HP 38G permits you to redisplay an entire calculation, edit it easily, then execute the *corrected* calculation.

Suppose you had typed 12 + 34 + 56, but had *not yet* pressed ENTER, when you realize that 34 should have been 74. Simply press ◄ (the *left* arrow key) as many times as necessary to move the blinking cursor left to 3, then press 7 DEL. On the other hand, if 34 should have been 384, move the cursor back to 4, and then type 8. The 8 is inserted at the cursor position and the other characters are pushed to the right. If the 34 should have been 3 only, move the cursor to 4, and press DEL to delete it.

Technology Tip: To move quickly to the *beginning* of an expression you are currently editing, press ■ ◄; to jump to the *end* of that expression, press ■ ►.

The history area may contain many previous entries, arranged from the most recent at the bottom to the oldest entry way above, even out of view. You may still edit one of these older expressions. First, press ON to clear the edit line. Next press ▲ and ▼ as necessary to move the highlight up and down through the history list until you reach the expression you want, then press [COPY] to bring it back to the edit line. Now you can change it. When you are finished, press ENTER to evaluate this new expression.

Technology Tip: To move quickly to the *oldest* entry at the top of the history area (which may be out of view), press ■ ▲; to jump back down to the edit line, press ■ ▼.

The HP 38G retains its history area even when it is turned off, so long as its batteries are good. When you want to clear the history area, press ■ CLEAR; this clears the edit line at the same time. Since the history area takes up calculator memory, it's a good idea to clear the history area frequently.

Technology Tip: When you need to evaluate a formula for different values of a variable, use the editing feature to simplify the process. For example, suppose you want to find the balance in an investment account if there is now $5000 in the account and interest is compounded annually at the rate of 8.5%. The formula for the balance is $P = \left(1 + \frac{r}{n}\right)^{nt}$, where P = principal, r = rate of interest (expressed as a decimal), n = number of times interest is compounded each year, and t = number of years. In our example, this becomes $5000(1 + .085)^t$. Here are the keystrokes for finding the balance after t = 3, 5, and 10 years (Figure 11.3).

Years	Keystrokes	Balance
3	5000 (1 + .085) x^y 3 ENTER	$6386.45
5	▲ ▲ [COPY] ◄ 5 DEL ENTER	$7518.28
10	▲ ▲ [COPY] ◄ 10 DEL ENTER	$11,304.92

Then to find the balance from the same initial investment but after 5 years when the annual interest rate is 7.5%, press the keys to change the last calculation above: ▲ ▲ [COPY] DEL DEL 5 ◄ ◄ ◄ ◄ ◄ 7 DEL ENTER.

Technology Tip: Press ▲ as many times as necessary to move the highlight up into the history area and onto one of the expressions you just entered. Then press [SHOW] for the HP 38G to display this expression in standard mathematical form, the way you see expressions in your textbook. This is especially helpful for checking your entry of a complicated formula. When you have finished examining the expression in standard form, press [OK] to return to the home screen. Figure 11.4 illustrates this with the last calculation.

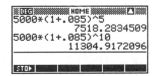

Figure 11.3: Editing expressions

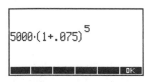

Figure 11.4: [SHOW]

11.1.3 Key Functions:
Most keys on the HP 38G offer access to more than one function, just as the keys on a computer keyboard can produce more than one letter ("g" and "G") or even quite different characters ("5" and "%"). The primary function of a key is indicated on the key itself, and you access that function by a simple press on the key.

To access the *second* function indicated *above* a key, first press ■ and *then* press the key. For example to calculate 5^2, press 25 ■ x^2 ENTER.

When you want to use an uppercase letter at the *lower right* corner of a key, first press A...Z. The symbol α appears at the top of screen. Then press the key. For example, to use the letter K in a formula, press A...Z K. If you need several letters in a row, press and hold A...Z, and press all the letters you want. For a lowercase letter, press ■ a...z and then the letter. So press ■ a...z K for the letter k.

11.1.4 Order of Operations:
The HP 38G performs calculations according to the standard algebraic rules. Working outwards from inner parentheses, calculations are performed from left to right. Powers and roots are evaluated first, followed by multiplications and divisions, and then additions and subtractions.

Enter these expressions to practice using your HP 38G.

Expression	Keystrokes	Display
$7 - 5 \cdot 3$	7 – 5 * 3 ENTER	–8
$(7 - 5) \cdot 3$	(7 – 5) * 3 ENTER	6
$120 - 10^2$	120 – 10 ■ x^2 ENTER	20
$(120 - 10)^2$	(120 – 10) ■ x^2 ENTER	12100
$\dfrac{24}{2^3}$	24 / 2 x^y 3 ENTER	3

$\left(\dfrac{24}{2}\right)^3$	(24 / 2) x^y 3 ENTER	1728
$(7 - -5) \cdot -3$	(7 – –x 5) * –x 3 ENTER	–36

11.1.5 Algebraic Expressions and Memory: Your calculator can evaluate expressions such as $\dfrac{N(N+1)}{2}$ *after* you have entered a value for N. Suppose you want $N = 200$. Press 200 [STO ▶] A...Z N ENTER to store the value 200 in memory location N. Whenever you use N in an expression, the calculator will substitute the value 200 until you make a change by storing *another* number in N. Next enter the expression $\dfrac{N(N+1)}{2}$ by typing A...Z N * (A...Z N + 1) / 2 ENTER. For $N = 200$, you will find that $\dfrac{N(N+1)}{2} = 20100$.

The contents of any memory location may be revealed by typing just its letter name and then ENTER. And the HP 38G retains memorized values even when it is turned off, so long as its batteries are good.

11.1.6 Repeated Operations with Ans: The result of your *last* calculation is always stored in memory location Ans and replaces any previous result. This makes it easy to use the answer from one computation in another computation. For example, press 30 + 15 ENTER so that 45 is the last result displayed. Then press ■ ANSWER / 9 ENTER and get 5 because 45 ÷ 9 = 5.

With a function like division, you press the / key *after* you enter an argument. For such functions, whenever you would start a new calculation with the previous answer followed by pressing the function key, you may press just the function key. So instead of ■ ANSWER / 9 in the previous example, you could have pressed simply / 9 to achieve the same result. This technique also works for these functions: + – * x^y x^2 x^{-1}.

Here is a situation where this is especially useful. Suppose a person makes $5.85 per hour and you are asked to calculate earnings for a day, a week, and a year. Execute the given keystrokes to find the person's incomes during these periods (Figure 11.5).

Pay Period	Keystrokes	Earnings
8-hour day	5.85 * 8 ENTER	$46.80
5-day week	* 5 ENTER	$234
52-week year	* 52 ENTER	$12,168

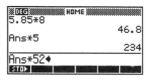

Figure 11.5: Using Ans

11.1.7 The MATH Menu: Operators and functions associated with a scientific calculator are available either immediately from the keys of the HP 38G or by the ■ keys. You have direct access to common arithmetic operations (■ x^2, $\sqrt{x}$, ■ x^{-1}, x^y), trigonometric functions (SIN, COS, TAN), and their inverses (■ ASIN, ■ ACOS, ■ ATAN), exponential and logarithmic functions (■ LOG, ■ 10^x, ■ LN, ■ e^x), and a famous constant (■ π).

A significant difference between the HP 38G graphing calculators and most scientific calculators is that the HP 38G requires the argument of a function *after* the function, as you would see in a formula written in your textbook. For example, on the HP 38G you calculate $\sqrt{16}$ by pressing the keys $\sqrt{x}$ 16 in that order.

Here are keystrokes for basic mathematical operations. Try them for practice on your HP 38G.

Expression	Keystrokes	Display
$\sqrt{3^2 + 4^2}$	$\sqrt{x}$ (3 ■ x^2 + 4 ■ x^2) ENTER	5
$2\frac{1}{3}$	2 + 3 ■ x^{-1} ENTER	2.33333333333
$|-5|$	■ ABS –x 5) ENTER	5
log 200	■ LOG 200) ENTER	2.30102999566
$2.34 \cdot 10^5$	2.34 * ■ 10^x 5 ENTER	234000

Additional mathematical operations and functions are available from the MATH menu (Figure 11.6). Press MATH to see the various options. You will learn in your mathematics textbook how to apply many of them. To leave the MATH menu and take no other action, press [CANCL].

Figure 11.6: MATH menu

The *factorial* of a non-negative integer is the *product* of *all* the integers from 1 up to the given integer. The symbol for factorial is the exclamation point. So 4! (pronounced *four factorial*) is $1 \cdot 2 \cdot 3 \cdot 4 = 24$. You will learn more about applications of factorials in your textbook, but for now use the HP 38G to calculate 4! The factorial command is located in the MATH menu's Prob submenu. To compute 4! press these keystrokes: 4 MATH ▲ ► ▼ [OK] ENTER.

On the HP 38G it is also possible to do calculations with complex numbers. A complex number will appear as an ordered pair (a, b) where a is the real part and b is the complex part. For example, press $\sqrt{x}$ –x 1 ENTER. This is the imaginary number i, which is returned as (0, 1).

You can enter a complex number $a + bi$ on the HP 38G either as (a, b) or $a + bi$. To enter i on your calculator, press ▮ a...z I. For example, to divide $2 + 3i$ by $4 - 2i$, press (2 + 3 ▮ a...z I) / (4 – 2 ▮ a...z I) ENTER or (2 , 3) / (4, –x 2) ENTER (Figure 11.7). The result is $0.1 + 0.8i$.

Commands for manipulating complex number are found in the MATH menu's COMPLEX submenu. For example, to find the complex conjugate of $4 + 5i$ press MATH ▮ ▲ ▼ ► ▼ [OK] (4 + 5 ▮ a...z I) ENTER (Figure 11.7).

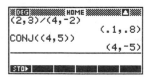

Figure 11.7: Complex number calculations

11.2   Functions and Graphs

11.2.1 Evaluating Functions: Suppose you receive a monthly salary of $1975 plus a commission of 10% of sales. Let x = your sales in dollars; then your wages W in dollars are given by the equation $W = 1975 + .10x$. If your January sales were $2230 and your February sales were $1865, what was your income during those months?

Here's how to use your HP 38G to perform this task. Press the LIB key to display the APLET LIBRARY (Figure 11.8) where the calculator organizes its built-in applications. Move the highlight, if necessary, to FUNCTION and press [START] or ENTER.

You may enter as many as ten different functions here for the HP 38G to use at one time. If there is already a function F1, use the up or down keys to move the highlight to F1 and press DEL to delete whatever was there.

Technology Tip: Press ▮ CLEAR to erase all ten functions in FUNCTION SYMBOLIC VIEW.

Technology Tip: To move quickly to the *first* function (which may be out of view), press ▮ ▲; to jump down to the *last* function, press ▮ ▼.

Enter the expression $1975 + .10x$ for F1 by pressing these keys: 1975 + .10 X,T,θ ENTER (Figure 11.9). (The X,T,θ key lets you enter a variable X without having to use the A...Z key.) Now press HOME to return to the main calculations screen.

Figure 11.8: APLET LIBRARY

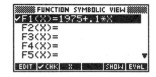

Figure 11.9: FUNCTION SYMBOLIC VIEW

Technology Tip: The function key [X] in the FUNCTION SYMBOLIC VIEW screen, as you see at the bottom of Figure 11.9, serves the same purpose as X,T,θ.

Set 2230 to the variable x by pressing 2230 [STO ▶] X,T,θ. Then press ■ : to allow another expression to be entered on the same command line. Like your textbook the HP 38G uses standard function notation. So press the following keystrokes to evaluate F1 and find January's wages: A...Z F 1 (X,T,θ) ENTER.

It is not necessary to repeat all these steps to find the February wages. Simply copy the previous input line to the edit line, change 2230 to 1865, and press ENTER (Figure 11.10).

Figure 11.10: Using function notation

You may also have the HP 38G create a table of values for a function. Press ■ NUM to set up the numerical tables. For NUMTYPE, choose Build Your Own (Figure 11.11); the other options are not important now. Then press NUM, enter 2230 for x, and press [OK] or ENTER. Continue to enter additional values for x and the calculator automatically completes the table with corresponding values of F1 (Figure 11.12).

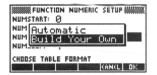

Figure 11.11: FUNCTION NUMERIC SETUP

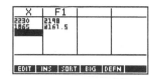

Figure 11.12: Numerical table

Technology Tip: The HP 38G does not require multiplication to be expressed between variables, so xxx means x^3. It is often easier to press two or three x's together than to search for the square key or the powers key. Of course, expressed multiplication is also not required between a constant and a variable. Hence to enter $2x^3 + 3x^2 - 4x + 5$ in the HP 38G, you might save keystrokes and press just these keys: 2 X,T,θ X,T,θ X,T,θ + 3 X,T,θ X,T,θ − 4 X,T,θ + 5

11.2.2 Functions in a Graph Window: On the HP 38G, you can easily generate the graph of a function. The ability to draw a graph contributes substantially to our ability to solve problems.

Once you have started the HP 38G's Function aplet and entered an expression in the FUNCTION SYMBOLIC VIEW, just press PLOT to see its graph.

For example, here is how to graph $y = -x^3 + 4x$. First press LIB, select Function, and then press [START].

Next press SYMB and delete anything that may be there by pressing ■ CLEAR [YES]. Then, with the highlight on the line with F1, press –x x,т,θ xʸ 3 + 4 x,т,θ ENTER (Figure 11.13). Now press PLOT to draw the graph (Figure 11.14).

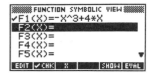

Figure 11.13: FUNCTION SYMBOLIC VIEW

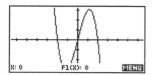

Figure 11.14: Graph of $y = -x^3 + 4x$

Technology Tip: Press [MENU] a couple of times to clear the screen of everything except the graph. Tap any function key to restore the menu option.

The graph window on the HP 38G may look like the one in Figure 11.14. Since the graph of $y = -x^3 + 4x$ extends infinitely far left and right and also infinitely far up and down, the HP 38G can only display a piece of the actual graph. This displayed rectangular part is called a *viewing rectangle*. You can easily change the viewing rectangle to enhance your investigation of a graph.

The viewing rectangle for the HP 38G in Figure 11.14 shows the part of the graph that extends horizontally form –6.5 to 6.5 and vertically from –3.1 to 3.2. Press ■ PLOT to see information about your viewing rectangle. Figure 11.15 shows the FUNCTION PLOT SETUP screen that corresponds to the viewing rectangle in Figure 11.14. This is the *standard* viewing rectangle for the HP 38G.

Figure 11.15: Standard viewing rectangle

The values following XRNG are the minimum and maximum *x*-values of the viewing rectangle; the values following YRNG are the minimum and maximum *y*-values.

XTICK and YTICK set the spacing between tick marks on the axes.

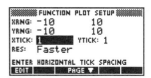

Figure 11.16: Custom window

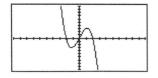

Figure 11.17: Graph of $y = -x^3 + 4x$

Use the arrow keys to move around this list. Enter a new value and press ENTER or [OK] to move along to the next item. You may also press [EDIT] to edit a highlighted entry as you would edit an expression in the home screen. Remember to use the −x key, not the − key (which is subtraction), when you want to enter a negative value. The following figures show different plot setups and the corresponding viewing rectangles.

Figure 11.18: Custom window

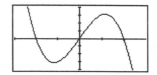

Figure 11.19: Graph of $y = -x^3 + 4x$

To set the range quickly to standard values (Figure 11.15), press ■ CLEAR when you are in FUNCTION PLOT SETUP. Press PLOT to redraw the graph.

Sometimes you may wish to display grid points corresponding to tick marks on the axes. This and other graph format options may be changed by pressing ■ PLOT [PAGE▼] to display the second page of the FUNCTION PLOT SETUP menu. Use the arrow keys to move the highlight cursor to GRID and press [✓CHK]; then press PLOT to redraw the graph. Figure 11.21 shows the same graph as Figure 11.19 but with the grid turned on. In general, you'll want the grid turned *off*, so do that by return to the second page of the FUNCTION PLOT SETUP menu by pressing ■ PLOT [PAGE▼], moving the highlight cursor back to GRID, and press [✓CHK] again.

Figure 11.20: FUNCTION PLOT SETUP

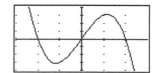

Figure 11.21: Grid turned on

11.2.3 Graphing Step and Piecewise–Defined Functions:
The greatest integer function, written $[[x]]$, gives the greatest *integer* less than or equal to a number x. On the HP 38G, the greatest integer function is called FLOOR and is located in the REAL submenu of the MATH menu. To calculate $[[6.78]] = 6$, enter the home screen and press MATH ▶ ▼ ▼ [OK] 6.78 ENTER. (It is not necessary to type the closing parenthesis.)

To graph $y = [[x]]$, we assume that the Function aplet is already set up. Press SYMB and enter the function as F1 by moving the highlight to F1 and pressing MATH ▶ ▼ ▼ [OK] X,T,θ ENTER. Press PLOT to draw the graph. Figure 11.22 shows this graph in a viewing rectangle from −5 to 5 in both directions.

The true graph of the greatest integer function is a step graph, more like the one in Figure 11.23. For the graph of $y = [[x]]$, a segment should *not* be drawn between every pair of successive points. You can change from the connected setting to the plotted setting by pressing ■ PLOT [PAGE▼] to display the second page of the FUNCTION PLOT SETUP menu. Use the arrow keys to move the highlight cursor to CONNECT and press [✓CHK]; then press PLOT to redraw the graph (Figure 11.23).

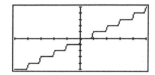

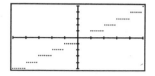

Figure 11.22: Connected graph of $y = [[x]]$ Figure 11.23: Plotted graph of $y = [[x]]$

In general, you'll want the graph to be connected, so do that by returning to the second page of the FUNCTION PLOT SETUP menu by pressing ■ PLOT [PAGE▼], moving the highlight cursor back to CONNECT, and press [✓CHK] again.

It is also possible to graph piecewise–defined functions on the HP 38G. We use the test functions located in the TEST submenu of the MATH menu. The test functions return either a 1 (for *true*) or 0 (for *false*). For example, in the home screen press 1 MATH ▼ ▼ ▼ ▶ [OK] 0 ENTER. The calculator returns 0 since the relation $1 < 0$ is not true.

Suppose we want to graph the piecewise–defined function $f(x) = \begin{cases} x^2 + 1, & x < 0 \\ x - 1, & x \geq 0 \end{cases}$. Assuming that Function

aplet is selected, press SYMB to enter FUNCTION SYMBOLIC VIEW and move the highlight onto F1. We

enter the piecewise function as $f(x) = (x^2 + 1)(x < 0) + (x - 1)(x \geq 0)$ by pressing (X,T,θ ■ x² + 1) (X,T,θ

MATH ▼ ▼ ▼ ▶ [OK] 0) + (X,T,θ – 1) (X,T,θ MATH ▼ ▼ ▼ ▶ ▶ ▼ [OK] 0) ENTER (Figure 11.24).

Use the standard viewing rectangle and press PLOT (Figure 11.25).

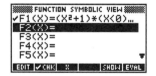

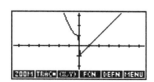

Figure 11.24: Inputting $f(x) = \begin{cases} x^2 + 1, & x < 0 \\ x - 1, & x \geq 0 \end{cases}$ Figure 11.25: Graph of $f(x) = \begin{cases} x^2 + 1, & x < 0 \\ x - 1, & x \geq 0 \end{cases}$

The graph in Figure 11.25 was drawn in the connected setting, so you can see a segment drawn from (0, 1) to (0, −1). Again, you can turn of the connected setting as described earlier if you do not wish to see this.

11.2.4 Graphing a Circle: Here is a useful technique for graphs that are not functions but can be "split" into a top part and a bottom part, or into multiple parts. Suppose you wish to graph the circle of radius 6 whose equation is $x^2 + y^2 = 36$. First solve for y and get an equation for the top semicircle, $y = \sqrt{36 - x^2}$, and for the bottom semicircle, $y = -\sqrt{36 - x^2}$. Then graph the two semicircles simultaneously.

Use the following keystrokes to draw this circle's graph, assuming that you have already started the HP 38G's Function aplet. Press SYMB and enter $\sqrt{36 - x^2}$ as F1 by scrolling to F1 and pressing √x (36 − X,T,θ ■ x^2) ENTER. Then store $-\sqrt{36 - x^2}$ as F2 by pressing −x √x (36 − X,T,θ ■ x^2) ENTER (Figure 11.26). Next press PLOT to graph both halves (Figure 11.27). Make sure that the viewing rectangle is set large enough to display a circle of radius 6.

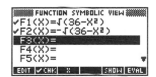

Figure 11.26: Two semicircles

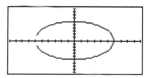

Figure 11.27: One view of the circle's graph

If your viewing rectangle was set to extend from −10 to 10 in both directions, your graph would look like Figure 11.27. Now this does *not* look a circle, because the units along the axes are not the same. You need what is called a "square" viewing rectangle.

The HP 38G's standard viewing rectangle is square but too small to display a circle of radius 6, so you should double the dimensions of the standard viewing rectangle. Change the range to extend horizontally from −13 to 13 and vertically from −6.2 to 6.2 (Figure 11.28). The graph for the better circle is shown in Figure 11.29.

Figure 11.28: Twice standard range

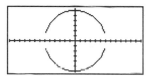

Figure 11.29: Better circle

Technology Tip: Another way to get an approximately square graph on the HP 38G is to change the viewing rectangle values so Ymax − Ymin is $\frac{1}{2}$ (Xmax − Xmin). For example, use the values in Figure 11.39 to get the corresponding graph in Figure 11.31. This method works because the dimensions of the HP 38G's display are such that the ratio of vertical to horizontal is approximately $\frac{1}{2}$.

Figure 11.30: $\frac{\text{vertical}}{\text{horizontal}} = \frac{14}{28} = \frac{1}{2}$

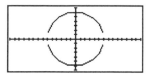

Figure 11.31: "Square" circle

The two semicircles in the figures do not meet because of an idiosyncrasy in the way the HP 38G plots a graph.

Instead of entering $-\sqrt{36-x^2}$ as F2, you could have entered –F1 as F2 and saved some keystrokes. First go back into FUNCTION SYMBOLIC VIEW by pressing SYMB. Then scroll F2 and press –x A...Z F 1 ([X]) ENTER (Figure 11.32). The graph should be as before.

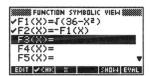

Figure 11.32: –F1 as F2

Technology Tip: The square viewing rectangle is also important when you want to judge whether two lines are perpendicular. The intersection of perpendicular lines will always *look* like a right angle in a square viewing rectangle.

11.2.5 TRACE: Graph the function $y = -x^3 + 4x$ from Section 11.2.2 using the standard viewing rectangle. Right after you plot, press ◄ and ► and see the cursor move from the center of the viewing rectangle. The coordinates that are displayed belong to points on the function's graph, so the *y*-coordinate is the calculated value of the function at the corresponding *x*-coordinate (Figure 11.33).

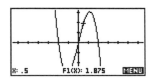

Figure 11.33: Tracing on $y = -x^3 + 4x$

Press [MENU] to display the various options for exploring this graph (Figure 11.34). Tap [TRACE] to turn the trace feature on and off. When the trace feature is off, use any of the arrow keys to move the cursor, called a *free-moving cursor* because it can move *anywhere* in the graph window. When the cursor is free-moving, press [(X,Y)] to display the coordinates of the cursor (Figure 11.35).

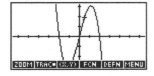

Figure 11.34: Plot menu

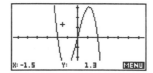

Figure 11.35: Free-moving cursor

Now plot a second function, $y = -.25x$, along with $y = -x^3 + 4x$. From the above graph window, press SYMB, and enter the second function as F2 by highlighting F2 and pressing −x .25 x,т,θ ENTER (Figure 11.36). Finally, press PLOT to draw both functions (Figure 11.37).

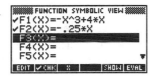

Figure 11.36: Two functions

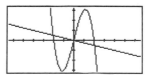

Figure 11.37: Graph of $y = -.25x$ and $y = -x^3 + 4x$

Notice that in Figure 11.36 there are check marks left of *both* F1 and F2. This means that *both* functions will be graphed. In the FUNCTION SYMBOLIC VIEW screen, move the cursor onto F1 and press [✓CHK]. The check mark next to F1 should no longer be there (Figure 11.38). Now press PLOT and see that only F2 is plotted (Figure 11.39).

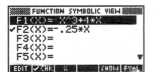

Figure 11.38: Only Y2 active

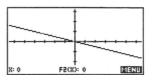

Figure 11.39: Graph of $y = .25x$

Many different functions can be stored in SYMBOLIC FUNCTION VIEW list and any combination of them may be graphed simultaneously. You can make a function active or inactive for graphing by scrolling onto the function pressing the [✓CHK]. Now go back to SYMBOLIC FUNCTION VIEW and do what is needed in order to graph Y1 but not Y2.

Now activate both functions so that both graphs are plotted. Press PLOT and the trace cursor appears on the graph of $y = -x^3 + 4x$ because it is higher up on the list of active functions. You know the cursor is on F1 because F1(X) is displayed on the bottom of the screen. Press the up ▲ or down ▼ arrow key to move the cursor vertically to the graph of $y = -.25x$. Now F2(X) is displayed on the bottom of the screen. Next press the left and right arrow keys to trace along the graph of $y = -.25x$. When more than one function is plotted, you can move the trace cursor vertically from one graph to another in this way.

Technology Tip: By the way, trace the graph of $y = -.25x$ and press and hold either ◄ or ►. Eventually you will reach the left or right edge of the window. Keep pressing the arrow key and the HP 38G will allow you to continue the trace by panning the viewing rectangle. Check the FUNCTION PLOT SETUP screen (press ▉ PLOT) to see that the XRNG is automatically updated.

Technology Tip: Jump quickly to the left-most point (or to the right-most point) of a function in the current view by pressing ▉ ◄ (or ▉ ►) while tracing on its graph.

The HP 38G has a display of 131 horizontal columns of pixels and 64 vertical rows, so when you trace a curve across a graph window, you are actually moving from Xmin to Xmax in 130 equal jumps, each called Δx. You would calculate the size of each jump to be $\Delta x = \dfrac{\text{Xmax} - \text{Xmin}}{130}$. Sometimes you may want the jumps to be friendly numbers like 0.1 or 0.25 so that, when you trace along the curve, the x-coordinates will be incremented by such a convenient amount. Just set your viewing rectangle for a particular increment Δx by making Xmax = Xmin + 130 · Δx. For example, if you want Xmin = −5 and Δx = 0.3, set Xmax = −5 + 130 · 0.3 = 19.

On the HP 38G, to center your window around a particular point, say (h, k), and also have a certain Δx, set Xmin = $h − 65 · \Delta x$ and make Xmax = $h + 65 · \Delta x$. Likewise, make Ymin = $k − 31 · \Delta y$ and make Ymax = $h + 32 · \Delta x$. For example, to center a window around the origin (0, 0), with both horizontal and vertical increments of 0.25, set the range so that Xmin = 0 − 65 · 0.25 = −16.25, Xmax = 0 + 65 · 0.25 = 16.25, Ymin = 0 − 31 · 0.25 = −7.75 and Ymax = 0 + 32 · 0.25 = 8.

The HP 38G's standard viewing rectangle is already a friendly viewing rectangle, centered at the origin (0, 0) with $\Delta x = \Delta y = 0.1$.

See the benefit by first plotting $y = x^2 + 2x + 1$ in a window that extends from −10 to 10 in both directions. Trace near its y-intercept, which is (0, 1), and move towards its x-intercept, which is (−1,0). Then initialize the range to the standard viewing rectangle and trace again near the intercepts.

11.2.6 ZOOM: Plot again the two graphs, for $y = -x^3 + 4x$ and $y = -.25x$. There appears to be an intersection near $x = 2$. The HP 38G provides several ways to enlarge the view around this point.

You can change the viewing rectangle directly by pressing ▉ PLOT and editing the values of XRNG and YRNG. Figure 11.40 shows a new viewing rectangle for the range extending from 1.5 to 2.5 horizontally and from −2.5 to 2.5 vertically. Trace along the graphs until coordinates of a point that is close to the intersection are displayed.

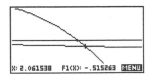

Figure 11.40: Closer view

A more efficient method for enlarging the view is to draw a new viewing rectangle with the cursor. Start again with a graph of the two functions $y = -x^3 + 4x$ and $y = -.25x$ in the standard viewing rectangle. (Press ■ PLOT ■ CLEAR to set the standard window. Then press PLOT.)

Now imagine a small rectangular box around the intersection point, near $x = 2$. Press [MENU] [ZOOM] ▼ to select Box... (Figure 11.41). Press [OK] to draw a box to define this new rectangle.

Use the arrow keys to move the cursor, which is now free-moving and whose coordinates are displayed at the bottom of the window, to one corner of the new viewing rectangle you imagine (Figure 11.42).

Figure 11.41: Zoom menu

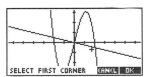

Figure 11.42: One corner selected

Press [OK] or ENTER to fix the corner where you moved the cursor. Use the arrow keys again to move the cursor to the diagonally opposite corner of the new rectangle (Figure 11.43). If this box looks all right to you, press ENTER. The rectangular area you have enclosed will now enlarge to fill the graph window (Figure 11.44).

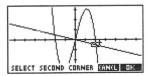

Figure 11.43: Box drawn

Figure 11.44: New viewing rectangle

You may cancel the zoom any time *before* you press this last ENTER. Press [CANCL] and start over.

The HP 38G has a split screen feature that enables you to see two views of a graph simultaneously. First re-draw the graph in the standard viewing rectangle (press ■ PLOT ■ CLEAR PLOT). Then press ■ VIEWS [OK]. The graph is plotted twice (Figure 11.45). Press [MENU] [ZOOM] and draw a box as you did before. The graph on the right side is now the zoomed graph (Figure 11.46).

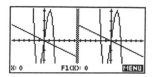

Figure 11.45: Split screen

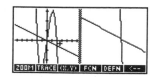

Figure 11.46: Split screen with zoom

The plot on the left side is called the *base* plot and is the plot which you act on. If you press [<--], the base plot will be converted to the scale of the plot on the right. To unsplit the plot, press PLOT and the base plot is displayed in the whole screen.

The HP 38G can quickly magnify a graph around the cursor's location. Return once more to the standard viewing rectangle for the graph of the two functions $y = -x^3 + 4x$ and $y = -.25x$. Trace along the graphs to move the free-moving cursor as close as you can to the point of intersection near $x = 2$ (Figure 11.47). Then press [MENU] [ZOOM] ▼ ▼ [OK] and the calculator draws a magnified graph, centered at the cursor's position (Figure 11.48). The range values are changed to reflect this new viewing rectangle. Look in FUNCTION PLOT SETUP screen to check.

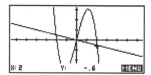

Figure 11.47: Before a zoom in

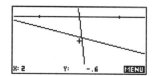

Figure 11.48: After a zoom in

As you see in the zoom menu (Figure 11.41), the HP 38G can zoom in (select In) or zoom out (select Out). Zoom out to see a larger view of the graph, centered at the cursor position. You can change the horizontal and vertical scale of the magnification by going into the zoom menu and moving the highlight up or down to Set Factors...; press [OK] and you can edit XZOOM and YZOOM, the horizontal and vertical magnification factors.

The default zoom factor is 4 in both directions (press ■ CLEAR in the SET ZOOM FACTORS screen to reset). It is not necessary for XZOOM and YZOOM to be equal. Sometimes, you may prefer to zoom in one direction only. In this case you can choose X-Zoom In, X-Zoom Out, Y-Zoom In, or Y-Zoom Out from the zoom menu. Press [CANCL] to leave the zoom menu without taking any action.

Technology Tip: The HP 38G remembers the window it displayed before a zoom, so if you should zoom in too much and lose the curve, you can press [Zoom] ▲ [OK] to select Un-zoom and go back to the previous window.

Technology Tip: The HP 38G can automatically select the necessary *vertical* range for a function. For auto scaling, press ■ VIEWS and select Auto Scale. Take care, because sometimes when you are graphing two functions together, the calculator will auto scale for one function in such a way that the other function will no longer be visible. For example, plot the two functions $y = -x^3 + 4x$ and $y = -.25x$ in the HP 38G's standard viewing rectangle, then auto scale and trace along both functions.

11.2.7 Relative Minimums and Maximums: Graph $y = -x^3 + 4x$ once again in the standard viewing rectangle. This function appears to have a relative minimum near $x = -1$ and a relative maximum near $x = 1$. You may zoom and trace to approximate these extreme values.

First trace along the curve near the local minimum. Notice by how much the x-values and y-values change as you move from point to point Trace along the curve until the y-coordinate is as *small* as you can get it, so that you are as close as possible to the local minimum, and zoom in. Now trace again along the curve and, as you move from point to point, see that the coordinates change by smaller amounts than before. Keep zooming and tracing until you find the coordinates of the local minimum point as accurately as you need them, approximately $(-1.15, -3.08)$.

Follow a similar procedure to find the local maximum. Trace along the curve until the y-coordinate is as *great* as you can get it, so that you are as close as possible to the local maximum, and zoom in. The local maximum point on the graph of $y = -x^3 + 4x$ is approximately $(1.15, 3.08)$.

The HP 38G has a feature that can automatically find the maximums and minimums for functions. After graphing $y = -x^3 + 4x$, trace your cursor near a minimum or maximum. From the plot menu, press [FCN]. Then press ▲ [OK] to select the Extremum option (Figure 11.49).

Figure 11.49: Extremum option

The calculator will calculate the minimum or maximum at where the cursor was closest. Figures 11.50 and 11.51 show the minimum and maximum of $y = -x^3 + 4x$.

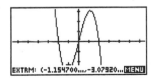

Figure 11.50: Minimum of $y = -x^3 + 4x$

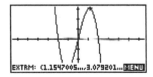

Figure 11.51: Maximum of $y = -x^3 + 4x$

11.3   Solving Equations and Inequalities

11.3.1 Intercepts and Intersections: Tracing and zooming are also used to locate an *x*-intercept of a graph, where a curve crosses the *x*-axis. For example, the graph of $y = x^3 - 8x$ crosses the *x*-axis three times (Figure 11.52). After tracing over to the *x*-intercept point that is farthest to the left, zoom in (Figure 11.53). Continue this process until you have located all three intercepts with as much accuracy as you need. The three *x*-intercepts of $y = x^3 - 8x$ are approximately −2.828, 0, and 2.828.

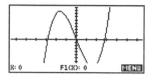

Figure 11.52: Graph of $y = x^3 - 8x$

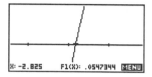

Figure 11.53: Near an *x*-intercept of $y = x^3 - 8x$

Technology Tip: As you zoom in, you may also wish to change the spacing between tick marks on the *x*-axis so that the viewing rectangle shows scale marks near the intercept point. Then the accuracy of your approximation will be such that the error is less than the distance between two tick marks. Change the *x*-scale from the FUNCTION PLOT SETUP screen. Move the cursor down to XTICK and enter an appropriate value.

The *x*-intercept of a function's graph is a *root* of the equation $f(x) = 0$, and the HP 38G can automatically search for the roots. First plot the function and then trace to a point near an *x*-intercept. From the plot menu press [FCN] [OK] to select the Root option (Figure 11.54). The value of the nearest root will be displayed (Figure 11.55); repeat this process to find the other roots.

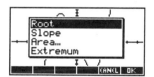

Figure 11.54: Root option

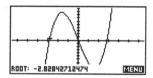

Figure 11.55: A root of $y = x^3 - 8x$

TRACE and ZOOM are especially important for locating the intersection points of two graphs, say the graphs of $y = -x^3 + 4x$ and $y = -.25x$. Trace along one of the graphs until you arrive close to an intersection point. Then press ▲ or ▼ to jump to the other graph. Notice that the *x*-coordinate does not change, but the *y*-coordinate is likely to be different (Figures 11.56 and 11.57).

When two *y*-coordinates are as close as they can get, you have come as close as you now can to the point of intersection. So zoom in around the intersection point, then trace again until the two *y*-coordinates are as close as possible. Continue this process until you have located the point of intersection with as much accuracy as necessary.

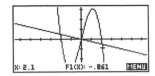

Figure 11.56: Trace on $y = -x^3 + 4x$

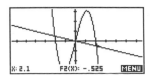

Figure 11.57: Trace on $y = -.25x$

While the graphs are displayed, the HP 38G can search for points of intersection automatically. Trace along one graph near an intersection. From the plot menu press [FCN] ▼ [OK] to select the Intersection option (Figure 11.58). Move the highlight, if necessary, to the name of the second function (Figure 11.59) and press [OK]. Figure 11.60 shows one of the intersection points.

Figure 11.58: Intersection option

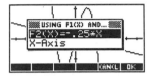

Figure 11.59: Choosing the second function

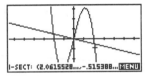

Figure 11.60: An intersection of $y = -x^3 + 4x$ and $y = -.25x$

11.3.2 Solving Equations by Graphing: Suppose you need to solve the equation $24x^3 - 36x + 17 = 0$. First graph $y = 24x^3 - 36x + 17$ in a window large enough to exhibit *all* its x-intercepts, corresponding to all the equation's roots. Then use trace and zoom or the Root option, to locate each one. In fact, this equation has just one solution, approximately $x = -1.414$ (Note that when using the ROOT option, you may get an extremum instead. If that is the case, move the cursor near the root and then use the ROOT option.

Remember that when an equation has more than one root, it may be necessary to change the viewing rectangle a few times to locate all of them.

Technology Tip: To solve an equation like $24x^3 + 17 = 36x$, you may first transform it into standard form, $24x^3 - 36x + 17 = 0$, and proceed as above. However, you may also graph the *two* functions $y = 24x^3 + 17$ and $y = 36x$, then zoom and trace to locate their point of intersection or use the Intersection option.

11.3.3 Solving Systems by Graphing: The solutions to a system of equations correspond to the points of intersection of their graphs (Figure 11.61). For example, to solve the system $y = x^3 + 3x^2 - 2x - 1$ and $y = x^2 - 3x - 4$, first graph them together in a suitable window. Then use zoom and trace or the Intersection option, to locate their point of intersection, approximately $(-2.17, 7.25)$.

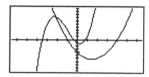

Figure 11.61: Graph of $y = x^3 + 3x^2 - 2x - 1$ and $y = x^2 - 3x - 4$

If you use zoom and trace, you must judge whether the two current y-coordinates are sufficiently close for $x = -2.17$ or whether you should continue to zoom and trace to improve the approximation.

The solutions of the system of two equations $y = x^3 + 3x^2 - 2x - 1$ and $y = x^2 - 3x - 4$ correspond to the solutions of the single equation $x^3 + 3x^2 - 2x - 1 = x^2 - 3x - 4$, which simplifies to $x^3 + 2x^2 + x + 3 = 0$. So you may also graph $y = x^3 + 2x^2 + x + 3$ and find its x-intercepts (roots) to solve the system.

11.3.4 Solving Inequalities by Graphing: Consider the inequality $1 - \dfrac{3x}{2} \geq x - 4$. To solve it with your HP 38G, graph the two functions $y = 1 - \dfrac{3x}{2}$ and $y = x - 4$ and locate their point of intersection, at $x = 2$ (Figure 11.62). The inequality is true when the graph of $y = 1 - \dfrac{3x}{2}$ lies *above* the graph of $y = x - 4$, and that occurs when $x < 2$. So the solution is the half-line $x \leq 2$, or $(-\infty, 2]$.

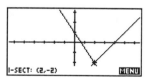

Figure 11.62: Solving $1 - \dfrac{3x}{2} \geq x - 4$

11.4 Trigonometry

11.4.1 Degrees and Radians: The trigonometric functions can be applied to angles measured either in radians or degrees, but you should take care that the HP 38G is configured for whichever measure you need. Press ■ MODES to see your current settings. Press [CHOOS] and move your cursor over the measure you want and then press [OK] or ENTER.

It's a good idea to check the angle measure setting before executing a calculation that depends on a particular measure. You may change a mode setting at any time and not interfere with pending calculations. Try the following keystrokes to see this in action.

Expression	Keystrokes	Display
$\sin 45°$	■ MODES [CHOOS]	
	(move cursor to Degrees) [OK]	
	HOME SIN 45 ENTER	.707106781187
$\sin \pi°$	SIN ■ π ENTER	5.48036651488E−2
$\sin \pi$	■ MODES [CHOOS] ▼ [OK]	
	HOME SIN ■ π ENTER	−2.06761537357E−13
$\sin 45$	SIN 45 ENTER	.850903524534
$\sin \dfrac{\pi}{6}$	SIN ■ π / 6) ENTER	.5

The first line of keystrokes sets the HP 38G in degree mode and calculates the sine of 45 *degrees*. While the calculator is still in degree mode, the second line keystrokes calculates the sine of π *degrees*, approximately 3.1415°. The third line changes to radian mode just before calculating the sine of π *radians*. The fourth line calculates the sine of 45 *radians* (the calculator remains in radian mode).

Note that for sin π the HP 38G does not return 0, but a very small number which we take as being 0 since it is so small.

The HP 38G makes it possible to convert a number in degrees to radians and vice versa in a calculation. The commands for these are located in the REAL sub-menu of the MATH menu. If your calculator is set for radians, we use the DEG→RAD command. Execute these keystrokes to calculate $\tan 45° + \sin \frac{\pi}{6}$ as shown in Figure 11.63: TAN MATH ► ▼ [OK] 45)) + SIN ■ π / 6) ENTER. If your calculator is set to degrees, we use the RAD→DEG command. Execute these keystrokes to calculate $\tan 45° + \sin \frac{\pi}{6}$ as shown in Figure 11.64: TAN 45) + SIN MATH ► ▲ ▲ ▲ ▲ ▲ [OK] (■ π / 6)) ENTER

Figure 11.63: Using DEG→RAD

Figure 11.64: Using RAD→DEG

11.4.2 Graphs of Trigonometric Functions: When you graph a trigonometric function, you need to pay careful attention to the choice of graph window and your angle setting. To change your angle setting, press ■

SYMB. (We assume that you have activated the Function aplet.) Then select the appropriate angle measure as you did before. For example, graph $y = \dfrac{\sin 30x}{30}$ in the standard viewing window in radians. Trace along the curve to see where it is. Zoom in to a better window, or use the period and amplitude to establish a better window.

Technology Tip:. The viewing window of a trigonometric graph on the HP 38G can be set to a trigonometric scale Press ■ VIEWS ▲ [OK]. The XRNG now goes from $-\frac{63\pi}{24}$ to $\frac{63\pi}{24}$ if the calculator is set for radians and from –472.5 to 472.5 for degrees. The YRNG goes from –3.1 to 3.2. This setting allows for convenient tracing. Your cursor moves in horizontal increments of $\frac{\pi}{24}$ in radians and 12° in degrees.

11.5   Scatter Plots

11.5.1 Entering Data: The table shows the total prize money (in millions of dollars) awarded at the Indianapolis 500 race from 1981 to 1989. (*Source*: Indianapolis Motor Speedway Hall of Fame.)

Year	1981	1982	1983	1984	1985	1986	1987	1988	1989
Prize ($million)	$1.61	$2.07	$2.41	$2.80	$3.27	$4.00	$4.49	$5.03	$5.72

We'll now use the HP 38G to construct a scatter plot that represents these points and to find a linear model that approximates the given data.

Press LIB to display the APLET LIBRARY. Move the highlight to Statistics and press [START] or ENTER. A data columns screen like the one in Figure 11.65 should appear. If the second right menu item at the bottom of the screen displays 1VAR as in Figure 11.65, press [1VAR], so it changes to 2VAR.

Next press SYMB to set the columns which make up our data. In this case we use the settings for S1 as shown in Figure 11.66; if not, press [C] 1 for the independent variable and [C] 2 for the dependent variable. The default setting for Fit1 should be the linear regression model m*X + b. (If not, press ■ SYMB, move the cursor to the setting for S1FIT, press [CHOOS], and select Linear. Return to STATISTICS SYMBOLIC VIEW by pressing SYMB.) Select S1 by checkmarking it by pressing [✓CHK], if it is not already checked.

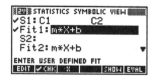

Figure 11.65: Statistics   Figure 11.66: STATISTICS SYMBOLIC VIEW

Press NUM to return to the data columns screen. If necessary, clear all columns by pressing ■ CLEAR and selecting All Columns. Instead of entering the full year 198x we will enter only x. Now begin entering the x's in C1 by having the cursor in the first row of C1 and pressing 1 ENTER. As your cursor moves down the column, continue to enter all the other years (pressing ENTER after each entry). Then move the cursor to the first row of C2 by pressing ▶ and enter the corresponding prize data (Figure 11.67).

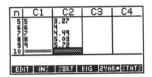

Figure 11.67: Entering the data

To edit an entry in a column move the cursor to the value for any entry you wish to edit; then press [EDIT], edit the expression as you would in your home screen, and press ENTER. To change an entry move the cursor onto the entry, type a new entry, and press ENTER. To insert or delete data, move the cursor to the value for any entry you wish to add or delete. Press [INS] and a new entry in the list is created; press DEL and the entry is deleted. (You can scroll to the bottom and top of a column by pressing ■ ▲ and ■ ▼, respectively.)

11.5.2 Plotting Data: Once all the data points have been entered, press ■ [PLOT] ▼ ▼ [CHOOS] ▼ [OK] to select a mark for S1 that is more visible. The screen should be set as in Figure 11.68. The default setting for the viewing rectangle can be set by pressing ■ CLEAR. Although, this will change the mark back to a dot. Set the viewing rectangle so XRNG and YRNG is from –2 to 10.

To get the scatter plot of the data press PLOT. Figure 11.69 shows the scatter plot in the default setting.

Figure 11.68: STATISTICS PLOT SETUP 1

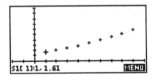

Figure 11.69: Scatter plot

11.5.3 Regression Line: The HP 38G can draw and calculate the equation for the line that best fits all the data. From the plot of the data press [MENU]. Then press [FIT] and this line is drawn (Figure 11.70).

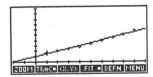

Figure 11.70: Linear regression line

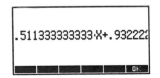

Figure 11.71: Equation of line

To see the equation of the line, press SYMB, move the cursor to entry for Fit1, and press [SHOW]. The equation will be displayed (Figure 11.71).

11.5.4 Exponential Growth Model: The table shows the world population (in millions) from 1980 to 1992.

Year	1980	1985	1986	1987	1988	1989	1990	1991	1992
Population (millions)	4453	4850	4936	5024	5112	5202	5294	5384	5478

As in the linear regression example, press NUM (assuming that the Statistics aplet is selected) and enter the data in C1 and C2. Use 0 for 1980, 5 for 1985, and so on.

Press ■ SYMB and change the setting for S1FIT to Exponential. Now press ■ VIEWS and select Auto Scale. The scatter plot should appear with or without the exponential growth model in a viewing rectangle automatically determined by the calculator. If the model is not there, press [MENU] [FIT]. You may need to set a suitable YTICK. To do this press ■ PLOT [PAGE▼] and enter a suitable YTICK, for example, 100. Redraw the graph by pressing PLOT (Figure 11.72)

To see the equation of the exponential growth model, press SYMB, move the cursor to entry for Fit1, and press [SHOW]. The equation will be displayed (Figure 11.73). Press [OK] when you are done.

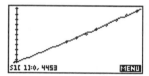

Figure 11.72: Graph of exponential growth model

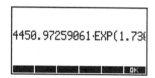

Figure 11.73: Exponential growth model

11.6   Matrices

11.6.1 Making a Matrix: The HP 38G can work with 10 different matrices. Here's how to create this 3×4

matrix $\begin{bmatrix} 1 & -4 & 3 & 5 \\ -1 & 3 & -1 & -3 \\ 2 & 0 & -4 & 6 \end{bmatrix}$ in your calculator.

Press ■ MATRIX to see the MATRIX CATALOG (Figure 11.74).

Technology Tip: To delete a single matrix, move the highlight there and press DEL; to delete every matrix in the catalog, press ■ CLEAR. To move quickly to the *first* matrix (which may be out of view), press ■ ▲; to jump down to the *last* matrix, press ■ ▼.

Move the highlight to M1 and press [NEW] to create a new matrix with that name. Make this a Real matrix because all its elements are real numbers.

Figure 11.74: MATRIX CATALOG

Press [OK]. Tap [GO] until [GO→] is displayed so that the cursor will automatically move to the right after each entry. Input the top row by pressing 1 ENTER –x 4 ENTER 3 ENTER 5 ENTER. Now press ▼to move down to the second row and back to the first column, then press –x 1 ENTER 3 ENTER –x 1 ENTER –x 3 ENTER 2 ENTER 0 ENTER –x 4 ENTER 6 ENTER to finish creating this matrix (Figure 11.75).

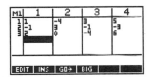

Figure 11.75: Creating a new matrix

Technology Tip: You may delete any row (or column) by moving the highlight into that row (or column) and pressing DEL. Insert a row above (or column to the left) of the highlight by pressing [INS].

Press ■ MATRIX again and see that matrix M1 is now listed as a 3×4 real matrix. If you wish to change matrix M1, press [EDIT] and use the arrow keys to move directly to any element.

11.6.2 Matrix Math: Matrix calculations can be performed in the home screen. To calculate the scalar multiplication 2 M1 press 2 A...Z M 1 ENTER. The resulting matrix is displayed in a form where each element *and* row is separated by a comma (Figure 11.76).

Figure 11.76: 2 Mat A

To add two matrices, say M1 and M2, create M2 (with the same dimensions as M1) in the same manner as you created M1. Then enter the home screen and then press A...Z M 1 + A...Z M 2 ENTER. If you want to store the answer as a specific matrix, say M3, then press [STO ▶] A...Z M 3. Subtraction is performed in a similar manner.

Now enter the matrix $\begin{bmatrix} 2 & 0 & 3 \\ 1 & -5 & -1 \end{bmatrix}$ as M3. For matrix multiplication of M3 by M1, press A...Z M 3 ∗ A...Z M 1 ENTER. If you tried to multiply M1 by M3, your calculator would signal an error because the dimensions of the two matrices do not permit multiplication in this way.

The *transpose* of a matrix is another matrix with the rows and columns interchanged. To calculate the transpose of Mat A , use the command TRN found in the MATRIX sub-menu of the MATH menu. In the home screen, press MATH ▲ ▲ ▲ ► ▲ [OK] A...Z M 1 ENTER.

11.6.3 Row Operations: The HP 38G does not have row operations built in.

11.6.4 Determinants and Inverses: Enter this 3×3 square matrix as M1: $\begin{bmatrix} 1 & -2 & 3 \\ -1 & 3 & 0 \\ 2 & -5 & 5 \end{bmatrix}$. To calculate its determinant $\begin{vmatrix} 1 & -2 & 3 \\ -1 & 3 & 0 \\ 2 & -5 & 5 \end{vmatrix}$, go in the home screen and press MATH ▲ ▲ ▲ ► ▼ ▼ ▼ [OK] A...Z M 1 ENTER. You should find that the determinant is 2 (Figure 11.77). You can also compute the determinant by first holding down the A...Z key and pressing D E T, and then pressing (A...Z M 1 ENTER.

Since the determinant of the matrix is not zero, it has an inverse matrix. Press MATH ▲ ▲ ▲ ► ► ► [OK] A...Z M 1 ENTER to calculate the inverse (Figure 11.77).

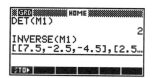

Figure 11.77: Determinant of Mat A

Now let's solve a system of linear equations by matrix inversion. Consider the system $\begin{cases} x - 2y + 3z = 9 \\ -x + 3y = -4 \\ 2x - 5y + 5z = 17 \end{cases}$.

The coefficient matrix for this system is the matrix $\begin{bmatrix} 1 & -2 & 3 \\ -1 & 3 & 0 \\ 2 & -5 & 5 \end{bmatrix}$ which was entered in the previous example.

HP 38G Series Calculator

Enter the matrix $\begin{bmatrix} 9 \\ -4 \\ 17 \end{bmatrix}$ as M2. Then in the home screen, press MATH ▲ ▲ ▲ ▶ ▶ ▶ [OK] A...Z M 1) *

A...Z M 2 ENTER to get the answer as shown in Figure 11.78.

Figure 11.78: Solution matrix

The solution is $x = 1$, $y = -1$, and $z = 2$.

11.7   Sequences

11.7.1 Iteration with the Ans key: The Ans feature enables you to perform iterations to evaluate a function repeatedly. As an example, calculate $\dfrac{n-1}{3}$ for $n = 27$. Then calculate $\dfrac{n-1}{3}$ for $n =$ the answer to the previous calculation. Continue to use each answer as n in the *next* calculation. Here are keystrokes to accomplish this iteration on the HP 38G calculator (Figure 11.79). Notice that when you use Ans in place of n in a formula, it is sufficient to press ENTER to continue an iteration.

Iteration	Keystrokes	Display
1	27 ENTER	27
2	(■ ANSWER – 1) / 3 ENTER	8.66666666667
3	ENTER	2.55555555556
4	ENTER	.51851851852
5	ENTER	–.16049382716

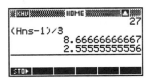

Figure 11.79: Iteration

Press ENTER several more times and see what happens with this iteration. You may wish to try it again with a different starting value.

11.7.2 Arithmetic and Geometric Sequences: Use iteration with the Ans variable to determine the n-th term of a sequence. For example, find the 18th term of an *arithmetic* sequence whose first term is 7 and whose common difference is 4. Enter the first term 7, then start the progression with the recursion formula, ■ ANSWER + 4 ENTER. This yields the 2nd term, so press ENTER sixteen more times to find the 18th term. For a *geometric* sequence whose common ratio is 4, start the progression with ■ ANSWER × 4 ENTER.

You can also define the sequence recursively with the HP 38G by pressing LIB and starting Sequence in the APLET LIBRARY. Once again, let's find the 18th term of an *arithmetic* sequence whose first term is 7 and whose common difference is 4. You need to enter the first *two* terms and a formula for the n-th term. For the sequence U1 set the first term $u_1 = 7$ and the second term $u_2 = 11$. Make $u_n = u_{n-1} + 4$ by moving the highlight to U1(N) and pressing [U1] [(N–1)] + 4 [OK] (Figure 11.80).

Now press NUM to leave this menu and go to a numerical table. To find the 18th term u_{18} of this sequence, either scroll to the line where $n = 18$ or else enter 18 directly into the first column (Figure 11.81).

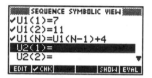

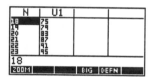

Figure 11.80: SEQUENCE SYMBOLIC VIEW

Figure 11.81: $a_{18} = 75$

Technology Tip: Up to ten different sequences may be entered in SEQUENCE SYMBOLIC VIEW. To move quickly to the *first* sequence, press ■ ▲; to jump down to the *last* sequence, press ■ ▼.

Technology Tip: While you are scrolling through the numerical table, press [DEFN] to toggle between displaying the formula and the numerical value at the bottom of the screen.

Of course, you could use the *explicit* formula for the n-th term of an arithmetic sequence, $t_n = a + (n-1)d$. First enter values for the variables a, d, and n in the home screen; then evaluate the formula by pressing A...Z A + (A...Z n – 1) A...Z D ENTER. For a geometric sequence $t_n = a \cdot r^{n-1}$, enter the values for the variables a, r, and n in the home screen; then evaluate the formula by pressing A...Z A A...Z R x^y (A...Z N – 1) ENTER.

11.7.3 Finding Sums of Sequences: You can use the Σ command found in the Loop sub-menu of the MATH menu to find the sum of a sequence. For example, suppose you want to find the sum $\sum_{n=1}^{12} 4(0.3)^n$. In the home screen press MATH ▲ ▲ ▲ ▲ ► ▲ [OK] X,T,θ MATH ▼ ▼ ► [OK] 1 , 12 , 4 * .3 x^y X,T,θ ENTER (Figure 11.82).

Now calculate the sum starting at $n = 0$ by editing the expression. You should obtain a sum of approximately 5.712848. In general, the command is in the form Σ (*variable = initial value, final value, expression*).

Figure 11.82: $\displaystyle\sum_{n=1}^{12} 4(0.3)^n$

11.8   Parametric and Polar Graphs

11.8.1 Graphing Parametric Equations: The HP 38G plots parametric equations as easily as it plots functions. First press LIB, select Parametric, and then press [START]. Be sure, if the independent parameter is an angle measure, that the angle measure has been set to whichever you need, radians or degrees. To set the appropriate angle measure press ■ SYMB and select the angle measure you want.

You can now enter the parametric functions. For example, here are the keystrokes needed to graph the parametric equations $x = \cos^3 t$ and $y = \sin^3 t$. First check that angle measure is in radians and then return to PARAMETRIC SYMBOLIC VIEW by pressing SYMB. Clear or uncheck any existing functions and then select a location for the equation of x, say X1(T). Press (COS X,T,θ)) x^y 3 ENTER (SIN X,T,θ)) x^y 3 ENTER (Figure 11.83).

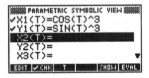

Figure 11.83: $x = \cos^3 t$ and $y = \sin^3 t$

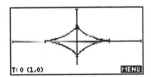

Figure 11.84: Parametric graph of $x = \cos^3 t$ and $y = \sin^3 t$

Press ■ PLOT ■ CLEAR to get the default range settings. Notice that the values of TRNG go from 0 to 12 in steps of .1. In order to provide a better viewing rectangle and set the XRNG and YRNG to go from −2 to 2. Press PLOT to draw the graph (Figure 11.84).

You may zoom and trace along parametric graphs just as you did with function graphs. As you trace along this graph, notice that the cursor moves in the counterclockwise direction as T increases.

11.8.2 Rectangular-Polar Coordinate Conversion: It is not possible to directly convert between rectangular and polar coordinate systems on the HP 38G.

11.8.3 Graphing Polar Equations: The HP 38G graphs polar functions in the form $r = f(\theta)$. First press LIB, select Polar, and then press [START]. Be sure that the angle measure has been set to whichever you need, radians or degrees. To set the appropriate angle measure press ■ SYMB and select the angle measure you want. Here we will use radian measure.

Return POLAR SYMBOLIC VIEW by pressing SYMB and clear or unselect any existing graphs. Here we will use radian measure. For example, to graph $r = 4 \sin\theta$, select a location for the function, say R1(θ) and press 4 SIN X,T,θ ENTER.

Press ■ PLOT ■ CLEAR to get the default range settings. Notice that the values of θRNG go from 0 to 2π. Choose a good viewing rectangle and an appropriate interval and increment for θ. In Figure 11.85, the viewing window is roughly "square" and extends from −8 to 8 horizontally and from −4 to 4 vertically. (Refer back to the Technology Tip in Section 11.2.4.) Press PLOT to draw the graph.

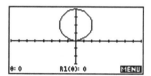

Figure 11.85: Polar graph of $r = 4 \sin\theta$

Trace along this graph to see the polar coordinates of the cursor's location displayed at the bottom of the window. Zooming works just the same as before.

11.9   Probability

11.9.1 Random Numbers: The command Random generates a number between 0 and 1. You will find this command in the Prob. sub-menu of the MATH menu. In the home screen press MATH ▲ ► ▼ ▼ ▼ [OK] ENTER to generate a random number. Press ENTER to generate another number; keep pressing ENTER to generate more of them.

If you need a random number between, say, 0 and 10, then press 10 MATH ▲ ► ▼ ▼ ▼ [OK] ENTER. To get a random number between 5 and 15, press 5 + 10 MATH ▲ ► ▼ ▼ ▼ [OK] ENTER.

11.9.2 Permutations and Combinations: To calculate the number of permutations of 12 objects taken 7 at a time, $_{12}P_7$, press MATH ▲ ▶ ▼ ▼ [OK] 12 , 7 ENTER (Figure 11.86). Thus $_{12}P_7 = 3,991,680$.

For the number of combinations of 12 objects taken 7 at a time, $_{12}C_7$, press MATH ▲ ▶ [OK] 12 , 7 ENTER (Figure 11.86). Thus $_{12}C_7 = 792$.

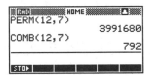

Figure 11.86: $_{12}P_7$ and $_{12}C_7$

11.9.3 Probability of Winning: A state lottery is configured so that each player chooses six different numbers from 1 to 40. If these six numbers match the six numbers drawn by the State Lottery Commission, the player wins the top prize. There are $_{40}C_6$ ways for the six numbers to be drawn. If you purchase a single lottery ticket, your probability of winning is 1 in $_{40}C_6$. Press 1 / MATH ▲ ▶ [OK] 40 , 6 ENTER to calculate your chances, but don't be disappointed.

11.10   Programming

11.10.1 Entering a Program: The HP 38G is a programmable calculator that can store sequences of commands for later replay. Here's an example to show you how to enter a useful program that solves quadratic equations by the quadratic formula.

Press ■ PROGRAM to access the PROGRAM CATALOG. The HP 38G has space for many programs, each called by a name you give it. The names of all your programs are listed alphabetically in the PROGRAM CATALOG. Create a new program now, so press [NEW].

Now enter a descriptive title, say Quadratic. Notice that there is a function key [A...Z] in addition to the calculator's A...Z key. Press [A...Z], which acts as a letters lock, and type all the letters you want. Remember to press [A...Z] when you are finished and want to restore keys to their primary functions. To lock in lower case letters, press ■ [A...Z]. To unlock from lowercase, press [A...Z] again. So press A...Z Q ■ [A...Z] U A D R A T I C (Figure 11.87) and then press [OK] or ENTER to move to the program editor. The function key [A...Z] is also available in the program editor.

Figure 11.87: Entering a title

Any command you could enter directly in the HP 38G's home screen can be entered as a line in a program. There are also special programming commands.

Some programming commands are obtained by pressing MATH [CMDS], then moving the highlight to a command category, then pressing ▶ and moving the highlight to a specific command. To simplify these directions, we use the notation MATH [CMDS/*category*/*command*] in the program listing below.

You may also type a command directly, instead of searching for it in the PROGRAM COMMANDS. For example, instead of all the keystrokes for MATH [CMDS/*Branch*/*If*] [OK], just [A...Z] I F [A...Z].

The instruction manual for your HP 38G gives detailed information about programming. Refer to it to learn more about programming and how to use other features of your calculator.

Enter the program Quadratic by pressing the keystrokes given below. Individual program commands are separated by :. You may interrupt program input at any stage. To return later for more editing, press ▮ PROGRAM, move the highlight down to this program's name, and press [EDIT].

Program Line	*Keystrokes*
INPUT A; "Coefficient"; "A";"Enter number"; 1:	MATH [CMDS/*Prompt*/*INPUT*] [OK] [SPACE] A...Z A ▮ ; ▮ CHARS ▲ [OK] A...Z C ▮ [A...Z] O E F F I E N T [a...z] ▮ CHARS ▲ [OK] ▮ ; ▮ CHARS ▲ [OK] A...Z A ▮ CHARS ▲ [OK] ▮ ; ▮ CHARS ▲ [OK] A...Z E ▮ [A...Z] N T E R [SPACE] N U M B E R [a...z] ▮ CHARS ▲ [OK] ▮ ;1 ▮ : ENTER

waits for you to input a value that will be assigned to the variable A

INPUT B; "Coefficient"; "B";"Enter number"; 1:	MATH [CMDS/*Prompt*/*INPUT*] [OK] [SPACE] A...Z B ▮ ; ▮ CHARS ▲ [OK] A...Z C ▮ [A...Z] O E F F I E N T [a...z] ▮ CHARS ▲ [OK] ▮ ; ▮ CHARS ▲ [OK] A...Z B ▮ CHARS ▲ [OK] ▮ ; ▮ CHARS ▲ [OK] A...Z E ▮ [A...Z] N T E R [SPACE] N U M B E R [a...z] ▮ CHARS ▲ [OK] ▮ ;1 ▮ : ENTER
INPUT C; "Coefficient"; "C";"Enter number"; 1:	MATH [CMDS/*Prompt*/*INPUT*] [OK] [SPACE] A...Z C ▮ ; ▮ CHARS ▲ [OK] A...Z C ▮ [A...Z] O E F F I E N T [a...z] ▮ CHARS ▲ [OK] ▮ ; ▮ CHARS ▲ [OK] A...Z C ▮ CHARS ▲ [OK] ▮ ; ▮ CHARS ▲ [OK] A...Z E ▮ [A...Z] N T E R [SPACE] N U M B E R [a...z] ▮ CHARS ▲ [OK] ▮ ;1 ▮ : ENTER
$B^2 - 4AC$ ▶ D	A...Z B ▮ x^2 – 4 A...Z A A...Z C [STO▶] A...Z D ▮ : ENTER

calculates the discriminant and stores its value as D

CASE   MATH [CMDS/*Branch*/*CASE*] [OK] ENTER

    begins a series of tests on the discriminant

IF D>0   MATH [CMDS/*Branch*/*IF*] [OK] [SPACE]
                              A...Z D MATH [MTH/*Tests*/>] [OK] 0 [SPACE]

    tests to see if the discriminant is positive

THEN MSGBOX   MATH [CMDS/*Branch*/*THEN*] [OK] [SPACE]
"Two real roots: "   MATH [CMDS/*Prompt*/*MSGBOX*] [OK] [SPACE]
(−B + √D)/(2A) ", "   ■ CHARS ▲ [OK] A...Z T ■ [A...Z] W O [SPACE]
(−B − √D)/(2A): END   R E A L [SPACE] R O O T S [a...z] ■ : [SPACE]
                              ■ CHARS ▲ [OK] (−x A...Z B + √x A...Z D) / (2 A...Z A)
                              ■ CHARS ▲ [OK] , [SPACE] ■ CHARS ▲ [OK]
                              (−x A...Z B − √x A...Z D) / (2 A...Z A) ■ :
                              [SPACE] MATH [CMDS/*Branch*/*END*] [OK] ENTER

    displays the two real roots

IF D==0   MATH [CMDS/*Branch*/*IF*] [OK] [SPACE]
                              A...Z D MATH [MTH/*Tests*/==] [OK] 0 [SPACE]

    tests to see if the discriminant is zero

THEN MSGBOX   MATH [CMDS/*Branch*/*THEN*] [OK] [SPACE]
"Double root: "   MATH [CMDS/*Prompt*/*MSGBOX*] [OK] [SPACE]
−B/(2A): END   ■ CHARS ▲ [OK] A...Z D ■ [A...Z] O U B L E [SPACE]
                              R O O T [a...z] ■ : [SPACE] ■ CHARS ▲ [OK]
                              −x A...Z B / (2 A...Z A) ■ : [SPACE]
                              MATH [CMDS/*Branch*/*END*] [OK] ENTER

    displays the double root

IF D<0   MATH [CMDS/*Branch*/*IF*] [OK] [SPACE]
                              A...Z D MATH [MTH/*Tests*/<] [OK] 0 [SPACE]

    tests to see if the discriminant is negative

THEN MSGBOX   MATH [CMDS/*Branch*/*THEN*] [OK] [SPACE]
"Complex conjugates: "   MATH [CMDS/*Prompt*/*MSGBOX*] [OK] [SPACE]
−B /(2A)" ± "√−D/(2A)"i":   ■ CHARS ▲ [OK] A...Z C ■ [A...Z] O M P L E X [SPACE]
END   C O N J U G A T E S [a...z] ■ : [SPACE]
                              ■ CHARS ▲ [OK] −x A...Z B / (2 A...Z A)
                              ■ CHARS ▲ [OK] [SPACE]
                              ■ CHARS ► ► ► ► ► ► ► [OK] [SPACE]
                              ■ CHARS ▲ [OK] (√x −x A...Z D) / (2 A...Z A)

■ CHARS ▲ [OK] ■ A...Z I ■ CHARS ▲[OK]
■ : [SPACE] MATH [CMDS/*Branch*/*END*] [OK] ENTER

displays the complex roots; since $D < 0$, we must use $-D$ as the radicand

END   MATH [CMDS/*Branch*/*END*] [OK]

marks the end of a CASE group of commands

When you have finished, press HOME to leave the program editor and enter the home screen.

If you want to erase a program from memory, press ■ PROGRAM, move the highlight to the program's name, and press DEL to delete the entire program.

11.10.2 Executing a Program: To run the program you have entered, press ■ PROGRAM, move the highlight to the program's name, and press [RUN]. Or in the home screen, type RUN Quadratic and press ENTER.

The program has been written to prompt you for values of the coefficients a, b, and c in a quadratic equation $ax^2 + bx + c = 0$. Input a value, then press ENTER to continue the program.

If you need to interrupt a program during execution, press ON.

11.11   Differentiation

11.11.1 Limits: Suppose you need to find this limit: $\lim\limits_{x\to 0} \dfrac{\sin 4x}{x}$. Start the Function aplet and plot the graph of $f(x) = \dfrac{\sin 4x}{x}$ in a convenient viewing rectangle that contains the point where the function appears to intersect the line $x = 0$ (because you want the limit as $x \to 0$). Your graph should support the conclusion that $\lim\limits_{x\to 0} \dfrac{\sin 4x}{x} = 4$ (Figure 11.88).

To test whether the conclusion that $\lim\limits_{x\to\infty} \dfrac{2x-1}{x+1} = 2$ is reasonable, evaluate $f(x) = \dfrac{2x-1}{x+1}$ for several large positive values of x (since you want the limit as $x \to \infty$). For example, evaluate $f(100)$, $f(1000)$, and $f(10,000)$. Another way to test the conclusion is to examine the graph of $f(x) = \dfrac{2x-1}{x+1}$ in a viewing rectangle that extends over large values of x. See, as in Figure 11.89 (where the viewing rectangle extends horizontally from 0 to 100), whether the graph is asymptotic to the horizontal line $y = 2$. Enter $\dfrac{2x-1}{x+1}$ for F1(X) and 2 for F2(X).

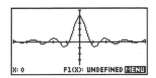

Figure 11.88: Checking $\lim_{n \to 0} \dfrac{\sin 4x}{x} = 4$

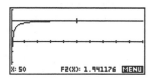

Figure 11.89: Checking $\lim_{r \to \infty} \dfrac{2x-1}{r+1} = 2$

11.11.2 Numerical Derivatives: The HP 38G has a command in the Calculus sub-menu of the MATH menu that can compute the derivative of a function f at x. For example, to find $f'(2.5)$ when $f(x) = x^3$, enter the home screen and press 2.5 [STO ▶] x,T,θ ENTER MATH ▼ ▼ ▼ ▼ ▼ ▶ [OK] x,T,θ (x,T,θ x^y 3 ENTER as shown in Figure 11.90. The format of this command is ∂ *variable (expression, variable, value,* Δx*).*

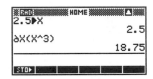

Figure 11.90: Using ∂

Technology Tip: It is sometimes helpful to plot both a function and its derivative together. In Figure 11.92, the function $f(x) = \dfrac{5x-2}{x^2+1}$ and its numerical derivative (actually, an approximation to the derivative given by the symmetric difference) are graphed on viewing window that extends from –6 to 6 vertically and horizontally. In the FUNCTION SYMBOLIC VIEW, enter $\dfrac{5x-2}{x^2+1}$ for F1(X). Move the highlight to F2(X) and press MATH ▼ ▼ ▼ ▼ ▼ ▶ [OK] x,T,θ (A...Z F 1 (x,T,θ) ENTER (Figure 11.91). Then press PLOT. Note the calculator may take some time in producing the graph.

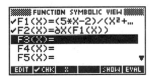

Figure 11.91: Entering $f(x)$ and $f'(x)$

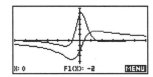

Figure 11.92: Graphs of $f(x)$ and $f'(x)$

Once the derivative is stored as F2(X) you can recall it in the home screen to evaluate. For example to compute $f'(-2.3)$, first press A...Z F 2 (–x 2.3) ENTER. This returns the unevaluated expression for the derivative of f at –2.3. Then recall and evaluate it by pressing ▲ [COPY] ENTER (Figure 11.93).

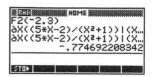

Figure 11.93: Evaluating $f'(-2.3)$

Technology Tip: To approximate the *second* derivative $f''(x)$ of a function $y = f(x)$ or to plot the second derivative, first enter the expression for F1(X) and its derivative for F2(X) as above. Then enter the second derivative by moving the highlight to F3(X) and pressing MATH ▼ ▼ ▼ ▼ ▼ ▶ [OK] X,T,θ (A...Z F 2 (X,T,θ) ENTER.

The HP 38G can compute the derivative of a point on a graph drawn in the graph mode by computing the slope. For example, graph the function $f(x) = \dfrac{5x-2}{x^2+1}$ in a viewing rectangle that extends from −6.5 to 6.5 horizontally and from −6.2 to 6.4 vertically. Then use the left arrow key to move the cursor so the x-coordinate is −2.3. Then press [MENU] [FCN], select Slope, and press [OK]. You can use the left and right arrow keys to move to a specific point, say $x = -2.3$. Figure 11.94 shows the slope (derivative) at that point to be about −0.774692.

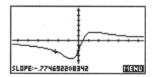

Figure 11.94: Slope of $f(x) = \dfrac{5x-2}{x^2+1}$ at $x = -2.3$

11.11.3 Newton's Method: With the HP 38G, you may iterate using Newton's method to find the zeros of a function. Recall that Newton's Method determines each successive approximation by the formula

$$x_{n+1} = x_n - \frac{f(x_n)}{f'(x_n)}.$$

As an example of the technique, consider $f(x) = 2x^3 + x^2 - x + 1$. In the FUNCTION SYMBOLIC VIEW, enter the function as F1(x). Set the range to the standard viewing rectangle and clear or uncheck any other graphs. Graph the function. A look at the graph suggests that it has a zero near $x = -1$, so return to the home screen and start the iteration by storing −1 as x (press −x 1 [STO▶] X,T,θ ENTER). Then, press these keystrokes: X,T,θ − A...Z F 1 (X,T,θ) / MATH ▼ ▼ ▼ ▼ ▼ ▶ [OK] X,T,θ (A...Z F 1 (X,T,θ)) [STO▶] X,T,θ ENTER ENTER (Figure 11.95) to calculate the first two iterations of Newton's method. Press ENTER repeatedly until two successive approximations differ by less than some predetermined value, say 0.0001. Note that each time you press ENTER, the HP 38G will use the *current* value of *x*, and that value is changing as you continue the iteration.

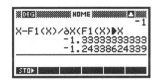

Figure 11.95: Newton's method

Technology Tip: Newton's Method is sensitive to your initial value for *x*, so look carefully at the function's graph to make a good first estimate. Also, remember that the method sometimes fails to converge!

You may want to write a short program for Newton's Method. See your calculator's manual for further information.

11.12   Integration

11.12.1 Approximating Definite Integrals: The HP 38G has a function which is accessed from the function analysis menu that will approximate a definite integral. For example, to find a numerical approximation to $\int_0^1 \cos x^2\, dx$ enter the home screen and press MATH ▼ ▼ ▼ ▼ ▼ ▶ ▼ [OK] 0 , 1 , COS X,T,θ ■ x²) , X,T,θ) ENTER (Figure 11.96).

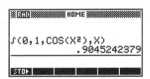

Figure 11.96: Using ∫

11.12.2 Areas: You may approximate the area under the graph of a function $y = f(x)$ between $x = A$ and $x = B$ with your HP 38G. For example, suppose you want to find the area under the graph of the function $y = \cos x^2$ between $x = 0$ and $x = 1$. The area is represented by the definite integral $\int_0^1 \cos x^2\, dx$. First graph $y = \cos x^2$ using the range that extends from −6.5 to 6.5 horizontally and from −3.1 to 3.2 vertically. The horizontal range is important, since we want "friendly" values to move the cursor from 0 to 1. Press [MENU] [FCN], select Area..., and the press [OK] when the *x*-coordinate of the cursor is 0. Then move the cursor so the *x*-coordinate is 1 and then press [OK] (Figure 11.97).

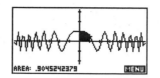

Figure 11.97: Graph and area

Technology Tip: Suppose that you want to find the area between two functions, $y = f(x)$ and $y = g(x)$ from $x = A$ and $x = B$. If $f(x) \geq g(x)$ for $A \leq x \leq B$, then graph the expression $f(x) - g(x)$ in the manner described above to find the required area.